A. Figà Talamanca (Ed.)

Harmonic Analysis and Group Representation

Lectures given at a Summer School of the
Centro Internazionale Matematico Estivo (C.I.M.E.),
held in Cortona (Arezzo), Italy,
June 24 - July 9, 1980

Springer

FONDAZIONE CIME ROBERTO CONTI

C.I.M.E. Foundation
c/o Dipartimento di Matematica "U. Dini"
Viale margagni n. 67/a
50134 Firenze
Italy
cime@math.unifi.it

ISBN 978-3-642-11115-0 e-ISBN: 978-3-642-11117-4
DOI:10.1007/978-3-642-11117-4
Springer Heidelberg Dordrecht London New York

Reprint of the 1st ed. C.I.M.E., Ed. Liguori, Napoli & Birkhäuser 1982
With kind permission of C.I.M.E.

Printed on acid-free paper

Springer.com

C O N T E N T S

CENTRO INTERNAZIONALE MATEMATICO ESTIVO

(C.I.M.E.)

NILPOTENT GROUPS AND ABELIAN VARIETIES

L. AUSLANDER AND R. TOLIMIERI

Lectures on

NILPOTENT GROUPS AND ABELIAN VARIETIES

by

L. Auslander and R. Tolimieri

Introduction

A. A. Albert, in an immense burst of creative energy succeeded in solving the "Riemann matrix problem." Although this is one of the great mathematical achievements of our century, there are few systematic accounts of Albert's work. Perhaps, C. L. Siegel's account [6] comes the closest to providing us with a view of this marvelous achievement. Albert's and Siegel's treatment are difficult because their arguments are based on matrix calculations. Because a coordinate system has been chosen, there is a hidden identification of a vector space with its dual and matrices play the role of both linear transformations and bilinear forms.

In these notes, we will present a way of using nilpotent groups to formulate the ideas of Abelian varieties and present part of the existence theorems contained in Albert's work. A full treatment of the existence part of Albert's work will appear in [4]. Our approach rests on nilpotent algebraic groups. This enables us to present a matrix-free treatment of the Riemann matrix problem. We hope this approach will reawaken admiration for, and interest in, Albert's achievement.

TABLE OF CONTENTS

1. Associative Algebras and Nilpotent Algebraic Groups

In these notes the word field will denote either the reals, $\mathbb{R}$, the complex, $\mathbb{C}$, or an algebraic number field, k, containing the rationals, $\mathbb{Q}$ and we let $[k,\mathbb{Q}] = h < \infty$. Further, all algebras and vector spaces are finite dimensional and all associative algebras have an identity.

Let $\mathcal{N}$ be an associative algebras over k of the form $k \cdot 1 \oplus \mathcal{R}$, where 1 is the identity of $\mathcal{N}$, and $\mathcal{R}$ is the radical of $\mathcal{N}$. Let $N(\mathcal{N})$ be the subset of the group of units of $\mathcal{N}$ of the form

$$N(\mathcal{N}) = \{1 + n \mid n \in \mathcal{R}\} .$$

Then $N(\mathcal{N})$ is a subgroup of the group of units, because

$$(1 + n_2)(1 + n_2) = 1 + n_1 + n_2 + n_1 n_2$$

and because $\mathcal{R}$ is an ideal if $n_1, n_2 \in \mathcal{R}$, then $n_1 + n_w + n_1 n_2 \in \mathcal{R}$. Since $\mathcal{R}$ is nilpotent, there exists k such that $\mathcal{R}^{k+1} = \{0\}$. By the binomial theorem,

$$(1 + n)^{-1} = 1 - n + \ldots + (-1)^k n^k .$$

Let $\mathcal{I} \subset \mathcal{R}$ be an ideal in $\mathcal{R}$. Let

$$G(\mathcal{I}) = \{(1 + n) \mid n \in \mathcal{I}\} .$$

Then $G(\mathcal{I})$ is easily seen to be a normal subgroup of $N(\mathcal{N})$ and $N(\mathcal{N})/G(\mathcal{I})$ is isomorphic to $N(\mathcal{N}/\mathcal{I})$.

Let the dimension of $N(\mathcal{N})$ as a k-vector space be m. For $1+n \in N(\mathcal{N})$, and $g \in \mathcal{N}$, the mapping

$$\rho(1 + n)(g) = (1 + n)g$$

is a representation of $N(\mathcal{N})$ in $GL(m,k)$. Further, there is a basis of $\mathcal{N}$ such that $\rho(N(\mathcal{N})) \subset U(m,k)$, where

$$U(m,k) = \begin{pmatrix} 1 & & * \\ & \ddots & \\ 0 & & 1 \end{pmatrix}$$

and $\rho(N(\mathcal{N}))$ is the set of zeros of a set of linear equations over k. We call $N(\mathcal{N})$ the k-algebraic group of N. It is easily verified that $N(\mathcal{N})$ is a nilpotent group.

The above has the following generalization.

Definition: Let G be a nilpotent group. We call G a k-nilpotent algebraic group if there exists an isomorphism $\rho:G \longrightarrow U(m,k)$ such that $\rho(G)$ is a k-algebraic variety in $U(m,k)$.

A nilpotent group G is called 2-step nilpotent if [G,G] is central. If G is a 2-step k-nilpotent algebraic group it is easily verified that G satisfies an exact sequence:

$$0 \to V_1 \to G \to V_2 \to 0,$$

where V_1 and V_2 are k-vector spaces such that $V_1 \supset [G,G]$ and V_1 is in the center of G. With these general definitions out of the way, we can discuss the special cases with which we will be concerned in these notes. Let V be a k-vector space and let $\Lambda(V)$ denote the exterior algebra over V. Then

$$\Lambda(V) = k \cdot 1 \oplus \mathscr{R}$$

where $\mathscr{R} = \sum_{i>0} \Lambda^i(V)$ is the radical of $\Lambda(V)$.. Hence, we may form the k-nilpotent algebraic group $N(\Lambda(V))$. It is clear that $\mathscr{I} = \sum_{i>2} \Lambda^i(V)$ is an ideal in $\Lambda(V)$. Hence we may form

$$\mathscr{F}_2(V) = N(\Lambda(V))/G(\mathscr{I}) = N(\Lambda(V)/\mathscr{I}).$$

Since $\mathscr{F}_2(V)$ is very important in the rest of this paper, we will present another more explicit description or "presentation" of $\mathscr{F}_2(V)$. As a set

$$\mathscr{F}_2(V) = V \times V \wedge V$$

and the group law of composition is given by

$$(v_1,w_1)(v_2,w_2) = (v_1 + v_2, w_1 + w_2 + v_1 \wedge v_2)$$

where $v_2 \epsilon V$ and $w_2 \epsilon V \wedge V$ for $\alpha = 1,2$.

$\mathscr{F}_2(V)$ is a 2-step k-nilpotent algebraic group with center (0,w), $w \epsilon V \wedge V$, and it is called the free 2-step k-nilpotent group over V. The reason for the name, "free," is the following: If G is a 2-step k-nilpotent algebraic group and

$$f: \mathscr{F}_2(V)/[\mathscr{F}_2(V), \mathscr{F}_2(V)] \longrightarrow G/[G,G]$$

is k-linear, then there exists a homomorphism

$$F: \mathscr{F}_2(V) \longrightarrow G$$

such that the kernel of F is a k-algebraic group and the following diagram is commutative:

$$\begin{array}{ccc} \mathcal{F}_2(V) & \xrightarrow{F} & G \\ \downarrow & & \downarrow \\ \mathcal{F}_2(V)/[\mathcal{F}_2(V),\mathcal{F}_2(V)] & \xrightarrow{f} & G/[G,G] \end{array}$$

The nilpotent algebraic groups $N(\Lambda(V))$ and $\mathcal{F}_2(V)$ exhibit a property that will have enormous implications in our later work. We observe that the representations of $N(\Lambda(V))$ or $\mathcal{F}_2(V)$ arising from the associative algebra structure can be defined by linear equations whose coefficients are in $\mathbb{Q}$ This will enable us to consider $N(\Lambda(V))$ and $\mathcal{F}_2(V)$ as $\mathbb{Q}$-nilpotent algebraic groups. We will now discuss how this can be done.

Let G be a k-nilpotent algebraic group and assume that a set of equations defining G can be chosen to have coefficients in $K \subset k$. We will then say that G is defined over K. Now let V be an m-dimensional k-vector space. If $[k:K] = h$, then we may consider V as an mh dimensional K vector space that we will denote by $V(K)$. Clearly, k linear transformation of V gives rise to a K linear transformation of $V(K)$. Thus we have an isomorphism

$$r(K):GL(m,k) \longrightarrow GL(mh,K) .$$

We will call $r(K)$ the isomorphism of reducing the field from k to K. It is easily seen that if G is a k-algebraic group defined over K, then $r(K)(G)$ will be a K-algebraic group. We will call $r(K)(G)$ the K-algebraic group obtained by reducing the field of G.

Again, let G be a k-algebraic group defined over K, $K \subset k$. Consider $G(K) \subset G$ consisting of those points in $GL(m,k)$, all of whose coefficients are in K. Then $G(K)$ will be a K-algebraic group in $GL(m,K)$. If all the k points of $G(K) = G$, we will call $G(K)$ a K-form of the k-algebraic group G. It should be remarked, that G may have non-isomorphic K-forms.

An example may help the reader understand all this better. Let k be a totally real algebraic number field over $\mathbb{Q}$ and let $[k:\mathbb{Q}] = h$. Consider the k-algebraic subgroup G of $GL(2,k)$ defined by

$$G = \begin{pmatrix} 1 & x \\ 0 & 1 \end{pmatrix} \qquad x \in k$$

A set of defining equations for G are given by $x_{11} = x_{22} = 1$ and $x_{21} = 0$ where

$$\begin{pmatrix} x_{11} & x_{12} \\ x_{21} & x_{22} \end{pmatrix} = GL(2,k)$$

Clearly, G may also be considered as the k-points of the $\mathbb{Q}$ algebraic group

$$\begin{pmatrix} 1 & x \\ 0 & 1 \end{pmatrix} \subset GL(2,\mathbb{Q}) \quad x \in \mathbb{Q}$$

We will now give an explicit map for $r(\mathbb{Q})$. Let r denote the regular representation of k over $\mathbb{Q}$. Then

$$r(\mathbb{Q})\begin{pmatrix} x_{11} & x_{12} \\ x_{21} & x_{22} \end{pmatrix} = \begin{pmatrix} r(x_{11}) & r(x_{12}) \\ r(x_{21}) & r(x_{22}) \end{pmatrix}$$

where we view the right hand matrix in $GL(2h,\mathbb{Q})$. Thus $r(\mathbb{Q})(G) = G(\mathbb{Q}) \subset GL(2h,\mathbb{Q})$ is a $\mathbb{Q}$-algebraic group. Let $G(\mathbb{Q})_{\mathbb{R}}$ denote the group of $\mathbb{R}$-points of $G(\mathbb{Q})$. Then $G(\mathbb{Q})_{\mathbb{R}} \subset GL(2h,\mathbb{R})$. Since k is totally real, there exists $A \in GL(h,\mathbb{R})$ such that

$$A^{-1}r(k)A = \begin{pmatrix} \chi_1(k) & & 0 \\ & \ddots & \\ 0 & & \chi_h(k) \end{pmatrix} = D(k) k \in k$$

and $\chi_i: k \longrightarrow \chi_i(k)$, $i = 1,\dots,h$, is an isomorphism of k into $\mathbb{R}$. Indeed, $\chi_1,\dots,\chi_n$ are distinct isomorphisms. Now

$$\begin{pmatrix} A^{-1} & 0 \\ 0 & A^{-1} \end{pmatrix} G(\mathbb{Q})_{\mathbb{R}} \begin{pmatrix} A & 0 \\ 0 & A \end{pmatrix} = \begin{pmatrix} I & D \\ 0 & I \end{pmatrix}$$

where D is as above.

Now, $N(\Lambda(V))$ and $\mathcal{F}_2(V)$ are easily seen to be defined over $\mathbb{Q}$ and so both may be considered - by reducing the field - as $\mathbb{Q}$- algebraic groups. Hence, we have the identity map $V(\mathbb{Q}) \longrightarrow V$ lifts to a morphism

$$\mathcal{F}_2(V(\mathbb{Q})) \longrightarrow \mathcal{F}_2(V).$$

There are certain homomorphisms of $\mathcal{F}_2(V)$ that will play an essential role in our theory. We will now establish a language with which to carry out this discussion. We begin by listing some standard notation that we will follow. If V and W are k-vector spaces, we use $\mathrm{Hom}(V,W)$ to denote the k-vector space of k-linear maps and $V^* = \mathrm{Hom}(V,k)$, the dual vector space. For $T \in \mathrm{Hom}(V,W)$, we have $T^* \in \mathrm{Hom}(W^*,V^*)$ and we will identify V^{**} with V.

Let $\mathrm{Bil}(V)$ denote the vector space of bilinear forms on $V \times V$. For $B \in \mathrm{Bil}(V)$, define $L(B) \in \mathrm{Hom}(V,V^*)$ by $(L(B)(u))(v) = B(u,v)$, $u,v \in V$. Since $L(B) \in \mathrm{Hom}(V,V^*)$, we have $L(B)^* \in \mathrm{Hom}(V,V^*)$. Clearly, B is alternating if and only if $L(B)^* = -L(B)$, and B is symmetric if and only if $L(B)^* = L(B)$. The set of alternating forms will be denoted by $\mathrm{Alt}(V)$, $\mathrm{Sym}(V)$ will denote the set of symmetric forms, and $\mathrm{Bil}(V) = \mathrm{Alt}(V) \oplus \mathrm{Sym}(V)$. If $L(B)$ is nonsingular, we say that B is non-singular and the space of non-singular bilinear forms will be denoted by $\mathrm{Bil}^\times(V)$. Analogously, we will use the notation $\mathrm{Alt}^\times(V) = \mathrm{Alt}(V) \cap \mathrm{Bil}^\times(V)$ and $\mathrm{Sym}^\times(V) = \mathrm{Sym}(V) \cap \mathrm{Bil}^\times(V)$.

Let $\mathcal{F}_2(V)$ denote the free 2-step k-nilpotent group over V. The dual space to $V \wedge V$ is $V^* \wedge V^*$, and we have the commutative diagram

$$\begin{array}{ccc} V \times V & \xrightarrow{\ \wedge\ } & V \wedge V \\ {\scriptstyle A}\downarrow & \swarrow {\scriptstyle l(A)} & \\ k & & \end{array}$$

where $A \in \mathrm{Alt}(V)$ and $l(A) \in V^* \wedge V^*$; this enables us to identify $\mathrm{Alt}(V)$ and $V^* \wedge V^*$.

Now, for $A \in \mathrm{Alt}(V)$, we may define a group structure $N(A)$ on the set $V \times k$ whose law of multiplication is

$$(v_1,k_1)(v_2,k_2) = (v_1 + v_2, k_1 + k_2 + A(v_1,v_2))$$

where $v_1,v_2 \in V$ and $k_1,k_2 \in k$. Then $N(A)$ has (o,k), $k \in k$ in its center and $N(A)$ modulo its center is Abelian. Hence, $N(A)$ is a 2-step k-nilpotent algebraic group. Define the surjection

$$P : \mathcal{F}_2(V) \longrightarrow N(A)$$

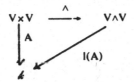

by $P(v,k) = (v, l(A)w)$, $(v,w) \in \mathcal{F}_2(V)$. If $i : V \longrightarrow V$ is the identity mapping, the following diagram is commutative:

$$
\begin{array}{ccc}
\mathcal{F}_2(V) & \xrightarrow{\;P\;} & N(A) \\
\downarrow & & \downarrow \\
\mathcal{F}_2(V)/[\mathcal{F}_2(V),\mathcal{F}_2(V)] & \longrightarrow & N(A)/[N(A),N(A)]
\end{array}
$$

We will call such morphisms of $\mathcal{F}_2(V)$ *polarizations and denote the set of polarizations by* $P(V)$. Clearly, we may identify $P(V)$ with $\mathrm{Alt}(V)$ as above. If $A \in \mathrm{Alt}^{\times}(V)$, all $N(A)$ are isomorphic and we will call $N(A)$ a k-Heisenberg group. The corresponding polarizations will be denoted by $P^{\times}(V)$. If $\dim V = 2m$, we will sometimes use $N_{2m+1}(k)$ to denote $N(A)$ and call $N_{2m+1}(k)$ the $2m+1$ k-Heisenberg group. Fixing an isomorphism of the center $\mathfrak{z}$ of $N_{2m+1}(k)$ with k, as when we present $N_{2m+1}(k)$ as $N(A)$, will be called an *orientation* of $N_{2m+1}(k)$.

The presentation $N(A)$ of the Heisenberg $N_{2m+1}(k)$ has the additional property of determining an isomorphism which we will denote by $A: V \longrightarrow V^*$ or, if P corresponds to A, by $P: V \longrightarrow V^*$. This follows from the fact that A is non-degenerate.

2. The Jacobi Variety of a Riemann Surface and Abelian Varieties

In this lecture we will need two special examples of a general phenomena; accordingly, we will begin with the general case and then specialize to the examples of interest to us.

Let M be a compact manifold and let $H^*(M,\mathbb{R})$ and $H^*(M,\mathbb{Z})$ be the cohomology rings of M with real and integer coefficients, respectively. If $\mathscr{R} = \sum_{i>0} H^i(M,\mathbb{R})$, then $\mathscr{R}$ is the radical of $H^*(M,\mathbb{R})$ and $H^*(M,\mathbb{R}) = \mathbb{R} \oplus \mathscr{R}$. Hence, we may form the nilpotent algebraic group $N(H^*(M,\mathbb{R}))$, which we will henceforth denote by N(M). Now the Lie algebra of N(M) is the Lie algebra associated with $\mathscr{R}$ by

$$[x,y] = xy - yx \qquad x,y \in \mathscr{R}.$$

Since, for $x \in H^i(M,\mathbb{R})$ and $y \in H^j(M,\mathbb{R})$, we have

$$xy = (-1)^{ij}\, yx \in H^{i+j}(M,\mathbb{R}) .$$

It follows that, if $x \in H^{2i}(M,\mathbb{R})$, then

$$[x,y] = xy - yx = 0 ,$$

and so $\sum H^{2i}(M,\mathbb{R}) \subset \mathscr{R}$ is in the center of $\mathscr{R}$ as a Lie algebra. Further, if $x,y \in \sum H^{2i+1}(M,\mathbb{R})$. Then $[x,y] \in \sum H^{2i}(M,\mathbb{R})$. Thus $\mathscr{R}$ is a 2-step nilpotent Lie algebra and so N(M) is a 2-step nilpotent Lie group.

By the standard theory of cohomology rings, there is a natural injection

$$i: H^*(M,\mathbb{Z}) \longrightarrow H^*(M,\mathbb{R})$$

such that $i(H^*(M,\mathbb{Z}))$ is a lattice in the vector space $H^*(M,\mathbb{R})$. Let $\mathscr{R}(\mathbb{Z}) = i(H^*(M,\mathbb{Z})) \cap \mathscr{R}$. Then we may argue as before and obtain that

$$\Gamma(M) = \{1 + n \mid n \in \mathscr{R}(\mathbb{Z})\}$$

is a subgroup of N(M). It is then easily verified that $\Gamma(M)\backslash N(M)$ is a compact manifold, called a nilmanifold. Hence, we have functorally assigned to every compact manifold M, the compact nilmanifold $\Gamma(M)\backslash N(M)$. By [1], there exists a unique $\mathbb{Q}$-nilpotent algebraic group $N_{\mathbb{Q}}(M)$ such that $\Gamma(M) \subset M_{\mathbb{Q}}(M) \subset N(M)$. We will call $N_{\mathbb{Q}}(M)$ the topological rational form of N(M). (It may happen that N(M) has other rational forms not isomorphic to $N_{\mathbb{Q}}(M)$).

The groups N(M) and P(M) constructed above have an additional structure that we will now discuss.

As a set N(M) = X × Y , where

$$X = \{1 + n \mid n \in \sum H^i(M,\mathbb{R}),\ i \text{ odd}\}$$

$$Y = \{1 + n \mid n \in \sum H^i(M,\mathbb{R}),\ i \text{ even},\ i > 0\}$$

where X and Y are vector spaces. If (x,y) ∈ X × Y, then the multiplication in N(M) is given by

$$(x_1,y_1)(x_2,y_2) = (x_1 + x_2, y_1 + y_2 + B(x_1,x_2))$$

where B : X × X ⟶ Y is skew symmetric. Such a presentation of a 2-step k-algebraic group will be called a grading. Notice that the presentation of $\mathcal{F}_2$(V) as V × V∧V in Section I was a graded presentation and N(A) was a graded presentation of the Heisenberg group.

The main purpose for introducing the graded structure of 2-step k-nilpotent algebraic groups is the following: If N = X × Y is a graded 2-step k-nilpotent algebraic group and if α : V ⟶ X is a morphism, then α has a *unique* extension to a morphism $\alpha^*:\mathcal{F}_2(V) \longrightarrow N$ that preserves gradings. This is because the composite mapping

$$V \times V \xrightarrow{\alpha \times \alpha} X \times X \xrightarrow{B} Y$$

is an alternating bilinear mapping on V × V and so we have a *unique* linear mapping ℓ(B):V∧V ⟶ Y that completes the commutative diagram

$$\begin{array}{ccc} V \times V & \xrightarrow{\wedge} & V \wedge V \\ {\scriptstyle B \circ \alpha \times \alpha}\ \big\downarrow & \swarrow {\scriptstyle \ell(B)} & \\ Y & & \end{array}$$

It follows that if A is any morphism of V then A determines a unique graded morphism of $\mathcal{F}_2$(V).

For the rest of this paper, we will restrict ourselves to graded nilpotent algebraic groups and all morphisms will be grading preserving morphism. *Henceforth, we will drop the word graded, but it will be what assures the uniqueness of various morphisms that occur in the discussion.*

Let M be a complex manifold. Then as in [9], the complex structure on M determines an automorphism J(M) of $H^*(M,\mathbb{R})$. Further, if

$$f: M_1 \longrightarrow M_2$$

is a complex analytic mapping, then

$$f^* J(M_2) = J(M_1) f^* .$$

It is clear that the complex structure determines an automorphism of N(M), which we will also denote by J(M). It is important to note that J(M) may *not* induce an automorphism of Γ(M) or even of $N_{\mathbb{Q}}(M)$.

To illustrate this, let us see how all this works for the m complex dimensional torus. Let W be an m dimensional complex vector space and let L be a discrete subgroup of W such that W/L is compact. We will begin by discussing another way of looking at W. Clearly, W is also a real vector space $W(\mathbb{R})$ of real dimension 2m. Let $e_1,\ldots,e_m$ be a basis of W. Then $e_1, ie_1,\ldots,e_m, ie_m$, $i = \sqrt{-1}$, is a basis of $W(\mathbb{R})$. For w ∊ W, the mapping $J: W \longrightarrow iW$ defines an automorphism of $W(\mathbb{R})$ which in terms of the above basis is given by

$$J = \begin{pmatrix} J_0 & & 0 \\ & \ddots & \\ 0 & & J_0 \end{pmatrix} = m\, J_0 \text{ where } J_0 = \begin{pmatrix} 0 & 1 \\ -1 & 0 \end{pmatrix}$$

Notice that J has the property that $J^2 = -I$, where I is the identity mapping.

Let A be a real linear transformation of $W(\mathbb{R})$. When does A induce a *complex* linear transformation of W? We will now verify that the answer is when

$$JA = AJ .$$

By a straightforward computation, one verifies that

$$J_0 \begin{pmatrix} a & b \\ c & d \end{pmatrix} = \begin{pmatrix} a & b \\ c & d \end{pmatrix} J_0$$

if and only if a = d, b = −c. But since the regular representation r of $\mathbb{C}$ over $\mathbb{R}$ is given by

$$r : a + ib \longrightarrow \begin{pmatrix} a & b \\ -b & a \end{pmatrix}$$

we have that our assertion is true for m = 1. Relative to the basis $e_1, \ldots, e_m$, let

$$C = (C_{\alpha\beta}) \in Hom(W,W) \qquad \alpha,\beta = 1,\ldots,m .$$

Then relative to the basis $e_1, ie_1, \ldots, e_m, ie_m$ of $W(\mathbb{R})$, C is given by

$$C = (r(C_{\alpha\beta})) .$$

It then follows easily that JC = CJ.

Now assume that JA = AJ and write A as an m×m matrix whose entries are 2×2 matrices

$$A = (A_{\alpha\beta}) \qquad \alpha,\beta = 1,\ldots,m .$$

By a direct computation, we have that JA = AJ implies that

$$J_0 A_{\alpha\beta} = A_{\alpha\beta} J_0 \qquad \text{all } \alpha,\beta .$$

Hence each $A_{\alpha\beta} = r(C_{\alpha\beta})$, $C_{\alpha\beta} \in \mathbb{C}$ and we have that A gives a complex linear transformation of W.

Now let V be a 2m dimensional real vector space and let J be an $\mathbb{R}$-linear transformation such that $J^2 = -I$. From the pair (V,J), we will construct a complex m-dimensional vector space W such that $W(\mathbb{R}) = V$ and the automorphism $J : w \longrightarrow iw$ is the mapping J.

Let $e_1 \neq 0$, $e_1 \in V$ and let $f_1 = J(e_1)$. Let $L(e_1,f_1)$ denote the linear subspace of V spanned by e_1 and f_1. Then $L(e_1,f_1)$ is J invariant and since $J^4 = I$, there exists V_2 such that $JV_2 = V_2$ and

$$V = L(e_1,f_1) \oplus V_2 .$$

Identifying $L(e_1,f_1)$ with $\mathbb{C}$ as a real vector space by

$$ae_1 + bf_1 \longrightarrow a + bi$$

we can solve our problem by induction.

Henceforth (V,J) *will be called a complex vector space and* J *will be called the complex structure.*

Let us now consider the complex torus V/L, where (V,J) is our complex vector space. It is well known that the 1-forms $dx_1, dy_1, \ldots, dx_m, dy_m$ are a basis of $H^1(V/L,\mathbb{R})$, where $V = x_1e_1 + y_1f_1 + \ldots + x_me_m + y_mf_m$ and $Je_i = f_i$ and $Jf_i = -e_i$, $i = 1,\ldots,m$. If $V^* = H^1(V/L,\mathbb{R})$, then

$$H^*(V/L,\mathbb{R}) = \bigwedge(V^*) .$$

Viewing V/L as a Lie group, we may identify the tangent space to V/L at the identity with V and V^* may be identified with the dual space to V. Hence, J induces J^* on V^* such that $(J^*)^2 = -I$. Thus a complex structure on a torus V/L is equivalent to an automorphism J^* of $\mathscr{F}_2(V^*)$ such that modulo the center of $\mathscr{F}_2(V^*)$, $(J^*)^2 = -I$.

Now let S be a compact Riemann surface of genus $m > 0$. (Topologically, S is a 2-sphere with m handles.) The classical facts about the cohomology ring $H^*(S,\mathbb{R})$ easily imply that N(S) is isomorphic to $N_{2m+1}(\mathbb{R})$. The orientation of S then determines an orientation of $N_{2m+1}(\mathbb{R})$. Let J(S) be the automorphism of $N_{2m+1}(\mathbb{R})$ induced by the complex structure on S, then J(S) acts trivially on $\mathfrak{z}$, the center of N(S), and if J_1 denotes the action of J(S) on $N(S)/\mathfrak{z}$, then $(J_1)^2 = -I$. Finally

$$[g, J(S)\, g\,] > 0 \qquad\qquad g \notin \mathfrak{z},\ g \in N(S)$$

where $[a,b] = aba^{-1}b^{-1}$.

<u>Definition:</u> Let $N_{2m+1}(\mathbb{R})$ be an oriented $\mathbb{R}$-Heisenberg group and let J be an automorphism of $N_{2m+1}(\mathbb{R})$ satisfying all the above conditions. We will call J a positive definite CR structure.

We are almost ready to define the concept of a Jacobi variety, but it will be convenient to make a slight detour in order to first define the concept of a dual torus.

Let V be an n-dimensional $\mathbb{R}$-vector space and let L be a discrete subgroup of V such that V/L is compact or a torus. Let V^* be the dual vector space to V and let $L^* \subset V^*$ be the subset of V^* such that $\ell^* \in L^*$ if and only if $\ell^*(L) \subset \mathbb{Z}$, where $\mathbb{Z}$ denotes the integers. One verifies that L^* is a discrete subgroup of V^* such that V^*/L^* is a torus which we will denote by $(V/L)^*$. We call L^* the dual lattice to L and $V^*/L^* = (V/L)^*$ the dual torus to V/L. Since $(L^*)^* = L$, we have $((V/L)^*)^* = V/L$.

Notice that $L \subset V$ determines a unique rational vector space $V(\mathbb{Q})$ such that

$$L \subset V(\mathbb{Q}) \subset V$$

and $L^* \subset V^*(\mathbb{Q}) \subset V^*$. Clearly, $V^*(\mathbb{Q})$ can be identified with $(V(\mathbb{Q}))^*$. We have already described $H^*(V/L,\mathbb{R})$ as $\Lambda(V^*)$. Let $\mathcal{I}_2 = \sum_{i \leq 2} H^i(V/L, \mathbb{R})$ and consider $G(\mathcal{I}_2)$, the normal subgroup of $N(\Lambda(V^*))$ determined by the ideal $\mathcal{I}_2$. Then one verifies that

$$N(\Lambda(V^*))/G(\mathcal{I}_2)\Gamma(V/L)$$

is the dual torus to V/L.

Now form $N(S)/\Gamma(S)\mathfrak{z}$, where $\mathfrak{z}$ is the center of $N(S)$. Then $N(S)/\Gamma(S)\mathfrak{z}$ is a torus with a complex structure J^* determined by the positive definite CR structure on $N(S)$ and $N(S)/\Gamma(S)\mathfrak{z}$ determine a unique rational form for $N(S)/\mathfrak{z}$. Let V/L be the complex dual torus to $N(S)/\Gamma(S)\mathfrak{z}$. The complex torus $(V/L,J)$ is called the Jacobi variety of S.

The Jacobi variety of S is related to S by two important mappings. The first is at the cohomology level; the second is at the manifold level.

Consider $N(S)$. Since this is a graded nilpotent group, there exists an $A \in \mathrm{Alt}^{\times}(V^*)$ such that $N(S) = N(A)$, where $V^* = N(S)/\mathfrak{z}$ and $\mathfrak{z}$ is the center of $N(S)$. Let V/L, the dual torus of $N(S)/\Gamma(S)\mathfrak{z}$, be the Jacobi variety of S. Now $\Lambda(V^*) = H^*(V/L,\mathbb{R})$ and so we have a natural homomorphism $N(V/L) \longrightarrow \mathcal{F}_2(V^*)$ with kernel $G(\mathcal{I})$, where $\mathcal{I}$ is the ideal $\sum_{i \geq 3} H^i(V/\Gamma,\mathbb{R})$. *Since*

$$\mathcal{F}_2(V^*)/[\mathcal{F}_2(V^*),\mathcal{F}_2(V^*)] = N(S)/\mathfrak{z} = V^* .$$

There is a unique surjection $P : \mathcal{F}_2(V^*) \longrightarrow N(S)$ (of course, P is the polarization determined by A) such that

$$\begin{array}{ccc} \mathcal{F}_2(V^*) & \xrightarrow{P} & N(S) \\ & \searrow \quad \swarrow & \\ & V^* & \end{array}$$

is a commutative diagram. Let $J(V/L)$ be the automorphism of $\mathcal{F}_2(V^*)$ induced by J. Then $J(V/L)$ is the same automorphism of $\mathcal{F}_2(V^*)$ as that induced by the complex structure on V/L. The mapping P is the first mapping at the cohomology level that we sought.

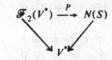

It is natural to ask if there exists a complex analytic mapping $f : S \longrightarrow V/L$ such that

$$f^{*}:N(\wedge(V^{*})) \longrightarrow N(S)$$

equals the composition $N(V/L) \longrightarrow \mathcal{F}_2(V^{*}) \xrightarrow{P} N(S)$. The answer is yes and the mapping of is called the Jacobi imbedding. Working this out in detail would take us too far afield from our main object so we will have to refer the reader to any of the many standard texts (for instance, [7]) for the proof of this result.

Now a complex torus that is a Jacobi variety has the remarkable property of having sufficiently many meromorphic functions to separate points.

Definition: A complex torus $(V/L,J)$ is called an Abelian variety if it has sufficiently many meromorphic functions to separate points.

Remarks: *Not every complex torus is an Abelian variety. Not every Abelian variety is a Jacobi variety.*

We will now formulate necessary and sufficient conditions for a complex torus to be an Abelian variety.

Let $(V/L,J)$ be a complex structure and let $\mathcal{F}_2(V^{*}) = N(\wedge(V^{*})/\sum_{i\geq 3} H^i(V/L,\mathbb{R}))$ and let J^{*} be the automorphism of $\mathcal{F}_2(V^{*})$ induced by J. Recall that $L \subset V(\mathbb{Q}) \subset V$ and let $V^{*}(\mathbb{Q})$ be the dual rational vector space to $V(\mathbb{Q})$ with $V^{*}(\mathbb{Q}) \subset V$. Then $\mathcal{F}_2(V^{*}(\mathbb{Q})) \subset \mathcal{F}_2(V^{*})$.

Definition: We call $P \in Pol^{\times}(V^{*})$ rational if the kernel P in $\mathcal{F}_2(V^{*})$ is the closure of a subspace of $\mathcal{F}_2(V^{*}(\mathbb{Q}))$.

We can now state the fundamental theorem of Abelian varieties. Again we will have to leave a proof to outside sources such as [7]. Nilpotent proofs can be found in [8] and [3].

Theorem: A necessary and sufficient condition for $(V/L,J)$ to be an Abelian variety is that there exists a rational polarization $P:\mathcal{F}_2(V^{*}) \longrightarrow N_{2m+1}(\mathbb{R})$, where $N_{2m+1}(\mathbb{R})$ is oriented, such that

1) The kernel of P is J^{*} invariant.

2) The automorphism that J^{*} induces on $N_{2m+1}(\mathbb{R})$ is a positive definite CR structure.

<u>Definition:</u> If (V/L,J) is an Abelian variety, J is called a Riemann matrix. If P satisfies the above theorem, (J,P) will be called a Riemann pair.

It should be remarked that for fixed J there may be many rational polarizations {P} such that (J,P) is a Riemann pair for $P \in \{P\}$. Also for each fixed P there may be many complex structures {J} such that (J,P) is a Riemann pair for $J \in \{J\}$.

3. Morphisms of Abelian Varieties and the Structure of Riemann Matrices

Let $(V_1/L_1,J_1)$ and $(V_2/L_2,J_2)$ be Abelian varieties and let $f: V_1/L_1 \longrightarrow V_2/L_2$ be a complex analytic mapping. Then $f^*: N(V_2/\Gamma_2) \longrightarrow N(V_1/L_1)$ and

$$f^* J_2 = J_1 f^* .$$

We will call f or, by abuse of language, f^*, a morphism of the Abelian varieties.

Let End($\mathbb{A}$) be the ring of morphisms of an Abelian variety $\mathbb{A} = (V/L,J)$. Let

$$\mathrm{End}(\mathbb{A}) \otimes_{\mathbb{Z}} \mathbb{Q} = \mathcal{M}(J)$$

and call $\mathcal{M}(J)$ the rational multiplier algebra of J. Notice that $\mathcal{M}(J)$ depends on the rational structure of V and on J, but not on the lattice L. Clearly, $\mathcal{M}(J)$ is actually a *representation* of a rational associative algebra. Since $\mathcal{M}(J) \supset \mathbb{Q}$ $\mathcal{M}(J)$ is never trivial.

Let (J,P) be a Riemann pair and let A be the alternating form corresponding to P. Then A determines an isomorphism $A: V(\mathbb{Q}) \longrightarrow V^*(\mathbb{Q})$. For $M \in \mathcal{M}(J)$, one verifies [6] that $A^{-1}M^* A \in \mathcal{M}(J)$ and hence

$$\sigma: M \longrightarrow \sigma(M) = A^{-1}M^*A$$

is an involution of $\mathcal{M}(J)$.. This involution, called the Rosati involution , is also positive; i.e., the trace MM^* is positive if $M \neq 0$.

Let us now state a lemma due to Poincare′ that will enable us to completely structure $\mathcal{M}(J)$.

<u>Poincare′ Lemma:</u> Let (V/L,J) be an Abelian variety and let $V_1(\mathbb{Q}) \subset V(\mathbb{Q})$ be such that $V_1 = V_1(\mathbb{Q}) \otimes_{\mathbb{Q}} \mathbb{R} \subset V$ is J invariant. Then there exists $V_2(\mathbb{Q}) \subset V(\mathbb{Q})$ such that

1) $V(\mathbb{Q}) = V_1(\mathbb{Q}) \otimes V_2(\mathbb{Q})$
2) $V_2 = V_2(\mathbb{Q}) \otimes_{\mathbb{Q}} \mathbb{R} \subset V$ is J invariant.
3) If L_2 is a lattice in $V_2(\mathbb{Q})$, then $(V_2/L_2, J \mid V_2)$ is an Abelian variety.

<u>Remark:</u> The existence of $V_1(\mathbb{Q})$ is equivalent to $\mathcal{M}(J)$ containing a proper projection operator or idempotent.

We may find $V_2(\mathbb{Q})$ as follows. Let (J,P) be a Riemann pair and let G be the subgroup of $N_{2m+1}(\mathbb{R})$ generated by $P(V_1)$. Let $\mathscr{C}(G)$ denote the centralizer of G. Then $\mathscr{C}(G)/\mathfrak{z}$, where $\mathfrak{z}$ is the center of $N_{2m+1}(\mathbb{R})$, will be V_2. A proof of these assertions can be found in [3], Chapter III.

Remark: Clearly, the Poincare' Lemma implies that $\mathscr{M}(J)$ is completely reducible and so is semi-simple.

We say that J is $\mathbb{Q}$-irreducible if $\mathscr{M}(J)$ has no non-trivial projections. Clearly, the Poincare' Lemma implies that

$$V = V_1 \oplus \dots \oplus V_n$$

where $J(V_i) = V_i$ and if $J_i = J \mid V_i$, then J_i is $\mathbb{Q}$-irreducible, all i. We say that J_i and J_j are $\mathbb{Q}$-equivalent if there exists $D \in \mathscr{M}(J)$ such that

$$D: V_i \longrightarrow V_j$$

and $DJ_i = J_j D$. We may group together all the $\mathbb{Q}$-equivalent J_i's and change the indexing to write

$$J = \sum_1^k m_i J_i \qquad m_i \in \mathbb{Z}^+ .$$

We call m_i the multiplicity of J_i. It follows that

$$\mathscr{M}(J) = \oplus \sum_{i=1}^{k} \mathscr{M}(m_i J_i) .$$

Now, it is easily seen that if J_i is irreducible, then $\mathscr{M}(J_i)$ is a representation ρ of a division algebra $\mathscr{D}_i$. Further, $\mathscr{M}(m_i J_i)$ is the $m_i \times m_i$ matrix algebra over $\rho(\mathscr{D}_i)$. Thus to determine all Riemann matrices J, Albert had to first solve the following algebraic problems.

1) Determine the set Δ of all rational division algebra with a positive involution.

2) For $\mathscr{D} \in \Delta$ determine the set of all positive involutions.

Since the solution to these algebraic problems have many good expositions [1], we will just pull the algebraic results out of the hat as we need them. We adopt the following language that has become customary in this subject. Let $\mathscr{D} \in \Delta$ and, let $\mathscr{k}$ be the center of $\mathscr{D}$, and let σ be a positive involution of $\mathscr{D}$. Then $\sigma(\mathscr{k}) = \mathscr{k}$ and $\sigma \mid \mathscr{k}$ is a positive involution. $\mathcal{D}$ is said to be of the *first kind* if $\sigma(k) = k$, all $k \in \mathcal{k}$ and if this fails, $\mathcal{D}$ is said to be of the *second kind*.

In these notes, we will not discuss the problem of $\mathbf{Q}$-irreducible Riemann matrices, but discuss the following simpler problem.

Main Problem: Let $\mathcal{D} \in \Delta$ and let ρ be a right $\mathbf{Q}$ representation of $\mathcal{D}$. Find all Riemann matrices J such that $\mathcal{M}(J) = \rho(\mathcal{D})$.

We will now outline our approach to this problem:

By the general representation theory, we know that every right representation ρ of $\mathcal{D}$ that could be a candidate for an irreducible J can be considered as pr, where r is the right regular representation of $\mathcal{D}$ over $\mathbf{Q}$ and $p \in \mathbf{Z}^+$. Assume ρ acts on the $\mathbf{Q}$-vector space V and form $\mathcal{F}_2(V)$, noting that ρ induces a representation of $\mathcal{D}$ as morphisms of $\mathcal{F}_2(V)$ that we will also denote by ρ. We next determine all $P \in \mathrm{Pol}^{\times}(V)$ such that if A is the alternating form corresponding to P then

$$A^{-1}\rho^{*}(d)A = \rho(\sigma(d)) \qquad d\in\mathcal{D},\sigma\in\{\sigma\}$$

where $\{\sigma\}$ is the set of positive involutions of $\mathcal{D}$. We let $\mathcal{A}(\mathcal{D},\rho,\sigma)$ denote the set of such polarizations. In other words, we first find the polarizations that can be candidates for a Rosati involution.

For each $P \in \mathcal{A}(\mathcal{D}, \rho,\sigma)$, we produce a Riemann matrix J(P) such that

1) $\mathcal{M}(J(P)) = \rho(\mathcal{D})$

2) (J(P),P) is a Riemann pair.

Finally, from J(P) we determine all Riemann matrices $\{J\}_P$ such that if $J \in \{J\}_P$:

1) $\mathcal{M}(J) \supset \rho(\mathcal{D})$

2) (J,P) is a Riemann pair.

4. Riemann Matrices Whose Multiplier Algebras are Totally Real Fields

The simplest examples of division algebras with positive involution are the totally real fields with the identity mapping as positive involution. Indeed, the identity mapping is the only positive involutions for totally real fields. Recall that $\mathcal{k}$ is totally real, $[\mathcal{k}:\mathbb{Q}] = h$, if and only if $\mathcal{k}$ has h distinct isomorphisms into $\mathbb{R}$; or, if and only if the regular representation r of $\mathcal{k}$ over $\mathbb{Q}$ is diagonalizable over $\mathbb{R}$.

Assume for the rest of this section that $\mathcal{k}$ is totally real, $[\mathcal{k}:\mathbb{Q}] = h$, and r is the regular representation of $\mathcal{k}$ over $\mathbb{Q}$. Up to rational equivalence, we may restrict ourselves to representations ρ of the form qr, $q \in \mathbb{Z}^{+}$; i.e., to multiples of the regular representation. Let $V(\mathbb{Q})$ be the $\mathbb{Q}$-vector space for the representation ρ. Then $\dim V(\mathbb{Q}) = hq$. But $V(\mathbb{Q})$ can also be considered as a $\mathcal{k}$-vector space by defining

$$kv = \rho(k)v \qquad k \in \mathcal{k},\ v \in V(\mathbb{Q}).$$

As a $\mathcal{k}$-vector space we will denote $V(\mathbb{Q})$ by $V(\mathcal{k})$. Of course, the $\mathcal{k}$-dimension of $V(\mathcal{k})$ is q.

We will now solve the problem of determining all polarizations of $\mathcal{F}_2(V(\mathbb{Q}))$ that induce the positive involution on $\mathcal{k}$. For this argument, it will be convenient to adopt the following notation: For $A \in \mathrm{Alt}(V)$, let $\pi(A) \in \mathrm{Pol}(V)$ be the polarization of $\mathcal{F}_2(V)$ corresponding to A.

Let $A \in \mathrm{Alt}^{\times}(V(\mathcal{k}))$ and let $\pi = \pi(A) \in \mathrm{Pol}^{\times}(V(\mathcal{k}))$. Let $t : \mathcal{k} \longrightarrow \mathbb{Q}$ be the trace mapping and set $B = t \circ A$. Then $B \in \mathrm{Alt}^{\times}(V(\mathbb{Q}))$ and $\pi' = \pi(B) \in \mathrm{Pol}^{\times}(V(\mathbb{Q}))$. Let $x,y \in V(\mathcal{k})$ and let $a \in \mathcal{k}$, then

$$\rho^{*}(a)B(x,y) = B(x,ay) = t(A(x,ay)) = t(aA(x,y))$$

and

$$B\rho(a)(x,y) = B(ax,y) = t(A(ax,y)) = t(aA(x,y)).$$

Thus

$$B^{-1}\rho^{*}(a)B = \rho(a)\ .$$

Let $\mathcal{A}(\mathcal{k},\rho,\sigma)$, where σ is the identity mapping, denote all polarization $\pi(B)$ such that

$$B^{-1}\rho^{*}(a)B = \rho(\sigma(a)) = \rho(a)\ .$$

Then the image of $\mathrm{Pol}^{\times}(V(\mathcal{k}))$ under t in $\mathrm{Pol}^{\times}(V(\mathbb{Q}))$ is contained in $\mathcal{A}(\mathcal{k},\rho,\sigma)$.

We will prove that $t(\mathrm{Pol}^{\times}(V(k)) = \mathscr{A}(k,\rho,\sigma)$.

Suppose $\pi' \in \mathrm{Pol}^{\times}(V(\mathbb{Q}))$, $\pi' = \pi'(B)$, $B \in \mathrm{Alt}^{\times}(V(\mathbb{Q}))$, and that $\pi' \in \mathscr{A}(k,\rho,\sigma)$ or

$$B^{-1}\rho^{*}(a)B = \rho(a) .$$

The equation is equivalent to $B(ax,y) = B(x,ay)$, $a \in k$, $x,y \in V(k)$. Then the mappings

$$\theta_{ij}: a \longrightarrow B(ar_i,v_j) \qquad a \in k$$

are $\mathbb{Q}$-linear mappings of k to $\mathbb{Q}$, where $v_1,\dots,v_q$ define a basis of $V(k)$. Since the trace form is non-singular, there exists $\xi_{ij} = k$ such that

$$\theta_{ij}(a) = t(\xi_{ij}a) \qquad i \leq i,j \leq q .$$

Since B is alternating, $\xi_{ij} = -\xi_{ji}$. Let $x = \sum_1^q a_i v_i$ and $y = \sum_1^q b_j v_j$ where $a_i,b_j \in k$. Then

$$B(v,y) = t\Big(\sum_{i,j} \xi_{ij}(a_i b_j - a_j b_i)\Big) .$$

Let

$$A = \sum \xi_{ij}(a_i b_j - a_j b_i) .$$

Then, $A \in \mathrm{Alt}^{\times}(V(k))$ and $B = t \circ A$ and we have proven that $t(\mathrm{Pol}^{\times}(V(k)) = \mathscr{A}(k\rho,\sigma)$.

Thus the commutative diagram

$$\begin{array}{ccc} \mathscr{F}_2(V(k)) & \xrightarrow{\ \pi\ } & N_{2q+1}(k) \\ \uparrow p & & \downarrow t \\ \mathscr{F}_2(V(\mathbb{Q})) & \longrightarrow & N_{2qh+1}(\mathbb{Q}) \end{array}$$

enables us to identify $\mathscr{A}(k,\rho,\sigma)$ and $\mathrm{Pol}^{\times}(V(k))$.

Let us now consider the four groups that appear in the above diagram. $\mathscr{F}_2(V(\mathbb{Q}))$ and $N_{2qh+1}(\mathbb{Q})$ are rational algebraic groups, and therefore we may consider the group of real points of these algebraic groups, which we will denote by $\mathscr{F}_2(V(\mathbb{R}))$ and $N_{2qh+1}(\mathbb{R})$, respectively. In Section 1, we saw that $\mathscr{F}_2(V(k))$ and $N_{2q+1}(k)$ are the k points of rational algebraic groups; or, equivalently, are defined over $\mathbb{Q}$ Hence by reducing the field to $\mathbb{Q}$ we may form $\mathbb{Q}$-algebraic groups which are isomorphic to $\mathscr{F}_2(V(k))$ and $N_{2q+1}(k)$ and form the group of real points of these $\mathbb{Q}$-algebraic groups, which we will denote by $\mathscr{F}_2(V(k))_{\mathbb{R}}$ and $N_{2q+1}(k)_{\mathbb{R}}$ respectively. Let $\chi_i : k \longrightarrow \mathbb{R}$ $i=1,\dots,h$ be distinct isomorphisms of k into $\mathbb{R}$

We claim that

$$N_{2q+1}(\mathbb{k})_{\mathbb{R}} = \prod_{1}^{h} N_{2q+1}(\mathbb{R},\chi_i)$$

where $N_{2q+1}(\mathbb{R},\chi_i)$ is isomorphic to $N_{2q+1}(\mathbb{R})$, $i=1,....,k$.

The above assertion may be seen as follows:

$$N_{2q+1}(\mathbb{k}) = \begin{pmatrix} 1 & k_1 \ldots k_q & k \\ & \ddots \quad 0 & k_{q+1} \\ & \ddots & \vdots \\ & & k_{2q} \\ 0 & & 1 \end{pmatrix} \quad k_\alpha \epsilon \mathbb{k},\ k \epsilon \mathbb{k},\ \alpha = 1,\ldots,2q.$$

If we reduce the field to $\mathbb{Q}$ we obtain

$$r(\mathbb{Q})(N_{2q+1}(\mathbb{k})) = \begin{pmatrix} r(1) & r(k_1)\ldots r(k_q) & r(k) \\ & \ddots \quad r(0) & r(k_{q+1}) \\ & \ddots & \vdots \\ r(0) & & r(k_{2q}) \\ & & r(1) \end{pmatrix}$$

where the right side above is a matrix of h+h matrices. Now, as in Section 1, let $A \in GL(h,\mathbb{R})$ be such that

$$A^{-1}r(k)A = D(k) \qquad k \epsilon \mathbb{k}$$

where

$$D(k) = \begin{pmatrix} \chi_1(k) & & 0 \\ & \ddots & \\ 0 & & \chi_h(k) \end{pmatrix}$$

Let

$$(q+2)A = \begin{pmatrix} A & & 0 \\ & \ddots & \\ 0 & & A \end{pmatrix}$$

where the matrix on the right side is a $(q+2) \times (q \times 2)$ matrix of $h \times h$ matrices. Then

$$(q + 2)A^{-1}\, r(\mathbb{Q})(N_{2q+1}(\Bbbk))(q + 2)A \;=\; \begin{pmatrix} D(1) & D(k_1)\ldots D(k_q) & D(k) \\ & D(0) & \vdots \\ & & D(k_{2q}) \\ D(0) & & D(1) \end{pmatrix}$$

It is now easily seen that we may rearrange rows and columns of the above matrix to obtain

$$\oplus \sum_{i=1}^{h} \begin{pmatrix} \chi_i(1) & \chi_i(k_1)\ldots\chi_i(k_q) & \chi_i(k) \\ \ddots & & \chi_i(k_{q+1}) \\ & \chi_i(0) & \vdots \\ & & \chi_i(k_{2q}) \\ \chi_i(0) & & \chi_i(1) \end{pmatrix}$$

This proves our assertion and explains that the notation $N_{2q+1}(\mathbb{R},\chi_i)$ stands for the 2q+1 $\mathbb{R}$-Heisenberg with $\Bbbk$ imbedded in $\mathbb{R}$ by the isomorphism χ_i.

Now $t{:}N_{2q+1}(\Bbbk) \longrightarrow N_{2qh+1}(\mathbb{Q})$ uniquely extends to a homomorphism $T{:}\prod N_{2q+1}(\mathbb{R},\chi_i) \longrightarrow N_{2qh+1}(\mathbb{R})$. Further, T restricted to the center of $N_{2q+1}(\mathbb{R},\chi_i)$ is an isomorphism for each i. If $N_{2qh+1}(\mathbb{R})$ is oriented, we may use T to orient each $N_{2q+1}(\mathbb{R},\chi_i)$ by requiring that the induced isomorphism on each center is orientation preserving. We will then say that $\prod N_{2q+1}(\mathbb{R},\chi_i)$ is coherently oriented.

We are now going to find all complex structures J on $V(\mathbb{R})$ such that for $P \in \mathscr{A}(\Bbbk,\rho,\sigma)$, (J,P) is a Riemann pair and $\rho(k)\,J = J\rho(k)$, $k \in \Bbbk$.

Assume J satisfies the two conditions above. Since J is an automorphism of $\mathscr{F}_2(V\mathbb{R}))$ that commutes with $\rho(\Bbbk)$, it follows that J determines an automorphism of $\mathscr{F}_2(V(\Bbbk))_{\mathbb{R}}$ By hypothesis, J preserves the real closure of the kernel of $P = t \circ \pi \circ p$. Hence, J preserves the kernel of $\mathscr{F}_2(V(\Bbbk))_{\mathbb{R}} \longrightarrow N_{2q+1}(\Bbbk)_{\mathbb{R}}$ and so J induces an automorphism J_1 of $N_{2q+1}(\Bbbk)_{\mathbb{R}}$ Because $\Bbbk$ operates on each $N_{2q+1}(\mathbb{R},\chi_i)$ modulo its center as $\chi_i(\Bbbk)$ (i.e., as a diagonal matrix) and since χ_i, $i=1,\ldots,h$, are all distinct, it follows that J_1 must leave each $N_{2q+1}(\mathbb{R},\chi_i)$ invariant. If J is a positive definite CR structure relative to $N_{2qh+1}(\mathbb{R})$, it follows that $J_1 \mid N_{2q+1}(\mathbb{R},\chi_i)$, all i, is a positive definite CR structure. Hence, we have that if J satisfies our two conditions, then J may be written as $\prod^{h} J_i$ where each J_i is a positive definite CR structure for $N_{2q+1}(\mathbb{R},\chi_i)$, $i=1,\ldots,h$.

Now let J_i be a complex structure automorphism of $N_{2q+1}(\mathbb{R},\chi_i)$ that determines a positive definite CR structure on $N_{2q+1}(\mathbb{R},\chi_i)$ $i=1,\dots,h$. Let $J_1 = \prod^h J_i$. Then J_1 is an automorphism of $\prod^h N_{2q+1}(\mathbb{R},\chi_i)$ that acts trivially on the center of this group. Hence, $T \circ J_1$ is an automorphism of $N_{2hq+1}(\mathbb{R})$ that is a positive definite complex structure. Clearly, there exists an automorphism J of $\mathcal{F}_2(V(\mathbb{R}))$ that commutes with $\rho(\mathfrak{k})$ and lifts $T \circ J_1$.

Thus, if $\mathcal{Y}$ is the set of positive definite CR structures in $N_{2q+1}(\mathbb{R})$, we have that the set of all complex structures J on $V(\mathbb{R})$ such that for $P \in \mathcal{A}(\mathfrak{k},\rho,\sigma)$, (J,P) is a Riemann pair and $\rho(k)J = J\rho(k)$, $k \in \mathfrak{k}$ is of the form $\prod^h \mathcal{Y}$.

5. The Involution Problem for Division Algebras Of the First Kind (Part I)

Let $\mathscr{D}$ be a division algebra over $\mathbb{Q}$ that has a positive involution of the first kind. Then by the algebraic theory $\mathscr{D}$ may be described as follows: Let k be the center of $\mathscr{D}$. Then k is a totally real number field and we may assume $[k,\mathbb{Q}] = h$. If $\mathscr{D}$ does not equal k, then $\mathscr{D}$ is a quaternion division algebra $\mathbb{K} = \mathbb{K}(a,b)$ over k defined as a 4-dimensional k-vector space with k basis 1, i, j, k satisfying

$$i^2 = a \quad j^2 = b \quad k^2 = -ab$$

$$ij = -ji = k; \; jk = -kj = -bi; \; ki = -ik = -aj$$

where $a,b \in k$.

Let L be any subfield of $\mathbb{K}$ and consider $\mathbb{K}$ as a left L-vector space. The right regular representation of $\mathbb{K}$ over L is given by

$$r_L : \mathbb{K} \longrightarrow \mathrm{Hom}_L(\mathbb{K},\mathbb{K})$$

as L-vector spaces, where $r_L(\delta_1)\delta_2 = \delta_2\delta_1$, $\delta_1,\delta_2 \in \mathbb{K}$. Let

$$N_L(\delta) = \det r_L(\delta) .$$

In particular, we define the norm of δ, $N(\delta)$, $\delta \in \mathbb{K}$, by $N(\delta) = N_L(\delta)$ where L is the maximal commutative subfield $k(i)$, $i^2 = a \in k$. Explicitly, if $\delta = x + yi + zj + tk \in \mathbb{K}$, then $N(\delta) = x^2 - ay^2 - bz^2 + abt^2$. The algebraic theory tells us that because $\mathbb{K}$ is a division algebra $N(\delta) \neq 0$, unless $\delta = 0$. If we identify k with $k \cdot 1 \subset \mathbb{K}$, we have that $N : \mathbb{K}^\times \longrightarrow k^\times$ is a homomorphism, where $\mathbb{K}^\times$ is the multiplicative subgroup of $\mathbb{K}$. We next note that $N(\delta) = \delta\sigma(\delta)$, $\delta \neq 0$, implies that $\sigma(\delta) = \delta^{-1} N(\delta)$. Since $N(\delta)$ is central in $\mathbb{K}$, we have

$$\sigma(\delta_1\delta_2) = \delta_2^{-1}\delta_1^{-1} N(\delta_1\delta_2) = \sigma(\delta_2)\sigma(\delta_1) .$$

Explicitly, if $\delta = x + iy + jz + kt$, then

$$\sigma(\delta) = x - iy - jz - kt .$$

We will use r to denote the right regular representation of $\mathbb{K}$ over k. Similarly, we introduce the left regular representation

$$l:\mathbb{K} \longrightarrow \mathrm{Hom}(\mathbb{K},\mathbb{K})$$

by $l(\delta_1)\delta_2 = \delta_1\,\delta_2$, $\delta_1, \delta_2 \in \mathbb{K}$. Clearly,

$$r(\delta_1)\, l(\delta_2) = l(\delta_2)\, r(\delta_1) \quad \delta_1,\delta_2 \in \mathbb{K}$$

It is well known that $\mathrm{Hom}(\mathbb{K},\mathbb{K}) = r(\mathbb{K}) \otimes_{k} l(\mathbb{K})$.

Define $T \in \mathrm{Sym}^{\times}(\mathbb{K}) \subset \mathrm{Hom}(\mathbb{K},\mathbb{K}^{*})$ by

$$T(\delta)(\delta) = T(\delta,\delta) = N(\delta) = \sigma(\delta)\delta .$$

One verifies that $\sigma(\delta_1)\delta_2 + \sigma(\delta_2)\delta_1 = N(\delta_1 + \delta_2) - N(\delta_2) - N(\delta_1)$ and hence that $T(\delta_1,\delta_2) = \frac{1}{2}(\sigma(\delta_1)\delta_2 + \sigma(\delta_2)\delta_1) \in k$.

The following two formulas lie at the heart of all we will do in the rest of this section.

(1) $T^{-1}\, r^{*}(\delta)\, T = r(\sigma(\delta))$

(2) $T^{-1}\, l^{*}(\delta)\, T = l(\sigma(\delta))$.

We will now verify these formulas. Since

$$T(\delta\delta_1,\delta\delta_2) = N(\delta)T(\delta_1,\delta_2)$$

we have $l^{*}(\delta)T\, l(\delta) = N(\delta)T = TN(\delta)$ and formula (2) is verified.

Next, notice that

$$T(\sigma(\delta_1), \sigma(\delta_2)) = \sigma(T(\delta_1,\delta_2)) = T(\delta_1,\delta_2) .$$

Hence,

$$\begin{aligned} T(\delta_1\delta,\delta_2\delta) &= T(\sigma(\delta)\sigma(\delta_1), \sigma(\delta)\sigma(\delta_1)) \\ &= N(\sigma(\delta))\, T(\sigma(\delta_1), \sigma(\delta_2)) \\ &= N(\sigma(\delta))\, T(\delta_1,\delta_2) \end{aligned}$$

But $N(\sigma(\delta)) = \sigma(N(\delta)) = N(\delta)$. Hence,

$$T(\delta_1\delta,\delta_2\delta) = N(\delta)T(\delta_1,\delta_2)$$

which easily implies 1.

If $\mathscr{k}$ is a totally real field and $\chi_1,\dots,\chi_h$ are distinct isomorphisms of $\mathscr{k}$ into $\mathbb{R}$ we say that $k \in \mathscr{k}$ is *totally positive* if $\chi_i(K) > 0$, $i=1,\dots,h$. We now distinguish two subsets of quaternions. The *totally positive quaternions*, $\mathbb{K}^+$, are those $\mathbb{K}(a,b)$ where -a and -b are totally positive. The totally indefinite quaternions, $\mathbb{K}^-$, are those $\mathbb{K}(a,b)$ for which a,-b are totally positive.

For $\mathbb{K}^+$, σ is the only positive involution; but for $\mathbb{K}^-$, σ is not a positive involution. However, there exists $u \in \mathbb{K}^-$ such that

$$\tau(x) = u^{-1}\sigma(x)\,u \qquad x \in \mathbb{K}$$

is a positive involution. The complete set of positive involutions of $\mathbb{K}^-$ can be described as follows. If τ_0 is a fixed positive involution of $\mathbb{K}^-$, then every other positive involution of $\mathbb{K}^-$ can be written as

$$\tau(x) = w^{-1}\,\tau_0(x)w$$

where $w \in \mathbb{K}^-$ is such that

α) $\tau_0(w) = w$

β) The right regular representation of w has positive eigenvalues.

In this section we will solve the following problem: Given $(\mathbb{K},\tau)$ determine the set $\mathscr{A}(\mathbb{K},r,\tau)$ of $A \in \mathrm{Alt}^\times(\mathbb{K})$ such that

$$3) \quad r(\delta)^* A = A\, r(\tau(\delta)) \qquad \delta \in \mathbb{K}$$

We have to treat the two cases $\mathbb{K}^+$ and $\mathbb{K}^-$ separately and, since the solution for $\mathbb{K}^+$ is most easily motivated, we will begin with $(\mathbb{K}^+,\sigma)$.

Consider equation 1. It shows that $T \in \mathrm{Sym}^\times(\mathbb{K}^+)$ satisfies 3. But if $B \in \mathrm{Bil}(\mathbb{K}^+)$ satisfies 3 and L commutes with $r(\delta)$, $\delta \in \mathbb{K}^+$, i.e., if $L = \ell(\delta)$, $\delta \in \mathbb{K}^+$, then BL also satisfies 3 because

$$r^*(\delta)BL = B\, r(\sigma(\delta))L = BL\, r(\sigma(\delta)) .$$

Hence, if B satisfies 3, then $B\ell(\mathbb{K})$ satisfies 3. Now let $\mathbb{K}_0^+ = \mathscr{k}i + \mathscr{k}j + \mathscr{k}k$. Clearly, $\sigma(\delta) = -\delta$ if and only if $\delta \in \mathbb{K}_0$. It is a consequence of equation 2, that $T\ \ell(\delta) \in \mathrm{Alt}^\times(\mathbb{K})$ for δ

$\in \mathbb{K}_0$. To see this, note that

$$(T\,\ell(\delta))^{*} = \ell^{*}(\delta)T = T\,\ell(\sigma(\delta)) = -T\,\ell(\delta).$$

Thus equations 1 and 2 imply that $\mathscr{A}(\mathbb{K}^+,r,\sigma) \supset T\,\ell(\mathbb{K}_0^+)$.

It is now our purpose to prove that

$$\mathscr{A}(\mathbb{K}^+,r,\sigma) = T\,\ell(\mathbb{K}_0^+).$$

The crucial idea in this proof is the idea of using the action of $r(\mathbb{K}^+)$ on $\mathrm{Bil}(\mathbb{K}^+,\mathbb{K}^+)$. This is suggested by the observation that

$$\begin{aligned} r(\delta)^{*}T\,\ell(\delta_0)r(\delta) &= r^{*}(\delta)T\,r(\delta)\ell(\delta_0) \\ &= T\,r(\sigma(\delta))r(\delta)\ell(\delta_0) \\ &= N(\delta)T\,\ell(\delta_0) \end{aligned}$$

Hence the space $T\ell(\mathbb{K}_0^+)$ has the property that if $X \in T\ell(\mathbb{K}_0^+)$, then $r_1(\delta)X = N(\delta)X$. This suggests that

$$\mathrm{Alt}(\mathbb{K}^+, r,\sigma) = W_1 \cap \mathrm{Alt}^{\times}(\mathbb{K}^+)$$

where $X \in W_1$ if and only if $r_1(\delta)X = N(\delta)X$. This is indeed the case because if $X \in W_1$

$$r(\delta)^{*}X\,r(\delta) = Xr(\sigma(\delta)\delta) = X\,r(\sigma(\delta))r(\delta)$$

or

$$r(\delta)^{*}X = X\,r(\sigma(\delta)).$$

The converse is obvious.

Thus our task becomes to find W_1. (Notice, that W_1 is $\ell_1(\mathbb{K}^+)$ invariant because $\ell_1 r_1 = r_1\ell_1$.) We will now describe the action of $r_1(\mathbb{K})$ on $\mathrm{Alt}(\mathbb{K})$, *where* $\mathbb{K}$ *is a general quaternion division algebra over* k.

We have already observed that $\mathbb{K}_0$ is the set of δ such that $\sigma(\delta) = -\delta$. since

$$\sigma(\sigma(\delta)\mu\,\delta) = -\sigma(\delta)\mu\,\delta \qquad \mu \in \mathbb{K}_0, \delta \in \mathbb{K}$$

we have $\sigma(\delta)\mathbb{K}_0\delta \subset \mathbb{K}_0$. This implies that

$$r_1(\delta)(T\,r(\mathbb{K}_0)) \subset T\,r(\mathbb{K}_0).$$

Now

$$r_1(\delta)(T\,\ell(\mathbb{K}_0)) = T\,r(\sigma(\delta))\ell(\mathbb{K}_0)\,r(\delta)$$
$$= N(\delta)\,T\,\ell(\mathbb{K}_0)$$

or

$$r_1(\delta)(T\,\ell(\mathbb{K}_0)) \subset T\,\ell(\mathbb{K}_0)\,.$$

We have already verified that $T\,\ell(\mathbb{K}_0) \subset \mathrm{Alt}(\mathbb{K})$. To verify that $T\,r(\mathbb{K}_0) \subset \mathrm{Alt}(\mathbb{K})$, let $\mu \in \mathbb{K}_0$, then

$$T\,r(\sigma(\mu)) = r^*(\mu)T = (T\,r(\mu))^*$$

and

$$T\,r(\sigma(\mu)) = -T\,r(\mu).$$

Hence, $T\,r(\mathbb{K}_0) \subset \mathrm{Alt}(\mathbb{K})$.

Let $W_3 = T\,r(\mathbb{K}_0)$. Then $\dim W_1 = \dim W_3 = 3$ and W_3 is an $r_1(\mathbb{K})$ invariant irreducible subspace of $\mathrm{Alt}(\mathbb{K})$. Since $\dim \mathrm{Alt}\mathbb{K} = 6$, and $W_1 \cap W_3 = 0$, we have

$$\mathrm{Alt}(\mathbb{K}) = W_1 \oplus W_3$$

and *we can conclude that* $\mathcal{A}(\mathbb{K}^+, r, \sigma) = T\,\ell(\mathbb{K}_0^+) = W_1$.

Because we will need it later and it falls into place naturally here, we will now describe the action of $r_1(\mathbb{K})$ on $\mathrm{Sym}(\mathbb{K})$. Clearly, $k\,T \subset \mathrm{Sym}(\mathbb{K}^*)$ and

$$r_1(\delta)(a\,T) = N(\delta)(a\,T) \qquad a \in k, \delta \in \mathbb{K}^*.$$

Let $\alpha \in \{i,j,k\}$. We claim that equations 1 and 2 imply

$$T\,l(\alpha)r(\mathbb{K}_0) \subset \mathrm{Sym}(\mathbb{K})\,.$$

This is because, for $k_0 \in \mathbb{K}_0$

$$(T\,r(k_0)\ell(\alpha))^* = \ell^*(\alpha)\,r^*(k_0)T$$
$$= -\ell^*(\alpha)\,T\,r(k_0)$$
$$= T\,\ell(\alpha)\,r(k_0)$$

Now let $W(\alpha) = T\,\ell(\alpha)\,r(\mathbb{K}_0)$. It is straightforward to verify that $r_1(\mathbb{K})\,W(\alpha) = W(\alpha)$ and that $W(\alpha)$ is a 3-dimensional $r_1(\mathbb{K})$ invariant, irreducible subspace of $Sym(\mathbb{K})$. Since dim $Sym(\mathbb{K}) = 10$, we have easily that

$$Sym(\mathbb{K}) = \mathcal{L}T \oplus W(i) \oplus W(j) \oplus W(k).$$

We now turn to the study of $T_1 \in Sym(\mathbb{K}^-)$ that satisfy equations

$$1_1) \quad T_1^{-1}\, r^*(\delta) T_1 = r(\tau(\delta))$$

$$2_1) \quad T_1^{-1}\, \ell^*(\delta) T_1 = \ell(\tau(\delta)),$$

where τ is a positive involution of $\mathbb{K}^-$. It follows from our previous discussion that $\tau(\delta) = u^{-1}\,\sigma(\delta)u$ where $\sigma(u) = -u$. We claim that if T is a solution of equations 1 and 2 and $T \in Sym(\mathbb{K}^-)$, then

$$T_1 = T\,\ell(u)\,r(u)$$

satisfies equations 1_1 and 2_1 and $T_1 \in Sym(\mathbb{K}^-)$. We will now verify this assertion.

Noting that

$$\begin{aligned}(T\,\ell(u)\,r(u))^* &= r^*(u)\,\ell^*(u)T \\ &= r^*(u)\,T\,\ell(\sigma(u)) \\ &= -r^*(u)\,T\,\ell(u) \\ &= T\,r(u)\,\ell(u)\end{aligned}$$

we have that $T_1 \in Sym(\mathbb{K}^-)$.

We will now verify that T_1 satisfies equation 1, that T_1 satisfies equation 2, can be verified in a similar manner.

$$\begin{aligned}T_1^{-1}\, r^*(\delta)T_1 &= (T\,\ell(u)\,r(u))^{-1}\, r^*(\delta)(T\,\ell(u)\,r(u)) \\ &= r(u)^{-1}\,\ell(u)^{-1}\,T^{-1}\,r^*(\delta)\,T\,\ell(u)\,r(u)\,.\end{aligned}$$

But

$$T^{-1}\, r^*(\delta)T = r(\sigma(\delta))$$

and $\ell(\delta_1)r(\delta_2) = r(\delta_2)\ \ \ell(\delta_1)$, and so

$$T_1^{-1}\, r^*(\delta)T_1 = r(u)^{-1}\, r(\sigma(\delta))\, r(u)$$
$$= r(u^{-1}\sigma(\delta)u) = r(\tau(\delta))$$

and equation 1_1 has been verified.

We will now see that $A = T_1\, \ell(\delta_0^{-1}) \in \mathscr{A}(\mathbb{K}^-, r, \tau)$, where $\tau = \delta_0^{-1}\, \sigma\, \delta_0$. First

$$A = T\, r(\delta_0)$$

so $A^* = r(\delta_0)^*\ T = T\, r^*(\sigma(\delta_0)) = -A$.

We next show that if $B \in \mathscr{A}(\mathbb{K}^-, r, \tau)$, then $B = kA$, $k \in k^\times$. Since $A^{-1}\, r(\delta)A = B^{-1}\, r(\delta)B$, we have BA^{-1} commutes with $r(\delta)$ all $\delta \in \mathbb{K}^-$. Hence, $BA^{-1} = \ell(\delta)$, $\delta \in \mathbb{K}^-$. But $B^* = -B$ and so $(\ell(\delta)A)^* = -\ell(\delta)A$. But $(\ell(\delta)A)^* = A^*\ell^*(\delta) = -A\ell^*(\delta) = -\ell(\sigma(\delta))A$. Hence, $\ell(\sigma(\delta)) = \ell(\delta)$ or $\sigma(\delta) = \delta$ and $\delta \in k$. $\mathscr{A}(\mathbb{K}^- r, \tau) = T_1\, \ell(\mathbb{K}_0^-)$.

We will close the study of this section by examining how $\mathscr{A}(\mathbb{K}^-, r, \tau)$ behaves under the action $r_1(\delta)$, $\delta \in \mathbb{K}^-$. Let $A \in \mathscr{A}(\mathbb{K}^-, r, \tau)$ and consider $B = r^*(\delta)A\ \ r(\delta)$. We will now prove that $B^{-1}\, r^*(d)B = r(d^\#)$ where $d^\# \in \mathbb{K}^-$. Further, if $\tau'(d) = d^\#$, then τ' is a positive involution of $\mathbb{K}^-$. For notational convenience in the proof of these assertions, we let $r(\delta) = M$.

We begin by observing that

$$\begin{aligned} B^{-1}r^*(d)B &= M^{-1}A^{-1}(Mr(d)M^{-1})^*AM \\ &= M^{-1}\{A^{-1}r^*(\delta d\delta^{-1})A\}M \\ &= M^{-1}r(\sigma(\delta d\delta^{-1}))M \\ &= r(\delta^{-1}\sigma(\delta d\delta^{-1})\delta). \end{aligned}$$

Hence, $d^\# = \sigma(\delta d\delta^{-1})\delta$. It is easily verified that if $\tau'(d) = d^\#$ then τ' is an involution. It remains to verify that it is positive or, equivalently, trace $r(d^\# d) >> 0$, where $>> 0$ denotes totally positive. By hypothesis, $\mathrm{Trace}(A^{-1}r^*(\delta)Ar(\delta)) > > 0$.

Now

$$\begin{aligned} \text{trace } r(d^* d) &= \text{trace } (M^{-1}A^{-1}(M^{*-1}r^*(\delta)M^*)A(Mr(d))) \\ &= \text{trace } (A^{-1}(M^{*-1}r^*(d)M^*)A(Mr(d)M^{-1})) \end{aligned}$$

Letting $C^* = M^{*-1}r^*(\delta)M^*$ we see that

$$\text{trace } r(d^\# d) = \text{trace } (A^{-1}C^*AC)$$

But the latter is totally positive by hypothesis and we have proven our assertion.

totally positive. By hypothesis,

$$\mathrm{Trace}(A^{-1}r^{*}(\delta)Ar(\delta)) >> 0 .$$

Now

$$\begin{aligned}\mathrm{trace}\ r(d^{*}d) &= \mathrm{trace}\ (M^{-1}A^{-1}(M^{*-1}r^{*}(\delta)M^{*})A(Mr(d))) \\ &= \mathrm{trace}\ (A^{-1}(M^{*-1}r^{*}(d)M^{*})A(Mr(d)M^{-1}))\end{aligned}$$

Letting $C^{*} = M^{*-1}r^{*}(\delta)M^{*}$ we see that

$$\mathrm{trace}\ \ r(d^{\#}d) = \mathrm{trace}\ (A^{-1}C^{*}AC)$$

But the latter is totally positive by hypothesis and we have proven our assertion.

6. The Involution Problem for Division Algebras of the First Kind (Conclusion

Let $\mu:\mathbb{K} \longrightarrow \mathrm{End}(V(\mathbb{Q}))$ be a right representation of $\mathbb{K}$ as a $\mathbb{Q}$-algebra. Since k is central in $\mathbb{K}$, we may view $V(\mathbb{Q})$ as a k-vector space which we will denote by V and μ induces a representation $\rho:\mathbb{K} \longrightarrow \mathrm{End}(V)$. Identifying the simple right $\mathbb{K}$-module over k with $\mathbb{K}$ itself, we have

$$V = \oplus\sum_{s=1}^{p} \mathbb{K}_s \qquad \mathbb{K}_s = \mathbb{K} \text{ all } s.$$

Clearly, $\dim \mathbb{K} = 4$, $\dim V = 4p$, and $\dim V(\mathbb{Q}) = 4ph$. By our usual convention, we have a representation $\rho_1(\mathbb{K})$ on $\mathrm{Hom}(V,V^*)$. We will view V^* as $\oplus\sum\mathbb{K}_s^*$.

Let B_{rs} denote the space of elements of $\mathrm{Hom}(V,V^*)$ which satisfy $B_{rs}(\mathbb{K}_r) \subset \mathbb{K}_s^*$ and $B_{rs}(\mathbb{K}_t) = 0$, $r \neq t$. Then

$$\mathrm{Hom}(V,V^*) = \oplus\sum_{r,s} B_{rs} \ .$$

Since $\rho(\mathbb{K})(\mathbb{K}_r) \subset \mathbb{K}_r$ and $\rho^*(\mathbb{K})\mathbb{K}_r^* \subset \mathbb{K}_r^*$, we have

$$\rho_1(\mathbb{K})(B_{rs}) \subset B_{rs} \ .$$

Clearly, we may identify B_{rs} with $\mathrm{Hom}(\mathbb{K}_r, \mathbb{K}_s^*)$ and, hence suppressing the indices, we have

$$B_{rs} \simeq \mathrm{Hom}(\mathbb{K}, \mathbb{K}^*)$$

and, with respect to this identification,

$$\rho_1(\mathbb{K}) \mid B_{rs} = r_1(\mathbb{K})$$

Clearly, $\mathrm{Alt} = \oplus\sum_{r,s} A_{rs}$, where $A_{rr} = \mathrm{Alt}(\mathbb{K}_r) \simeq \mathrm{Alt}(\mathbb{K})$ and A_{rs}, $r \neq s$, may be described as follows: As above,

$$B_{rs} \oplus B_{sr} = \mathrm{Hom}(\mathbb{K}, \mathbb{K}^*) \oplus \mathrm{Hom}(\mathbb{K},\mathbb{K}^*),\ r \neq s.$$

Then $A_{rs} = \{B-B^* \mid B \in B_{rs}\} \simeq \mathrm{Hom}(\mathbb{K},\mathbb{K}^*)$ and $\rho_1(\mathbb{K}) \mid A_{rs} = r_1(\mathbb{K})$. Thus

$$A_{rr} = W_1(r) \oplus W_3(r)$$

where $W_\alpha(r)$ corresponds to W_α, $\alpha = 1,3$, under identification $A_{rr} \simeq \mathrm{Alt}(\mathbb{K})$.

Further

$$A_{rs} = W_1(r,s) \oplus W_3(r,s)$$

where $W_\alpha(r,s)$ corresponds to W_α, $\alpha = 1,3$, where $A_{rs} \simeq \mathrm{Hom}(\mathbb{K},\mathbb{K}^*)$. It follows that Alt(V) $= W_\alpha(\rho) \oplus W_3(\rho)$ where

$$W_\alpha(\rho) = \oplus\sum_r W_\alpha(r) \oplus \sum_{r<s} W_\alpha(r,s) \qquad \alpha = 1,3 .$$

Clearly $W_1(\rho)$ is a $\rho_1(\mathbb{K})$ invariant subspace satisfying

$$\rho_1(\delta) \mid W_1(\rho) = N(\delta)I \qquad \delta \in \mathbb{K}$$

and $W_3(\rho)$ is a direct sum of 3-dim. $\rho_1(\mathbb{K})$ invariant irreducible subspaces. It is easy to verify that

$$\dim W_1(\rho) = \frac{4(p^2-p)}{2} + 3p$$

$$\dim W_3(\rho) = \frac{12(p^2-p)}{2} + 3p$$

A similar argument gives $\mathrm{Sym}(V) = W_1^{\#}(\rho) \oplus W_3^{\#}(\rho)$ where $W_1^{\#}(\rho)$ satisfies $\rho_1(\delta) \mid W_1^{\#}(\rho) = N(\delta)I$ and $W_3^{\#}(\rho)$ is the direct sum of 3 dimensional invariant irreducible $\rho_1(\mathbb{K})$ subspaces. It will be important for our later discussion to note that

$$\dim W_1^{\#}(\rho) = \frac{4(p^2-p)}{2} + p$$

As usual, we view $V \wedge V$ as the dual space to Alt(V). Then $V \wedge V = w_1 \oplus w_3$ where w_α is the annihilator of W_α, $\alpha = 1, 3$. Clearly, $\rho_1(\delta)X = N(\delta)X$, $X \in w_1$ and w_3 is the direct sum of 3-dimensional $\rho_1(\mathbb{K})$ invariant, irreducible subspaces.

We are now in a position to solve the involution problem for the quaternion algebra $\mathbb{K}(a,b)$. Recall that when $\mathbb{K} = \mathbb{K}^+$, a totally positive quaternion division algebra, we have $\mathscr{A}(\mathbb{K}^+,\rho,\sigma)$, σ, the positive involution of $\mathbb{K}^+$, consists of the $A \in \mathrm{Alt}(V)$ such that

$$A^{-1}\rho^*(\delta)A = \rho(\sigma(\delta))$$

or

$$\rho^*(\delta)A = A\rho(\sigma(\delta)).$$

Since $N(\delta) = \sigma(\delta)\delta$, we have

$$\rho^*(\delta)\, A\rho(\delta) = A\rho(\sigma(\delta))\, \rho(\delta) = AN(\delta) = N(\delta)A$$

Hence $A \in \mathscr{A}(\mathbb{K}^+, \mu_1 \sigma)$ if and only if $t \circ A$; where $A \in W_1(\rho)$, where t is the trace map of k.

Thus

$$\begin{array}{ccccc}
\mathscr{F}_2(V(\mathbb{Q})) & & \longrightarrow & & N_{4ph+1}(\mathbb{Q}) \\
\downarrow i & & & & \uparrow t \\
\mathscr{F}_2(V) & \longrightarrow & \mathscr{F}_2(V)/w_3 \simeq N(\mathbb{K}^+, \rho, 1) & \xrightarrow{\Pi} & N_{4p+1}(k)
\end{array}$$

In terms of matrices, we have $A \in \mathscr{A}(\mathbb{K}^+, \rho, \sigma)$ if and only if

$$A = \begin{pmatrix} T & & 0 \\ & \ddots & \\ 0 & & T \end{pmatrix} \left(\ell(\delta_{ij}) \right)$$

where $\delta_{ij} = -\sigma(\delta_{ji})$ $1 \leq i,j \leq p$

Now let $\mathbb{K}^-$ be a totally indefinite quaternion division algebra. Fix a positive involution τ of $\mathbb{K}^-$, which, by our previous discussion, is given by

$$\tau(\delta) = \delta_0^{-1}\, \sigma(\delta)\delta_0 \qquad \sigma(\delta_0) = -\,\delta_0 \in \mathbb{K}^-.$$

Let $\mathscr{A}(\tau) = \mathscr{A}(\mathbb{K}^-, \rho, \tau)$. Fix $\delta_1 \in \mathbb{K}^-$, $\delta_1 \neq 0$ consider $\rho_1(\delta_1)\, \mathscr{A}(\tau)$. By our previous discussion, if $u = \tau(\delta_1)\delta_1$, then $\tau(u) = u$, the mapping $\delta \longrightarrow u^{-1}\tau(\delta)u$ is a positive involution, and $\rho_1(\delta_1)\, \mathscr{A}(\tau) = \mathscr{A}(u^- \tau u)$.

We will now describe $\mathscr{A}(\tau)$. Let $W_1{}^{\#}\rho \subset \mathrm{Sym}(V)$ be as above. Let $S \in W_1{}^{\#}$ and let $A = S\rho(\delta_0)$. Then, since $\sigma(\delta_0) = -\,\delta_0$, we have $A^* = -A$ and

$$\rho^*(\delta)A = A\, \rho(\tau(\delta)).$$

Hence $W_1{}^{\#}\rho(\delta_0) \subset \mathscr{A}(\tau)$ and the converse if easily seen. We have therefore verified that $W_1{}^{\#}\rho(\delta_0) = \mathscr{A}(\tau)$. Let $w(\tau)$ be the annihilator of $\mathscr{A}(\tau)$, we have proven that

$$\mathscr{F}_2(V(\mathbb{Q}) \longrightarrow \mathscr{F}_2(V) \longrightarrow \mathscr{F}_2(V)/w(\tau) \xrightarrow{\Pi} N_{4p+1}(k) \xrightarrow{t} N_{4ph+1}(\mathbb{Q})$$

defines a monomorphism $\mathrm{Pol}(\mathscr{F}_2(V)/w(\tau))$ into $\mathrm{Pol}(V(\mathbb{Q}))$ whose image is $\mathscr{A}(\mathbb{K}^-, \mu, \tau)$.

We can refine this picture slightly by noting that if $N(\mathbb{K}^-,\rho,3) = \mathcal{F}_2(V)/W_3(\rho)$ then

$$\begin{array}{ccc} \mathcal{F}_2(V) & \longrightarrow & N(\mathbb{K}^-,\rho,3) \\ & \searrow & \downarrow \\ & & \mathcal{F}_2(V)/w(\tau) \end{array}$$

is a commutative diagram for all τ. This is because $W_3(\rho)$ is the smallest $\rho_1(\mathbb{K}^-)$ invariant subspace containing $W(\tau)$ for all τ.

Again, in terms of matrices, if $T(\tau) = T\ell(\delta_0)\, r(\delta_0)$. Then $A \in \mathcal{A}(\mathbb{K}^-,\rho,\tau)$ if and only if

$$A = \begin{pmatrix} T(\tau) & & 0 \\ & \ddots & \\ 0 & & T(\tau) \end{pmatrix} \begin{pmatrix} \ell(\delta_{ij}) \end{pmatrix}$$

where $\delta_{ij} = -\tau(\delta_{ji})\quad 1 \le i,j \le p$.

7. Existence of Riemann Matrices for Division Algebras of the First Kind

In Section 6, we obtained a description of $\mathcal{A}(\mathbb{K},\mu,\tau)$ where $\mathbb{K}$ is a quaternion division algebra over a totally real field k with positive involution τ and rational representation μ. For $A \in \mathcal{A}^{\times}(\mathbb{K},\mu,\tau)$, we need to describe the set $R(\mathbb{K},\mu,A)$ of all Riemann matrices J such that (J,A) is a Riemann pair and $\mu(\delta)J = J\mu(\delta)$, $\delta \in \mathbb{K}$ We will begin by showing how this problem can be reduced to finding for each $A \in \mathcal{A}^{\times}(\mathbb{K},\mu,\tau)$ one element of $R(\mathbb{K},\mu,A)$.

The first step in this reduction is to relate our current problem to the results in Section 4. By Section 4, since the Riemann matrix J that we are seeking commutes $\mathbb{K}$, it certainly commutes with k, and so we may structure J as follows: Let $\chi_1,\dots,\chi_h$ be the $h = [k:\mathbb{Q}]$ distinct isomorphisms of k into $\mathbb{R}$ and let $k(i) = \chi_i(k)$. Define $\mathbb{K}(\mathbb{R},i) = \mathbb{K}\otimes_{k(i)}\mathbb{R}$. Then, using the notation of Section 4, we have a representation ρ_i of $\mathbb{K}(\mathbb{R},i)$ on $\mathcal{F}_2(V(\mathbb{R}),i)$. Thus each $A \in \mathcal{A}^{\times}(\mathbb{K},\rho,\tau)$ determines A_i as in Section 4. Thus our problem may be formulated as follows. For $i=1,\dots,h$, find the set $\mathcal{J}_i(A)$ of all complex structures J_i on $\mathcal{F}_2(V(\mathbb{R}),i)$ such that (J_i,A_i) is a Riemann pair with the property that

$$\rho_i(\delta)\, J_i = J_i\rho_i(\delta) \qquad \delta \in \mathbb{K}(\mathbb{R},i)$$

then $R(\mathbb{K},\mu,A) = \prod_1^h \mathcal{J}_i(A)$.

Let G_i be the group in non-singular matrices commuting with ρ_i and let $G_i = C_i \cap \mathrm{Aut}\, N(A_i)$, where $\mathrm{Aut}\, N(A_i)$ denotes the automorphism group of the real Heisenberg group $N(A_i)$. Then $\mathcal{J}_i(A) \subset G_i$. Let $J_0 \in \mathcal{J}_i(A)$ and let $S \in G_i$ be such that

$$J_0^{-1}\, S\, J_0 = S^{-1}\,.$$

Since $J^2 = -I$, where I is the identity of G_i, we have, $(J^{-1}S)^2 = -J^{-1}S\, JS = -E$. Hence $J^{-1}S = R$ is a complex structure. Conversely, if $R \in G_i$ and $R^2 = -I$, the $JR = S$ is certainly in G_i. Since

$$\begin{aligned} R^2 = \;& J^{-1}S\, J^{-1}S = && -E \qquad \text{we have} \\ & J^{-1}S\, J^{-1} = && -S^{-1} \\ \text{or} \qquad & J^{-1}S\, J = S^{-1} && \end{aligned}$$

Let $G_i^{\#} \subset G_i$ be the subgroup generated by $g \in G_i$ such that $J_0^{-1}\, g\, J_0 = g$ or $J^{-1}g\, J_0 = g^{-1}$ and let G_i^0 be the subgroup of elements such that $J_0^{-1}g\, J_0 = g$. There is a natural mapping of $G_i^{\#}/G_i^0$ into the S such that $J_0\, S = R$ is a Riemann matrix. The subset of

$G_i^{\#}/G_i^{0}$ for which positivity holds is precisely $\mathcal{J}_i(A)$. This discussion suggests that Cartan symmetric spaces enter into the description of $\mathcal{J}_i(A)$. This is true and this is where the Siegel moduli enters into the picture. For a full discussion of this see [4].

We will now show how to produce $J_0 \in \mathcal{J}_i(A)$ where $A \in \mathcal{A}^{\times}(\mathbb{K}, \rho, \tau)$ and $A_i \in \mathcal{A}^{\times}(\mathbb{K} \oplus_{k(i)} \mathbb{R}, \rho_i, \tau)$ and $i=1,\ldots,h$. We will begin by looking at two special cases of our problem.

We will first produce for $A \in \mathcal{A}(\mathbb{K}^+, r, \sigma)$, and r the right regular representation, the desired J_0.

Let $T \in \mathrm{Sym}(\mathbb{K}^+)$ satisfy equations 1 and 2. Then we know we may write $A = T\ell(\delta_0)$, $\delta_0 \in \mathbb{K}^+$, $\sigma(\delta_0) = -\delta_0$. Clearly $T^{-1}A = \ell(\delta_0)$ is a linear transformation commuting with $r(\delta)$. To see that $\ell(\delta_0)$ defines an automorphism of N(A), we simply note that

$$\begin{aligned} \ell^*(\delta_0)A\ell(\delta_0) &= \ell^*(\delta_0)T\ell(\delta_0)\ell(\delta_0) \\ &= N(\delta_0)\, T\ell(\delta_0)\ . \end{aligned}$$

Of course $(\ell(\delta_0))^2$ may not be $-I$. Since k is totally real and σ is positive, for each i, $\chi_i(N(\delta_0)) > 0$. Hence in $\mathbb{K}^+(\mathbb{R},i)$, $\dfrac{\delta_0}{\sqrt{N(\delta_0)}}$ exists, and its square is $-I$. And so we have solved our problem in that special case.

Now let $A \in \mathcal{A}(\mathbb{K}^-, r, \tau)$, where $\tau(\delta) = \delta_0^{-1}\sigma(\delta)\,\delta_0$, and $\sigma(\delta_0) = -\delta_0$.

Let

$$A \in \mathcal{A}(\mathbb{K}^-, r, z)$$

where $\tau(\delta) = \delta_0^{-1}\sigma(\delta)\delta_0$ and $\sigma(\delta_0) = -\delta_0$. Then, let

$$S = T\ell(\delta_0)\, r(\delta_0) \text{ and } A = Tr(\delta_0).$$

Consider $A^{-1}S = \ell(\delta_0)$. Clearly, $\ell(\delta_0)$ commutes with $r(\delta_2)$, $\delta \in \mathbb{K}^1$. Again, $\ell(\delta_0)^2$ may not be -I, but we may complete the argument exactly as in the $\mathbb{K}^+$ case, above.

We will now produce a canonical form theorem that will reduce the general problem to the two special cases we have just verified.

Consider the matrix

$$C = (\ell(\delta_{ij})) \qquad 1 \leq i,j \leq k \qquad \delta_{ij} \in \mathbb{K}$$

where $\mathbb{K}$ is a division algebra with involution τ. Define

$$C^{\#} = (\ell(\tau(\delta_{ji}))) .$$

Since $(C^{\#})^{\#} = C$ and $(CD)^{\#} = D^{\#}C^{\#}$, we have if $A^{\#} = -A$ the $(C^{\#} AC)^{\#} = -C^{\#} AC$. We will now prove the following result. Given A such that $A^{\#} = -A$. Then there exists C such that

$$C^{\#}AC = \begin{pmatrix} \ell(\delta_1) & & 0 \\ & \ell(\delta_s) & \\ 0 & & K \end{pmatrix} \qquad \tau(\delta_i) = -\delta_i$$

where

$$K = \begin{pmatrix} \begin{pmatrix} 0 & \ell(1) \\ -\ell(1) & 0 \end{pmatrix} & 0 \\ 0 & \begin{pmatrix} 0 & \ell(1) \\ -\ell(1) & 0 \end{pmatrix} \end{pmatrix}$$

To prove this, begin by noting that if $a,b,c, \in \mathbb{K}$ are such that $\tau(a) = -a \neq 0$ and $\tau(c) = -c$, then

$$\begin{pmatrix} 1 & 0 \\ -\ell(\tau(\frac{b}{a})) & 1 \end{pmatrix} \begin{pmatrix} \ell(a) & \ell(b) \\ -\ell(\tau(b)) & \ell(c) \end{pmatrix} \begin{pmatrix} 1 & -\ell(\frac{b}{a}) \\ 0 & 1 \end{pmatrix} = \begin{pmatrix} \ell(a) & 0 \\ 0 & \ell(d) \end{pmatrix}$$

By repeated applications of this fact we may assume there exists C_1 such that

$$C^{\#}AC = \begin{pmatrix} \begin{pmatrix} \ell(\delta_1) & 0 \\ 0 & \ddots \; \ell(\delta_r) \end{pmatrix} & 0 \\ 0 & B_1 \end{pmatrix}$$

where $B_1^{\#} = -B_1$ and

$$B_1 = \begin{pmatrix} 0 & & * \\ & \ddots & \\ * & & 0 \end{pmatrix}$$

Now there exists C_2 in the entries $\ell(1)$ and $\ell(0)$ such that

$$C_2{}^{\#}B_1C_2 = \begin{pmatrix} \begin{pmatrix} 0 & \ell(\delta_1) \\ -\ell(\tau(\delta_1)) & 0 \end{pmatrix} & B_3 \\ B_4 & B_2 \end{pmatrix}$$

and there clearly exists C_3 such that

$$C_3{}^{\#}C_2{}^{\#}B_1C_2C_3 = \begin{pmatrix} \begin{pmatrix} 0 & \ell(1) \\ -\ell(1) & 0 \end{pmatrix} & B'_3 \\ B'_4 & B_2 \end{pmatrix}$$

If B'_3 or B'_4 are not both the zero matrix, we can rearrange the matrix such that the upper left hand corner looks like

$$D_1 = \begin{pmatrix} 0 & \ell(1) & \ell(\delta_1) \\ -\ell(1) & 0 & \ell(\delta_2) \\ -\ell(\tau(\delta_1)) & -\ell(\tau(\delta_2)) & 0 \end{pmatrix}$$

If $\delta_1 \neq 0$ then if $a = -\frac{\delta_2}{\delta_1}$ we have

$$\begin{pmatrix} \ell(1) & \ell(0) & \ell(0) \\ \ell(\tau(a)) & \ell(1) & \ell(0) \\ \ell(0) & \ell(0) & \ell(1) \end{pmatrix} D_1 \begin{pmatrix} \ell(1) & \ell(a) & \ell(0) \\ \ell(0) & \ell(1) & \ell(0) \\ \ell(0) & \ell(0) & \ell(1) \end{pmatrix}$$

$$= \begin{pmatrix} \ell(0) & \ell(1 & \ell(\delta_1) \\ -\ell(1) & \ell(\tau(a)-a) & \ell(0) \\ \ell(\tau(\delta_1)) & \ell(0) & \ell(0) \end{pmatrix}$$

If $\tau(a)-a \neq 0$, we may return to the above argument to reduce B_2 by 1 row and column and so be able to apply an induction to finish the argument. If $\tau(a)-a = 0$ we have reduced to the case

$$D_2 = \begin{pmatrix} \ell(0) & \ell(1) & \ell(\delta_1) \\ -\ell(1) & \ell(0) & \ell(0) \\ \ell(\tau(\delta_1)) & \ell(0) & \ell(0) \end{pmatrix}$$

and we have easily a matrix F such that

$$F^{\#}D_2F = \begin{pmatrix} 0 & \ell(1) & 0 \\ -\ell(1) & 0 & 0 \\ 0 & 0 & 0 \end{pmatrix}$$

A repeated use of the above argument proves our canonical form theorem.

Clearly if

$$B = \begin{pmatrix} \ell(\delta_1) & 0 & 0 \\ & \ell(\delta_k) & \\ 0 & 0 & K \end{pmatrix} \qquad \tau(\delta_i) = -\delta_i$$

Then there exists $\xi_1,\ldots,\xi_k \in \mathbb{R}$ such that if

$$B_1 = \begin{pmatrix} \ell(\chi_i(\delta_1)) & & \ell(0) \\ \xi_1 & \ell(\chi_i(\delta_k)) & \\ \ell(0) & \xi_k & K \end{pmatrix}$$

is an automorphism of N(B) for $\mathbb{K} = \mathbb{K}^+$ or $\mathbb{K}^-$. Since $C^{\#}AC = B$, $J_1 = (C^{\#-1})^{\#} B_1 C^{-1}$ is a Riemann matrix such that (J_1, A) is a Riemann pair in G. This proves the existence of a Riemann matrix.

REFERENCES

[1] A. A. Albert, Structure of Algebras, American Math. Society, 1939.

[2] L. Auslander, An exposition of the structure of solvmanifolds, Part I, Algebraic Theory. Bull. A.M.S. 79 (1973) 227-261.

[3] L. Auslander, Lectures on Nil-Theta Functions, C.B.M.S. Regional Conference Series, American Math. Soc. No. 34.

[4] L. Auslander and R. Tolimieri, A matrix-free treatment of the problem of Riemann matrices, to appear.

[5] L. Auslander and R. Tolimieri, Abelian harmonic analysis, Theta

[6] C. L. Siegel, Lectures on Riemann Matrices, Tata Institute, Bombay (1963).

[7] H. P. F. Swinnerton-Dyer, Analytic Theory of Abelian Varieties, London Math. Soc. (1974).

[8] R. Tolimieri, Heisenberg manifolds and Theta functions, Trans. A.M.S., 239 (1978) 293-319.

[9] A. Weil, Introduction a l'etude des varietes, Kahleriennes, Paris (1958).

[10] A. Weil, Sur certaines groupes d'operateurs unitaires, Acta Math., 111 (1964) 143-211.

CENTRO INTERNAZIONALE MATEMATICO ESTIVO

(C.I.M.E.)

UNITARY AND UNIFORMLY BOUNDED REPRESENTATIONS OF SOME SIMPLE LIE GROUPS

MICHAEL COWLING

0. INTRODUCTION

We denote by F the real or complex numbers (R or C) or the quaternions (Q). We consider R as a subfield of C, and C as a subfield of Q. If $z \in F$, then we may write

$$z = s + t\underline{i} + u\underline{j} + v\underline{k} ,$$

with s, t, u, and v in R. The conjugate $\bar{z}$ is now described thus:

$$\bar{z} = s - t\underline{i} - u\underline{j} - v\underline{k} .$$

Note that $z\bar{z} = \bar{z}z = |z|^2$. The real and imaginary parts of z in F are given by the formulae:

$$2\mathrm{Re}(z) = z + \bar{z} , \qquad 2\mathrm{Im}(z) = z - \bar{z} .$$

This is *not* the usual imaginary part in the complex case.

We shall consider the vector space F^{n+1}; in the quaternionic case, the scalars act on the left. We choose a basis $(e_0, e_1, \ldots , e_n)$ for F^{n+1} over F, and consider the sesquilinear form q on F^{n+1}, given by the formula:

$$q(z, \zeta) = (\textstyle\sum_{j=1}^{n} z_j \bar{\zeta}_j) - z_0 \bar{\zeta}_0 \qquad z, \zeta \in F^{n+1} .$$

We shall be interested in the group O(q) of all linear transformations of F^{n+1} which preserve this form. One of our principal aims here is to describe some aspects of the harmonic analysis of O(q). Along the way, we shall meet other groups, on which we shall describe some aspects of harmonic analysis, some of which will be directly pertinent to our study of O(q) and some of which will be complementary. These groups are compact, abelian, and nilpotent. Harmonic analysis on compact groups has been under intensive investigation during the last half-century, and on abelian groups for much longer, but harmonic analysis on nilpotent groups is less well known. We shall therefore dedicate the first three sections to this.

The main thrust of our development of harmonic analysis on nilpotent Lie groups owes much to E.M. Stein. In particular, much of what we present is already contained in work of R.A. Kunze and E.M. Stein ([KS1], [KS2], [KS3]), of A.W. Knapp and E.M. Stein ([KnS]), and of G.B. Folland and E.M. Stein ([FoS]).

We have simplified and even improved the methods of the above authors, but who soever is at home with the above works will find no real surprises.

The last five sections are dedicated to some semisimple groups of rank one. In the fourth section, we discuss some general properties of these groups, and indicate how they are related to certain geometric objects, which appear in various contexts and guises. This section may be considered as a study of some aspects of the geometry of semisimple Lie groups, together with a few easy consequences of a measure-theoretic nature.

In the fifth section, we consider certain unitary representations of these groups and their analytic continuations. In fact, we construct analytic families of representations $\pi_{\mu,\zeta}$ which, if the complex parameter ζ lies in the tube (strip) T —

$$T = \{\zeta \in C: |Re(\zeta)| < r\} \text{ —}$$

act isometrically in certain L^p-spaces, where $1/p = \zeta/2r + 1/2$. We also describe the intertwining operators $A(w, \mu, \zeta)$ which express the equivalence between $\pi_{\mu,\zeta}$ and $\pi_{\mu,-\zeta}$.

The sixth section contains some results on the L^1- and L^p-harmonic analysis of the group O(q), often abbreviated to G. Here the questions treated are about the Fourier transforms of L^p-functions and of certain convolution algebras on G.

In the penultimate section, we study the question of uniformly bounded representations, and touch on the problem of complementary series. Both these questions arise when one attempts to develop a calculus of analysis and synthesis of representations which would extend the direct integral theory for unitary representations to include the bounded Banach representations of G.

The last section is dedicated to the study of an extrapolation principle. We show that, metamathematically, the finite-dimensional representations of G determine its harmonic analysis.

We conclude this introduction with a disclaimer. This is not an attempt to provide an unbiased version of the representation theory of semisimple Lie groups. Rather, we have some hopes of offering the "commutative harmonic analyst" a vision of some aspects of the "noncommutative theory". Consequently, the references at the end are only those works cited in the text.

Horror of horrors, the labours of Harish-Chandra, without which this work would never have been written, have not been mentioned. It seems worthwhile to add G. Warner's tomes [War] to the list so that the reader may obtain a less slanted view of the literature and the subject.

It is a pleasure to thank Professor Alessandro Figà-Talamanca for his invitation to present this material at C.I.M.E., and to thank M. Cristina for her understanding during the preparation and writing up of it all.

1. ANALYSIS ON HEISENBERG GROUPS. I

We shall work on a group V, whose elements are of the form (x,y), where $x \in F^{n-1}$ and $y \in \mathrm{Im}(F)$. The product of (x',y') and (x,y) in V is given by the rule

$$(x',y')(x,y) = (x'+x,\ y'+y - 2\mathrm{Im}(x'x^*))\ ,$$

where we consider x as a row vector, and x^* is thus the conjugate of the corresponding column vector. In particular, we note that (0,0) is the identity of V, and that (-x,-y) is the inverse of (x,y) in V. If $F = R$, then $\mathrm{Im}(F)$ is trivial, and the group V is just the Euclidean space R^{n-1}; otherwise V is a two-step nilpotent group.

There are two important groups of automorphisms of V, which we shall now denote by M and A. An element of M is written m, or m(u,v), where $v \in F$ and $|v| = 1$, and u is an $(n-1)\times(n-1)$ matrix over F such that $xu = x$ for all x in F^{n-1}. We make the following definitions:

$$(1.1) \qquad R^m(x,y) = (v^{-1}xu,\ v^{-1}yv)$$

$$(1.2) \qquad D^s(x,y) = (sx,\ s^2y) \qquad s \in R^+ .$$

Then M "is" the group of rotations R^m, and A "is" the group of dilations D^s. The significance of these groups will become clear in §4. We also define a "norm" on V, which is suggested by the geometry of the situation (see Lemma 4.2):

$$N(x,y) = (|xx - y|^2)^{1/4}$$
$$= (|x|^4 + |y|^2)^{1/4} .$$

When $F = R$, this is the usual Euclidean norm, and in all cases, if $(x,y) \in V$,

$$N(D^s(x,y)) = sN(x,y) \qquad s \in R^+,$$
$$N(R^m(x,y)) = N(x,y) \qquad m \in M .$$

It is necessary to discuss certain singular integral operators on V. We shall here develop and extend work of A.W. Knapp and E.M. Stein [KnS], of A. Korányi and S. Vági [KoV], of G.B. Folland and E.M. Stein [FoS], and of many others.

The principal object of study will be certain families of homogeneous distributions. A function f on V is called homogeneous of degree d ($d \in C$) if

$$f \circ D^s = s^d f \qquad s \in R^+ ;$$

it is clear that, if we ignore the point (0,0), then any such function may be written in the form

$$f(x,y) = \Omega(x,y)N(x,y)^d \qquad (x,y) \in V,$$

where Ω is homogeneous of degree zero. We shall deal with families of homogeneous distributions K_ζ, where, for ζ in C,

$$K_\zeta(x,y) = \Omega(x,y)N(x,y)^{\zeta - r} \qquad (x,y) \in V,$$

and r, called the homogeneous dimension of V, is given by the rule $r = p + 2q$, where p and q are the dimensions of F^{n-1} and $\mathrm{Im}(F)$ as real vector spaces. To avoid some finicky provisos, we assume that $r \geq 3$ in what follows.

The first result we need about such distributions is of a technical nature, but is very important for our study. We write Σ for the unit sphere (ellipsoid) in V:

$$\Sigma = \{(x,y) \in V: N(x,y) = 1\} .$$

There is a unique smooth measure on Σ, $d\sigma$ say, such that for any f in $C_c(V)$,

$$\int_V dxdy\ f(x,y) = \int_\Sigma d\sigma \int_{R^+} ds\ s^{r-1}\ f(D^s\sigma) .$$

We abuse notation, and consider Ω indifferently as a function on Σ and as a function on V, homogeneous of degree 0.

LEMMA 1.1. Suppose that Ω is in $L^1(\Sigma)$. Then, if $\mathrm{Re}(\zeta) > 0$, $\Omega N^{\zeta - r}$ is locally integrable and defines a distribution on $C_c^\infty(V)$ by integration. The $C_c^\infty(V)'$-valued function $\zeta \mapsto N^{\zeta - r}$ is holomorphic in $\{\zeta \in C : \mathrm{Re}(\zeta) > 0\}$, and extends meromorphically to a $C_c^\infty(V)'$-valued function whose only possible poles are simple poles at the points $0, -1, -2, \ldots$.

Proof. We omit the easy calculation that, if $\mathrm{Re}(\zeta) > 0$, then $N^{\zeta - r}$ is locally integrable. We choose any f in $C_c^\infty(V)$, and develop f about (0,0) in a Taylor expansion. By grouping together all the terms of the same homogeneity, we may write f in the form

$$f = \Sigma_{j=0}^{J} f_j ,$$

where f_j is homogeneous of degree j if $0 \leq j < J$, and the error term f_J satisfies the condition that

$$f_J(x,y) = O(N(x,y)^J) \quad .$$

We take a $C_c^\infty(R)$-function ψ which takes the value 1 in a neighbourhood of 0. Clearly, if $\mathrm{Re}(\zeta) > 0$, then

$$\begin{aligned} &\int_V dxdy\ \Omega(x,y)N(x,y)^{\zeta-r}\ f(x,y) \\ &= \int_V dxdy\ \Omega(x,y)N(x,y)^{\zeta-r}\ \psi(N(x,y))\ f(x,y) \\ &+ \int_V dxdy\ \Omega(x,y)N(x,y)^{\zeta-r}\ [1 - \psi(N(x,y))]\ f(x,y) \quad . \end{aligned}$$

The second integral on the right hand side continues analytically into the whole complex plane. The first integral may be written thus:

$$\begin{aligned} &\int_V dxdy\ \Omega(x,y)N(x,y)^{\zeta-r}\ \psi(N(x,y))\ f(x,y) \\ &= \Sigma_{j=0}^{J} \int_V dxdy\ \Omega(x,y)N(x,y)^{\zeta-r}\ \psi(N(x,y))\ f_j(x,y) \\ &= \Sigma_{j=0}^{J-1} [\int_\Sigma d\sigma\ \Omega(\sigma)\ f_j(\sigma)][\int_{R^+} ds\ s^{\zeta+j-1}\ \psi(s)] \\ &+ \int_V dxdy\ \Omega(x,y)N(x,y)^{\zeta-r}\ \psi(N(x,y))\ f_J(x,y) \quad . \end{aligned}$$

The radial integral of the first J terms extends meromorphically to the whole complex plane, with poles at 0, -1, -2,... , while the integrand of the final term is integrable if $\mathrm{Re}(\zeta) > -J$ and so the last term extends analytically to $\{\zeta \in C: \mathrm{Re}(\zeta) > -J\}$. Since J is arbitrary, we have indeed a meromorphic continuation to the whole complex plane. We remark that the continuation thus obtained does not depend on the choice of the function ψ, by the uniqueness of meromorphic continuation. □

REMARK 1.2. Lemma 1.1 can be generalised to deal with families of distributions $\Omega_\zeta N^{\zeta-r}$, where Ω_ζ is a meromorphic function of ζ.

REMARK 1.3. It is clear from the proof that the distribution $\Omega N^{\zeta-r}$ is of order at most J in $\{\zeta \in C: \mathrm{Re}(\zeta) > -J\}$.

REMARK 1.4. We observe that, if Ω is even, in the sense that $\Omega(x,y) = \Omega(-x,y)$ for (x,y) in V, then the only possible poles of $\Omega N^{\zeta-r}$ are where ζ is even, while if Ω is odd, in the sense that $\Omega(x,y) = -\Omega(-x,y)$, then any poles of $\Omega N^{\zeta-r}$ lie in {-1, -3, -5,...} . This holds because, if the integrand is odd, then

$$\int_\Sigma d\sigma\ \Omega(\sigma)\ f_j(\sigma) = 0 \ .$$

We shall now examine the distributions K_ζ for certain values of ζ, viz $\zeta = i\eta$, $\eta \in R\setminus\{0\}$. We denote by $\langle\ ,\ \rangle$ the usual inner product on $L^2(V)$, by $\|\ \|_p$ the usual L^p-norm, and by $f * g$ the convolution of functions f and g:

$$f*g(u) = \int_V dv\; f(uv)\, g(v^{-1}) \qquad u \in V$$

(when this makes sense).

THEOREM 1.5. Suppose that $\Omega \in L^1(\Sigma)$, and that $f,g \in C_c^\infty(V)$. Then, if $\eta \in R\setminus\{0\}$,

$$|\langle K_{i\eta}*f,g\rangle| \leq C(\eta)\, \|\Omega\|_1\, \|f\|_2\, \|g\|_2 \quad ,$$

where, for some constant C,

$$C(\eta) \leq C[\,|\eta| + |\eta|^{-1}] \ .$$

Consequently K_ζ extends to a bounded convolution operator on $L^2(V)$.

Proof. By easy changes of variables, one finds that

$$\begin{aligned}\langle K_{i\eta}*f,g\rangle &= \langle K_{i\eta}, g*f^\sim\rangle \\ &= \int_\Sigma d\sigma\; \Omega(\sigma) \int_{R^+} ds\; s^{i\eta-1}\, (g*f^\sim)(D^s\sigma) \ ,\end{aligned}$$

where $f^\sim(v) = \bar{f}(v^{-1})$ $(v \in V)$. It will therefore suffice to show that, for any σ in Σ,

$$\left|\int_{R^+} ds\; s^{i\eta-1}\, (g*f^\sim)(D^s\sigma)\right| \leq C(\eta)\, \|f\|_2\, \|g\|_2 \ .$$

This is equivalent to showing that the distributions T^σ_ζ, given by the rule

$$T^\sigma_\zeta(f) = \int_{R^+} ds\; s^{\zeta-1}\, f(D^s\sigma) \ ,$$

enjoy the property that

$$|\langle T^\sigma_{i\eta}, g*f^\sim\rangle| \leq C(\eta)\, \|f\|_2\, \|g\|_2 \quad ,$$

i.e. that

$$|\langle T^\sigma_{i\eta}*f, g\rangle| \leq C(\eta)\, \|f\|_2\, \|g\|_2 \quad ,$$

or, by the converse of Hölder's inequality, that

$$\|T^\sigma_{i\eta}*f\|_2 \leq C(\eta)\, \|f\|_2$$

for all f in $C_c^\infty(V)$.

The distribution T^σ_ζ is supported in a subgroup H^σ of V:

$$H^\sigma = \{(sx,ty) : (s,t) \in R^2\} \quad ,$$

if $\sigma = (x,y)$ in V. Apart from the trivial cases where $x = 0$ or $y = 0$, the mapping $(s,t) \mapsto (sx,ty)$ is an isomorphism of R^2 onto H^σ. Roughly speaking, it is sufficient to prove that

$$(1.3) \qquad \|T^\sigma_{i\eta}*f\|_2 \leq C(\eta)\, \|f\|_2 \qquad f \in C_c^\infty(H^\sigma) \ .$$

In effect, this is because the cosets $H^\sigma v$ of H^σ foliate V smoothly, and convolution with T^σ_ζ acts on each of the cosets just as on H^σ itself; the uniform estimates thus obtained for what happens on all the leaves of the foliation imply estimates for what happens on V. (See the papers of A.P. Calderón and

A. Zygmund [CaZ], S. Saeki [Sae] and C.S. Herz [Her].)

The proof of inequality (1.3) is not trivial. Perhaps the easiest method involves a slight modification of a result of E.M. Stein and S. Wainger [SW1], which is based on work of J.G. van der Corput, described in A. Zygmund's book [Zyg] on p. 197. In our case, the proofs simplify, and we reproduce these for completeness.

Since the Fourier transform of the convolution $T^{\sigma}_{\zeta} * f$ is the product of the transforms of the factors, we may resolve the problem by estimating $(T^{\sigma}_{i\eta})^{\wedge}$. We set

$$\phi_{i\eta}(p,q) = \lim_{\zeta\to i\eta,\ \mathrm{Re}(\zeta)>0} \phi_{\zeta}(p,q)$$

and
$$\phi_{\zeta}(p,q) = \lim_{\alpha\to 0+} \int_0^{\infty} ds\ s^{\zeta-1} \exp(i[sp + s^2 q] - \alpha s^2) .$$

Since
$$\phi_{\zeta}(ap,a^2 q) = a^{-\zeta}\phi_{\zeta}(p,q) \qquad a \in R^+,\ (p,q) \in R^2$$

and
$$\phi_{\zeta}(-p,-q) = [\phi_{\bar\zeta}(p,q)]^{-} \qquad (p,q) \in R^2 ,$$

in our investigations we may suppose that $q = 1$.

First we consider the integral

$$\begin{aligned}
&\lim_{\alpha\to 0+} \int_0^1 ds\ s^{\zeta-1} \exp(i[sp + s^2] - \alpha s^2)\\
&= \int_0^1 ds\ s^{\zeta-1} \exp(i[sp + s^2])\\
&= \int_0^1 ds\ s^{\zeta-1} \exp(isp)\\
&\quad + \int_0^1 ds\ s^{\zeta-1} \exp(isp)\,[\exp(is^2) - 1] .
\end{aligned}$$

It is now easy to check that

$$\left| \lim_{\zeta\to i\eta,\ \mathrm{Re}(\zeta)>0} \lim_{\alpha\to 0+} \int_0^1 ds\ s^{\zeta-1} \exp(i[sp+s^2]-\alpha s^2)\right| \le C(1 + |\eta|^{-1}) . \tag{1.4}$$

Next, we consider the integral

$$\lim_{\alpha\to 0+} \int_1^{\infty} ds\ s^{\zeta-1} \exp(i[sp + s^2] - \alpha s^2).$$

Let ψ be that function for which

$$\int_s^{\infty} dt\ \psi(t) = s^{\zeta-1} \exp(-\alpha s^2) \qquad s \in R^+ ;$$

i.e.
$$\psi(t) = [(1 - \zeta)t^{\zeta-2} + 2\alpha t^{\zeta}] \exp(-\alpha t^2) .$$

By changing the order of integration, we find that

$$\begin{aligned}
&\int_1^{\infty} ds\ s^{\zeta-1} \exp(i[sp + s^2] - \alpha s^2)\\
&= \int_1^{\infty} dt\ \psi(t) \int_1^t ds\ \exp(i[sp + s^2]) .
\end{aligned}$$

We note that

$$\begin{aligned}
&\sup\{|\textstyle\int_1^t ds\ \exp(i[sp + s^2])| : t \ge 1,\ p \in R\}\\
&\le \sup\{|\textstyle\int_a^b ds\ \exp(is^2)| : a,b \in R\}
\end{aligned}$$

$$\leqslant C ,$$

and also that

$$\int_1^\infty dt\ |\psi(t)|$$
$$\leqslant \int_1^\infty dt\ |1 - \zeta|\ t^{\xi-2} \quad + \quad 2\alpha\int_1^\infty dt\ \exp(-\alpha t^2)$$
$$\leqslant (|1 - \zeta| / |1 - \xi|) \quad + \quad C\alpha^{1/2} \quad .$$

It now follows readily that

$$\lim_{\zeta\to i\eta,\ \mathrm{Re}(\zeta)>0} \lim_{\alpha\to 0+} \int_1^\infty ds\ s^{\zeta-1} \exp(i[\,sp+s^2] - \alpha s^2) \tag{1.5}$$
$$\leqslant C(1 + |\eta|) \quad .$$

Combining the estimates (1.4) and (1.5), we obtain the desired result, viz

$$\|(T^\sigma_{i\eta})^\frown\|_\infty \leqslant C(|\eta| + |\eta|^{-1}) \quad . \qquad \square$$

It should be noted that the above proof contains some gaps. For instance, the following question has been glossed over: the distribution $T^\sigma_{i\eta}$ is realised as an analytic continuation of distributions T^σ_ζ, with $\mathrm{Re}(\zeta) > 0$. While there is an obvious correspondence between the distributions T^σ_ζ as distributions on V and as distributions on H^σ when $\mathrm{Re}(\zeta) > 0$, this correspondence might conceivably break down when analytic continuation takes place. We leave it to the reader to ferret out and grapple with such difficulties.

Further study of the distributions K_ζ requires the use of differential operators. We shall identify the Lie algebra $\underline{V}$ of V with $F^{n-1} \times \mathrm{Im}(F)$; we write (X,Y) for the generic element of $\underline{V}$. It will be convenient to denote by $\underline{V}_1$ and $\underline{V}_2$ the subspaces of $\underline{V}$ of all elements of the forms (X,0) $(X \in F^{n-1})$ and and (0,Y) $(Y \in \mathrm{Im}(F))$ respectively. The exponential map is given by the formula

$$\exp t(X,Y) = (tX,tY) \in V \ .$$

The reason for these notations should become clear in §4.

If E is in $\underline{V}$, we write fE and Ef for the functions on V given by the rule:

$$fE(x,y) = [\,d/dt\ f((x,y)\exp(tE))]_{t=0}$$
$$Ef(x,y) = [\,d/dt\ f(\exp(tE)(x,y))]_{t=0} \quad ,$$

i.e. we identify $\underline{V}$ with a space of distributions supported at e, the identity of V. If E = (X,0), then we write Xf in place of Ef, and so on.

Let (X_j) and (Y_k) be orthogonal bases of F^{n-1} and $\mathrm{Im}(F)$ over R. Clearly, for (x,y) in V, we have, with the obvious notations,

(1.6) $\qquad (X_jf)(x,y)$

$$= [d/dt\, f(tX_j + x,\ y - 2\mathrm{Im}(X_j x^*))]_{t=0}$$
$$= \partial/\partial x_j\, f(x,y) - \Sigma_{k=1}^{q} \mathrm{Re}(2\mathrm{Im}(X_j x^*)Y_k^*)\ \partial/\partial y_k\, f(x,y)$$
$$= \partial/\partial x_j\, f(x,y) + 2\Sigma_{k=1}^{q} \mathrm{Re}(xX_j^* Y_k^*)\ \partial/\partial y_k\, f(x,y) \quad ,$$

and analogously,

(1.7) $$fX_j(x,y) = \partial/\partial x_j\, f(x,y) - 2\Sigma_{k=1}^{q} \mathrm{Re}(xX_j^* Y_k^*)\ \partial/\partial y_k\, f(x,y) \quad ,$$

and further $$fY_k = Y_k f = \partial/\partial y_k\, f \quad .$$

We now define distributions Δ and ¤ on V as follows: for f in $C_c^\infty(V)$,

(1.8) $$\Delta * f = -\Sigma_{j=1}^{p} X_j^2 f$$
$$¤ * f = -\Sigma_{k=1}^{q} Y_k^2 f \quad .$$

We have also that

(1.9) $$f * \Delta = -\Sigma_{j=1}^{p} fX_j^2$$
$$f * ¤ = -\Sigma_{k=1}^{q} fY_k^2 = ¤ * f \quad .$$

The important facts about the so-called sub-Laplacian Δ are that Δ is a homogeneous distribution of degree $-r-2$, and that Δ is hypoelliptic. In fact, we have the following proposition, due to G.B. Folland [Fol] and A. Kaplan [Kap].

PROPOSITION 1.6. Let k be the function N^{2-r}. Then, in the distributional sense,

$$k * \Delta = \Delta * k = c\delta \ ,$$

where c is a nonzero constant, and δ is the Dirac measure at (0,0).

Proof. We shall later prove a generalisation of this result (see Proposition 1.9 and Remark 1.10).

Proposition 1.6 can be proved by brutal calculations, but we prefer a longer but more elegant approach. The scope of the next two lemmata is to simplify later calculations.

LEMMA 1.7. The following equality holds:

$$\Sigma_{j=1}^{p} \Sigma_{k=1}^{q} \Sigma_{m=1}^{q} \mathrm{Re}(xX_j^* Y_k^*)\ \mathrm{Re}(xX_j^* Y_m^*)\ a_k b_m$$
$$= |x|^2\, \Sigma_{k=1}^{q} a_k b_k \quad .$$

Proof. Since Y_k is imaginary, $\mathrm{Re}(zY_k^*) = -\mathrm{Re}(Y_k z)$. Also, $\mathrm{Re}(cd^*)$ is the inner product of c and d in F^{n-1}, whence

$$\begin{aligned}
&\Sigma_{j=1}^{p}\Sigma_{k=1}^{q}\Sigma_{m=1}^{q}\operatorname{Re}(xX_j^*Y_k^*)\operatorname{Re}(xX_j^*Y_m^*)\,a_kb_m\\
&=\Sigma_{j=1}^{p}\Sigma_{k=1}^{q}\Sigma_{m=1}^{q}\operatorname{Re}(Y_kxX_j^*)\operatorname{Re}(Y_mxX_j^*)\,a_kb_m\\
&=\Sigma_{k=1}^{q}\Sigma_{m=1}^{q}\operatorname{Re}((Y_kx)(Y_mx)^*)\,a_kb_m\\
&=\Sigma_{k=1}^{q}\Sigma_{m=1}^{q}\operatorname{Re}(Y_k|x|^2Y_m^*)\,a_kb_m\\
&=|x|^2\,\Sigma_{k=1}^{q}a_kb_k\ . \qquad \square
\end{aligned}$$

We now recall that M acts on V (formula (1.1)):

$$R^m(x,y) = (v^{-1}xu,\ v^{-1}yv) \qquad (x,y) \in V,$$

where $m = m(u,v)$, $v \in F$, $|v| = 1$, and u is an isometric $(n-1)\times(n-1)$ matrix. We denote by $M^\sim$ the subgroup of M of those elements for which u is diagonal.

A function f on V is called polyradial if

$$f\circ R^m = f \qquad m \in M^\sim .$$

This means that for any (x,y) in V,

$$f(x_1,x_2,\ldots,x_{n-1},y) = f(\varepsilon_1x_1,\varepsilon_2x_2,\ldots,\varepsilon_{n-1}x_{n-1},\varepsilon_ny)$$

whenever

$$\begin{aligned}
&\varepsilon_j \in \{\pm1\} && 1 \leq j \leq n-1 && \text{if } F = R\\
&\varepsilon_j \in T && 1 \leq j \leq n-1,\ \ \varepsilon_n = 1 && \text{if } F = C\\
&\varepsilon_j \in U && 1 \leq j \leq n && \text{if } F = Q ,
\end{aligned}$$

where T and U are the multiplicative groups of the elements of C and Q of absolute value 1. A distribution T is called polyradial if

$$T(f) = T(f\circ R^m) \qquad m \in M^\sim,\ f \in C_c^\infty(V).$$

As the nomenclature would suggest, it is possible to approximate polyradial distributions by polyradial functions.

The next lemma uses an idea which the author learnt from A. Korányi.

LEMMA 1.8. The space of polyradial functions forms a commutative algebra under convolution.

Proof. We show first that the space is closed under convolution. If f and g are polyradial, then a change of variables shows that, for any u in V and any m in $M^\sim$,

$$\begin{aligned}
(f * g)(R^mu) &= \int_V dv\ f((R^mu)v)\ g(v^{-1})\\
&= \int_V dv\ f((R^mu)(R^mv))\ g((R^mv)^{-1})\\
&= \int_V dv\ f(R^m(uv))\ g(R^m(v^{-1}))\\
&= \int_V dv\ f(uv)\ g(v^{-1})\\
&= f * g(u) ,
\end{aligned}$$

because R^m is an automorphism of V.

To show that the algebra is commutative, we consider two cases. If $F \neq C$ then, given v in V, there exists m in M such that $R^m v = v^{-1}$. Consequently, a polyradial function f is equal to its reflection $f^\vee$ ($f^\vee(v) = f(v^{-1})$). Again, if f and g are polyradial, we may change variables to obtain that, for any u in V,

$$\begin{aligned}(f * g)(u) &= (f * g)(u^{-1}) \\ &= \int_V dv\ f(u^{-1}v)\ g(v^{-1}) \\ &= \int_V dv\ g^\vee(uv)\ f^\vee(v^{-1}) \\ &= g * f(u) \quad .\end{aligned}$$

Finally, if $F = C$, then the map $v \mapsto \bar{v}$ (i.e. $(x,y) \mapsto (\bar{x},\bar{y})$) is an automorphism (this is false if $F = Q$). For f on V, let f° be the function such that $f^\circ(v) = f(\bar{v})$. By similar calculations, we find that

$$(f * g) = (f * g)^{\circ\circ} = (f^\circ * g^\circ)^\circ \quad .$$

Further, if f and g are polyradial, then $f^\vee = f^\circ$ and $g^\vee = g^\circ$, whence

$$(f * g) = (f^\vee * g^\vee)^\circ = (g * f)^{\vee\circ} = g * f \ ,$$

since $g * f$ is also polyradial. □

A function f on V is called cylindrical (or bi-radial) if $f \circ R^m = f$ for all m in M. If $F = R$, then f(x) depends only on $|x|$ (i.e. is radial in the usual sense), while if $F = C$, then f(x,y) depends only on $|x|$ and y, and if $F = Q$, then f(x,y) depends only on $|x|$ and $|y|$. A distribution T is called cylindric if $T(f \circ R^m) = T(f)$ for all f in $C_c^\infty(V)$ and all m in M. Obviously, a cylindrical distribution may be approximated by cylindrical functions.

We now generalise Proposition 1.6, as promised.

PROPOSITION 1.9. The following equality is valid:

$$(\Delta * N^{\zeta-r})(x,y) = (r - \zeta)(\zeta - 2)\ |x|^2\ N(x,y)^{\zeta-r-4} \quad .$$

Proof. The proof proceeds in several stages. First, we note that it suffices to establish the result for ζ such that $\mathrm{Re}(\zeta) > r + 2$, in which case the evaluation of $\Delta * N^{\zeta-r}$ is by ordinary differentiation. The result will then extend to general ζ by meromorphic continuation.

The second step of the proof is to find a formula for Δ, valid for polyradial functions; the third stage is to specialise to the radial case. We now pass to the second step.

From the formulae (1.6) and (1.8), we have that

$$\begin{aligned}
&\Delta * f(x,y) \\
&= -\Sigma_{j=1}^{p} [\partial/\partial x_j + 2\Sigma_{k=1}^{q} \operatorname{Re}(xX_j^* Y_k^*)\, \partial/\partial y_k]^2 f(x,y) \\
&= -\Sigma_{j=1}^{p} \partial^2/\partial x_j^2 f(x,y) \\
&\quad -4\Sigma_{j=1}^{p} [\Sigma_{k=1}^{q} \operatorname{Re}(xX_j^* Y_k^*)\, \partial/\partial y_k]^2 f(x,y) \\
&\quad -4\Sigma_{j=1}^{p} \Sigma_{k=1}^{q} \operatorname{Re}(xX_j^* Y_k^*)\, \partial^2/\partial x_j \partial y_k f(x,y) \quad ,
\end{aligned}$$

since $\partial/\partial x_j \operatorname{Re}(xX_j^* Y_k^*) = \operatorname{Re}(X_j X_j^* Y_k^*) = 0$.

Lemma 1.7 implies that the second term is equal to

$$4|x|^2 \Sigma_{k=1}^{q} \partial^2/\partial y_k^2 f(x,y) \quad ,$$

and so we have the following expression for $\Delta * f$:

$$\begin{aligned}
(1.10) \qquad \Delta * f(x,y) &= -\Sigma_{j=1}^{p} \partial^2/\partial x_j^2 f(x,y) \\
&\quad -4|x|^2 \Sigma_{k=1}^{q} \partial^2/\partial y_k^2 f(x,y) - \Phi f(x,y) \quad ,
\end{aligned}$$

where $\Phi f(x,y) = 4\Sigma_{j=1}^{p} \Sigma_{k=1}^{q} \operatorname{Re}(xX_j^* Y_k^*)\, \partial^2/\partial x_j \partial y_k f(x,y)$.

Analogously, from formulae (1.7) and (1.9), it follows that

$$\begin{aligned}
(1.11) \qquad \Delta * f(x,y) &= -\Sigma_{j=1}^{p} \partial^2/\partial x_j^2 f(x,y) \\
&\quad -4|x|^2 \Sigma_{k=1}^{q} \partial^2/\partial y_k^2 f(x,y) + \Phi f(x,y)
\end{aligned}$$

(formulae (1.10) and (1.11) differ only in the sign of Φf). Now, if f is polyradial, we have from Lemma 1.8 and a standard limiting argument that $\Delta * f = f * \Delta$, for Δ is obviously cylindrical and a fortiori polyradial. We deduce from formulae (1.10) and (1.11) that $\Phi f = 0$, and

$$\begin{aligned}
(1.12) \qquad &\Delta * f(x,y) \\
&= -\Sigma_{j=1}^{p} \partial^2/\partial x_j^2 f(x,y) - 4|x|^2 \Sigma_{k=1}^{q} \partial^2/\partial y_k^2 f(x,y) \quad .
\end{aligned}$$

If f is cylindrical, then a further simplification is possible: we may pass to polar coordinates to evaluate $\Sigma_{j=1}^{p} \partial^2/\partial x_j^2$ and $\Sigma_{k=1}^{q} \partial^2/\partial y_k^2$. The key fact is familiar: for a radial function g on R^n,

$$\Sigma_{j=1}^{n} \partial^2/\partial x_j^2 g = [\partial^2/\partial\rho^2 + \{(n-1)/\rho\}\partial/\partial\rho]\, g \quad ,$$

where ρ is the radius.

The rest of the proof is now straightforward calculation, and is omitted. The reader may like to check the details. □

REMARK 1.10. We explain how Proposition 1.6 follows from Proposition 1.9. Let ε be a small positive quantity, and let Ω be defined by the formula

$$\Omega(x,y) = |x|^2 N(x,y)^{-2} \qquad (x,y) \in V \ .$$

Clearly $$\Delta * N^{2+\varepsilon-r} = \varepsilon(r - 2 - \varepsilon)\, \Omega N^{\varepsilon-r} ,$$

whence $$\Delta * N^{2-r} = (r-2)\mathrm{Res}(\Omega N^{\zeta-r}; \zeta = 0) ,$$

where the notation should not need explanation. A simple calculation, based on the same idea as Lemma 1.1 shows that

$$\lim_{\zeta \to 0} \int_V dxdy\ \zeta\ \Omega N^{\zeta-r}(x,y)\ f(x,y)$$
$$= [\int_\Sigma d\sigma\ \Omega(\sigma)]\ f(0,0) ,$$

and the Proposition 1.6 follows.

2. ANALYSIS ON HEISENBERG GROUPS. II

We have previously considered certain homogeneous distributions K_ζ of the form $\Omega N^{\zeta-r}$, where Ω is homogeneous of degree 0. In particular, we showed that for almost all values of ζ, K_ζ has a distributional interpretation, and that if $\zeta = i\eta$ ($\eta \in R\setminus\{0\}$) then K_ζ convolves $L^2(V)$ into itself (Th'm 1.5). In this section, we consider possible L^2-boundedness of K_0, and boundedness of K_ζ on more general potential-theoretic spaces. Our main tools are the analysis of the previous section, and the complex method of interpolation, developed by A.P. Calderon [Cal] and E.M. Stein [Ste]. Later, we consider some properties of the potential-theoretic spaces with which we deal.

First of all we consider the kernel K_0.

LEMMA 2.1. Let K_ζ be of the form $N^{\zeta-r}$, where $\Omega \in L^1(\Sigma)$. Define the mean value $MV(\Omega)$ of Ω by the formula

$$MV(\Omega) = \int_\Sigma d\sigma\ \Omega(\sigma) \quad .$$

Then there is the following dichotomy: if $MV(\Omega) = 0$, then K_0 is a distribution, and if $MV(\Omega) \neq 0$, then K_ζ has a pole at 0, and the residue there is $c\delta$, for some nonzero constant c.

Proof. We omit the proof, but refer to Lemma 1.1 and Remark 1.7 for the necessary techniques.

Lemma 2.1 suggests that, if $MV(\Omega) = 0$, then convolution with K_0 might be bounded on $L^2(V)$. In fact this is so, but only if some additional restriction be imposed on Ω. Several proofs of this fact are available. The first is due to A.W. Knapp and E.M. Stein [KnS] (see p. 494; the hypothesis of their theorem may be relaxed substantially using the same method of proof, but not to Ω in $C(\Sigma)$. We shall shortly present a second proof, which uses complex interpolation; the philosophy of this second proof is that if an operator exists in a weak sense (K_0) and if the ambient operators (K_ζ, with $\zeta \neq 0$) are bounded in some stronger sense, then so is the original operator. A third proof, offered later, relies on the fact that $N^{i-r} + N^{-i-r} = c\delta$ for some nonzero c. A fourth proof, similar to the third, uses the fact that $\Delta^{i} * \Delta^{-i} = \delta$, and relies on description of Δ^{i} and Δ^{-i} as operators of the form ΩN^{-2i-r} and $\tilde{\Omega} N^{2i-r}$, which description follows from work of G.B. Folland [Fo2] (see p. 184).

We now consider the distributions K_ζ for certain values of ζ with nonzero real parts.

LEMMA 2.2. Let n be a positive integer. Suppose that $K_\zeta = \Omega N^{\zeta-r}$, where $\Omega \in C^{2n}(\Sigma)$, and that $\zeta = 2n + i\eta$ ($\eta \in R\setminus\{0\}$). Then

$$\|\Delta^{n} * K_\zeta * f\|_2 \leq C(\Omega,\zeta)\ \|f\|_2 \qquad f \in C^\infty_c(V)\ ,$$

where $$C(\Omega,\zeta) \leq C(n,\Omega)\ (|\eta|^{-1} + |\eta|^{2n+1})\ .$$

Proof. It suffices to check that $\Delta^{n} * K_\zeta$ is a distribution of the form considered in Theorem 1.4, taking care of the growth of the constants all the while. We omit the details.

LEMMA 2.3. Suppose that $\zeta \mapsto \Omega_\zeta$ is a meromorphic $L^1(\Sigma)$-valued function, defined in the half-space H, where $H = \{\zeta \in C$: $Re(\zeta) < r - 2\}$. Set K_ζ equal to $\Omega_\zeta N^{\zeta-r}$ (in the sense of Remark 1.2), and let S be the set of poles of the distribution-valued function $\zeta \mapsto K_\zeta$. Let k be N^{2-r}.

If $Re(\zeta) \in (0, r-2)$, and if $\zeta \notin S$, then the integral defining the convolution $k * \Omega_\zeta N^{\zeta-r}$ converges (almost everywhere), and is a locally integrable function $\tilde{K}_\zeta$ of the following form:

$$\tilde{K}_\zeta = \tilde{\Omega}_\zeta N^{\zeta-r+2}\ ,$$

where $\tilde{\Omega}_\zeta$ is a $L^1(\Sigma)$-valued meromorphic function.

The $L^1(\Sigma)$-valued and $C^\infty_c(V)'$-valued functions $\tilde{\Omega}_\zeta$ and $\tilde{K}_\zeta$ extend meromorphically to H, and all poles of $\tilde{K}_\zeta$ lie in S. Distributionally,

$$k * K_\zeta = \tilde{K}_\zeta\ ,$$

and finally

$$\|\tilde{\Omega}_\zeta\|_1 \leq C(\zeta)\ \|\Omega_\zeta\|_1\ ,$$

where, if $\zeta = \xi + i\eta$,

$$C(\zeta) \leq C(\xi) + C(|\eta| + |\eta|^{-1})\ .$$

Proof. We shall not show that the integral defining the convolution $k * \Omega_\zeta N^{\zeta-r}$ converges, but this is straightforward. It is also clear that K_ζ is of the stated type if $\mathrm{Re}(\zeta) \in (0, r-2)$, for reasons of homogeneity. We shall consider the question of the meromorphic continuation.

We take ψ in $C^\infty_c(R)$ which is equal to 1 in a small neighbourhood of 0, and set ϕ equal to $\psi \circ N$. Now

$$k * K_\zeta = k * \phi K_\zeta + k * (1 - \phi)K_\zeta .$$

The first piece on the right hand side is the convolution of k with a compactly supported distribution, and so is a distribution. Since k is C^∞ away from (0,0), this piece is C^∞ off the support of ϕ. Provided that $\mathrm{Re}(\zeta) < r - 2$, the second piece "is" a locally integrable function. Both pieces continue meromorphically, with poles only where K_ζ has poles. The sum $K^\sim_\zeta$ is locally integrable off $\mathrm{supp}(\phi)$, which may be made arbitrarily small, whence $K^\sim_\zeta$ is in $L^1_{loc}(V\backslash\{(0,0)\})$.

By considerations of homogeneity, it must be true that, as distributions on $V\backslash\{(0,0)\}$,

$$K^\sim_\zeta = \Omega^\sim_\zeta N^{\zeta-r+2}$$

for an appropriate $L^1(\Sigma)$-valued function $\Omega^\sim_\zeta$. We now consider the two meromorphic $C^\infty_c(V)'$-valued functions $K^\sim_\zeta$ and $\Omega^\sim_\zeta N^{-r+2}$, defined as in Remark 1.2. These agree if $0 < \mathrm{Re}(\zeta) < r - 2$, and hence they agree everywhere. It is clear that $K^\sim_\zeta = k * K_\zeta$.

We conclude by noting that ϕK_ζ is a distribution with small support, say inside the set $\{(x,y)\colon N(x,y) < 1/2\}$, and of order at most $[\![-\xi]\!] + 1$, where $[\![.]\!]$ denotes the integral part function. If $N(x,y) = 1$, $k * \phi K_\zeta(x,y)$ can be estimated by the derivatives of k of order at most $[\![-\xi]\!] + 1$ in the anulus $\{(x,y)\colon 1/2 < N(x,y) < 3/2\}$. The estimate for $\|\Omega^\sim_\zeta\|_1$ uses this fact and some straightforward estimates for $k * (1 - \phi)K_\zeta$.

We may now improve somewhat Lemma 2.1.

PROPOSITION 2.4. If $\Omega \in C^2(\Sigma)$, and if $MV(\Omega) = 0$, then

$$\|\Omega N^{-r} * f\|_2 \leq C(\Omega)\,\|f\|_2 \qquad f \in C^\infty_c(V),$$

where $C(\Omega)$ depends on the derivatives of Ω of order at most 2.

Proof. We know that Δ is a positive unbounded operator with dense domain in $L^2(V)$, with no null space. The spectral calculus permits us to define $\Delta^{\zeta/2}$, as an unbounded operator on $L^2(V)$ with dense domain. The operators $\Delta^{i\eta/2}$, where $\eta \in R$, extend to isometries of $L^2(V)$. G.B. Folland [Fo2] has studied these operators.

We consider the family of distributions $\zeta \mapsto K^{\varphi}_{\zeta}$, where

$$K^{\varphi}_{\zeta} = \exp(\zeta^2)\,(\zeta^2 - 4)(\zeta + 1)\,\Omega N^{\zeta-r} ,$$

which is analytic in the strip $\{\zeta \in C\colon \mathrm{Re}(\zeta) \in [-3, 3]\}$, because any poles of $\Omega N^{\zeta-r}$ at -1 or -2 are annulled by the factor $(\zeta^2 - 4)(\zeta + 1)$, and by hypothesis K^{φ}_{ζ} has no pole at 0. On one hand, if $\xi = 2$, then from Lemma 2.2 and some juggling of estimates, we obtain that

$$\|\Delta * K^{\varphi}_{\zeta} * f\|_2 \leq C(\Omega)\,\|f\|_2 \qquad f \in C^{\infty}_c(V)$$

(the constant $C(\Omega)$ is now independent of η). Consequently

$$\|\Delta^{i\eta/2} * \Delta * K^{\varphi}_{\zeta} * f\|_2 \leq C(\Omega)\,\|f\|_2 \qquad f \in C^{\infty}_c(V),$$

and hence

$$|\langle \Delta^{\zeta/2} * K^{\varphi}_{\zeta} * f, g\rangle| \leq C(\Omega)\,\|f\|_2\,\|g\|_2 \qquad f,g \in C^{\infty}_c(V) .$$

On the other hand, if $\xi = -2$, then

$$\|k * K^{\varphi}_{\zeta} * f\|_2 \leq C(\Omega)\,\|f\|_2 \qquad f \in C^{\infty}_c(V) ,$$

by Lemma 2.3 and Theorem 1.2. Analogously, it follows that

$$|\langle \Delta^{\zeta/2} * K^{\varphi}_{\zeta} * f, g\rangle| \leq C(\Omega)\,\|f\|_2\,\|g\|_2 \qquad f,g \in C^{\infty}_c(V) .$$

Application of the Phragmén-Lindelöf "three lines theorem" (see, e.g., E.C. Titchmarsh [Tit], Section 5.65) implies that, if $\xi = 0$, then an analogous inequality holds. Considering the case where $\zeta = 0$, we obtain the inequality:

$$|\langle K^{\varphi}_{0} * f, g\rangle| \leq C(\Omega)\,\|f\|_2\,\|g\|_2 \qquad f,g \in C^{\infty}_c(V) ,$$

and the converse of Hölder's inequality completes the proof. □

REMARK 2.5. The above proof also shows that the norms of the convolution operators $K_{i\eta/2}$ remain bounded as η approaches 0.

A variation of the above proof establishes the following proposition.

PROPOSITION 2.6. Suppose that $\Omega \in C^{\infty}(\Sigma)$, and that $K_{\zeta} = \Omega N^{\zeta-r}$. Then, away from the poles of K_{ζ},

$$\|\Delta^{\zeta/2} * K_{\zeta} * f\|_2 \leq C(\Omega,\zeta)\,\|f\|_2 \qquad f \in C^{\infty}_c(V) .$$

At the poles of K_{ζ}, we have the formula

$$\|\Delta^{\zeta/2} * \mathrm{Res}(K_\zeta) * f\|_2 \leq C(\Omega,\zeta)\,\|f\|_2 \qquad f \in C_c^\infty(V) \quad .$$

where $\mathrm{Res}(K_\zeta)$ is the residue of K_ζ at ζ.

Proof. We omit the proof, which involves no new ideas.

The significance of the preceding proposition is that it shows that the operators K_ζ are no more singular than the operators $\Delta^{-\zeta/2}$. We can improve both this result and Proposition 2.4 by using a theorem which will be proved later (in § 8), namely, if ζ is not an even integer, then

(2.1) $$N^{\zeta-r} * N^{-\zeta-r} = c(\zeta)\delta \quad .$$

PROPOSITION 2.7. Suppose that $\Omega \in L^2(\Sigma)$, and that $MV(\Omega) = 0$. Then convolution with ΩN^{-r} extends from $C_c^\infty(V)$ to a bounded operator on $L^2(V)$.

Proof. In order to show that ΩN^{-r} convolves $L^2(V)$ into itself, it suffices to prove that

$$\|N^{i-r} * \Omega N^{-r} * f\|_2 \leq C(\Omega)\|f\|_2 \qquad f \in C_c^\infty(V) \quad ,$$

for then the equality above (2.1) and Theorem 1.2 would give the desired conclusion. Now let ϕ be a $C_c^\infty(V)$-function which is 1 near (0,0) and has small support; then

$$N^{i-r} * \Omega N^{-r}$$

$$= \{\phi N^{i-r} * \phi\Omega N^{-r}\} + \{(1-\phi)N^{i-r} * \phi\Omega N^{-r}\} + \{N^{i-r} * (1-\phi)\Omega N^{-r}\} \ .$$

The first term is a distribution with small support, the second is the convolution of a compactly supported distribution with a $C^\infty(V)$-function, and the third is in $L^2(V)$ by Theorem 1.2. It follows that there exists $\Omega^\sim$ in $L^2(V)$ such that

$$N^{i-r} * \Omega N^{-r} = \Omega^\sim N^{i-r} \ ,$$

by a homogeneity argument and the above discussion. Now Theorem 1.2 may be applied to complete the proof. □

REMARK 2.8. The afficionados of H^p-spaces will be pleased to know that Proposition 2.7 extends to those Ω which lie in a sort of $H^1(\Sigma)$ (in fact, let ψ be a $C_c^\infty(R)$-function with support in (1,4) which is equal to 1 on (2,3), and let ϕ be $\psi \circ N$: then we require that $\phi\Omega N^{-r}$ lie in $H^1(V)$). To show this, one uses the fact that N^{i-r} convolves $H^1(V)$ into $L^1(V)$, proved by R.R. Coifman and

G. Weiss [CoW] on p. 599. It is also possible to obtain L^p-estimates by suitably modifying Theorem 1.2 - indications of how to do this can be found in the paper of E.M. Stein and S. Wainger [SW2] (see p. 1253).

PROPOSITION 2.9. Convolution with the distribution $\Delta^{\zeta/2} * N^{\zeta-r}$ is a bounded invertible operator on $L^2(V)$ provided that ζ is not an even integer.

Proof. We have already seen that

$$\|\Delta^{\zeta/2} * N^{\zeta-r} * f\|_2 \leq C(\zeta)\,\|f\|_2 ,$$

unless $\zeta \in \{0, -2, -4, \ldots\}$. The invertibility of the operator is the new feature of this proposition.

From the formula (2.1), we deduce that

$$\Delta^{\zeta/2} * N^{\zeta-r} * N^{-\zeta-r} * \Delta^{-\zeta/2} = c(\zeta)\delta ,$$

and so we may write

$$(\Delta^{\zeta/2} * N^{\zeta-r})^{-1} = c(\delta)^{-1} N^{-\zeta-r} * \Delta^{-\zeta/2} .$$

In order to estimate the right hand side, we could use Lemma 1.7 to reduce the problem to that of estimating $\Delta^{-\zeta/2} * N^{-\zeta-r}$, which is resolved by applying Proposition 2.6, or we could reprove Proposition 2.6 for Δ acting on the right rather than on the left. Alternatively, we could observe that

$$[c(\zeta)^{-1} N^{-\zeta-r} * \Delta^{-\zeta/2}]^* = \bar{c}(\zeta)^{-1} \Delta^{-\omega/2} * N^{-\omega-r} ,$$

where $\omega = \delta^-$, and apply Proposition 2.6. In any case, the proposition is certainly proved. □

REMARK 2.10. In Section 8, we shall see that our proof of the equality (2.1) also yields Proposition 2.9 directly. However, there are other families of operators $\Omega N^{\zeta-r}$ (the intertwining operators - see below) for which, at least at the present time, an analogous inequality -

$$\Omega N^{\zeta-r} * \Omega' N^{-\zeta-r} = c'(\zeta)\delta \quad -$$

can be proved only by a method which does not imply the analogue of Proposition 2.9 except by the argument presented here (see A.W. Knapp and E.M. Stein [KnS] for further details).

Proposition 2.9 is a two-edged sword: on one hand, we may obtain results about the kernels $N^{\zeta-r}$ by using the group property of Δ, viz, $\Delta^{\zeta} * \Delta^{\omega} = \Delta^{\zeta+\omega}$, and on the other, we may use the kernels $N^{\zeta-r}$ as approximations to (the unknown kernel) $\Delta^{-\zeta/2}$. The following results, obtained in joint work with A.M. Mantero, will illustrate this principle. Our results extend work of R.S. Strichartz [Str] and N. Lohoué [Loh].

We shall need to use the Lorentz spaces $L^{p,q}$. The nonincreasing rearrangement of a measurable function f on V is the nonincreasing nonnegative right-continuous equimeasurable function f^* on R^+. In other words, f^* is nonincreasing and right-continuous, and further

$$m(\{v \in V: |f(v)| > \lambda\}) = m(\{s \in R^+: f^*(s) > \lambda\}) \qquad \lambda \in R^+,$$

m(S) denoting the measure of the set S. We follow R.A. Hunt [Hun] (see also E.M. Stein and G. Weiss [StW] pp. 188-205) and make the following definitions:

$$\|f\|^*_{pq} = \{(q/p) \int_0^\infty ds/s\ [s^{1/p} f^*(s)]^q\}^{1/q}$$

if $1 < p < \infty$, $1 < q < \infty$, while if $1 < p < \infty$, $q = \infty$,

$$\|f\|^*_{pq} = \sup\{[s^{1/p} f^*(s)]: s \in R^+\} \quad .$$

The space $L^{p,q}(V)$ is the space of those functions f on V for which $\|f\|^*_{pq} < \infty$. If $p = q$, then $L^{p,q}(V)$ is the usual Lebesgue space $L^p(V)$, while if $q = \infty$, then the space is often called "weak $L^p(V)$".

The next lemma is due to Hunt [Hun] (see p. 273).

LEMMA 2.11. Suppose that $t \in (1, \infty)$ and that $k \in L^{t,\infty}(V)$. Then there is a constant C(p,q) such that

$$\|k * f\|^*_{q2} \leq C(p,q) \|f\|^*_{p2} \qquad f \in C^\infty_c(V)$$

where $1/p + 1/t = 1/q + 1$, provided that $1 < p < q < \infty$.

Proof. Omitted.

We shall use a corollary of this lemma, which will be proved after the neccessary notation has been introduced.

We work with Hilbert spaces H^ξ: these are the completions of $C^\infty_c(V)$ with the norm $\|.\|^\xi$:

$$\|f\|^\xi = \langle \Delta^{-\xi/2} * f, f\rangle^{1/2}$$
$$= \|\Delta^{-\xi/4} * f\|_2 \quad .$$

It is clear that H and $H^{-\xi}$ are dual when equipped with the usual pairing:

$$\langle f,g\rangle = \int_V dv\ f(v)\ \bar{g}(v) \qquad f \in H^\xi,\ g \in H^{-\xi}.$$

COROLLARY 2.12. If $f \in C_c(V)$, then $f \in H^\xi$ whenever $0 < \xi < r$, and further

$$\|f\|^\xi \leq C(\xi) \|f\|^*_{p2} ,$$

where $1/p = 1/2 + \xi/2r$.

Proof. If $\xi = 0$, this result is trivial. We pick ξ very close to r, so that ξ is not an even integer. For f in $C_c(V)$, Proposition 2.9 implies that

$$\|\Delta^{-\xi/4} * f\|_2 \leq C(\xi) \|N^{\xi/2-r} * f\|_2 .$$

Since $N^{\xi/2-r}$ lies in $L^{t,\infty}(V)$, where $1/t = 1/2 + 1 - 1/p$ and $1/p = 1/2 + \xi/2r$, Lemma 2.11 implies the desired result. An application of complex interpolation now proves the corollary for general ξ. □

REMARK 2.13. Lemma 4.9 could also be proved by applying results of G.B. Folland [Fo2] (Proposition 3.17 on p. 184).

An easy corollary of the preceding corollary is proved by duality.

COROLLARY 2.14. If $-r < \xi < 0$, then H^ξ is contained in $L^{p,2}(V)$, where $1/p = 1/2 + \xi/2r$, and

$$\|f\|^*_{p2} \leq C(\xi) \|f\|^\xi .$$

Proof. Omitted.

We conclude this section by specialising another result of Hunt [Hun] (see p. 271) to fit our needs. We generalise a result of R.S. Strichartz [Str] and N. Lohoué [Loh], but the proof we offer is somewhat simpler than theirs.

LEMMA 2.15. Suppose that $1 < p, q, t < \infty$, and that $1/q = 1/t + 1/p$. If $f \in L^{p,2}$ and $m \in L^{t,\infty}$, then $mf \in L^{q,2}$ and

$$\|mf\|^*_{q2} \leq C(p,q) \|m\|^*_{t\infty} \|f\|^*_{p2} .$$

Proof. Omitted.

PROPOSITION 2.15. Let $M^d(V)$ be the space of all m in $C^\infty(V\setminus\{(0,0)\})$ which are homogeneous of degree d, where $d \in C$ and $Re(d) \leq 0$. If $-r < \xi_0 \leq \xi_1 < r$, and $\xi_0 - \xi_1 = 2Re(d)$, then pointwise multiplication by m in $M^d(V)$ defines a bounded operator from H^{ξ_0} to H^{ξ_1}.

Proof. We content ourselves with a brief sketch of the proof, and refer the reader to M. Cowling and A.M. Mantero [CoM] for further details. The proof involves five easy pieces.

First, if $\xi_0 \leq 0 \leq \xi_1$, then the result is an immediate consequence of Corollaries 2.12 and 2.14 and Lemma 2.15.

Next we tackle the case where $\xi_0 = \xi_1 = -2$. Let $(X_1, \ldots, X_p)$ be a basis of $\underline{V}_1$. To show that mf is in H^{-2} if f is, it suffices to prove that $X_j(mf)$ is in H^0. Since $X_j(mf) = (X_j m)f + m(X_j f)$, and since $X_j f$ lies in H^0, this can be done by the first argument.

Third, by complex interpolation and duality, the cases where $\xi_0 = \xi_1$ and $\xi_0 \in [-2, 2]$ are treated.

The fourth step is to consider the case where $\xi_0 = \xi_1 \leq 0$. This can be done by considering D(mf), where D is a right-invariant differential operator of order k and $2k + \xi_0 \in [0, 2)$.

Finally, duality and complex interpolation are applied to finish off the proof. □

3. REPRESENTATIONS OF HEISENBERG GROUPS.

We shall now develop the representation theory of V, using the Kirillov method (see A.A. Kirillov [Ki1] or [Ki2], or L. Pukanszky [Puk]). It would be possible to do things more directly, but the chosen path may have didactic advantages.

The group V is in fact a group of matrices. By $\underline{V}$ we denote the Lie algebra of all matrices of the form:

$$\begin{vmatrix} 0 & X & Y/2 \\ 0 & 0 & -X^* \\ 0 & 0 & 0 \end{vmatrix} \qquad X \in F^{n-1}, Y \in \mathrm{Im}(F)$$

(the (n+1)×(n+1) matrix is divided onto blocks of side 1 or (n-1): the central block is (n-1)×(n-1)). We abbreviate this matrix to $\underline{V}(X,Y)$ or just (X,Y). The group V is exactly the set of all exponentials of these matrices: we write V(x,y) or just (x,y) to indicate $\exp(\underline{V}(x,y))$, i.e.

$$V(x,y) = \exp\begin{vmatrix} 0 & x & y/2 \\ 0 & 0 & -x^* \\ 0 & 0 & 0 \end{vmatrix} = \begin{vmatrix} 1 & x & (y-xx^*)/2 \\ 0 & 1 & -x^* \\ 0 & 0 & 1 \end{vmatrix} .$$

This formulation gives rise to the multiplication with which we have been dealing.

Matters are arranged such that exp is a diffeomorphism of $\underline{V}$ onto V; the map $f \mapsto f\circ\exp$ is an isomorphism of $C_c^\infty(V)$ onto $C_c^\infty(\underline{V})$, and also of $S(V)$ onto $S(\underline{V})$ (these are the Schwartz spaces of functions which, together with all their derivatives, vanish at infinity faster than any polynomial grows). Further, the Haar measure of V corresponds to Lebesgue measure on $\underline{V}$, i.e.

$$\int_V dxdy\, f(x,y) = \int_{\underline{V}} dXdY\, f\circ\exp(X,Y) \qquad f \in C_c^\infty(V) .$$

The adjoint representation of V on $\underline{V}$ is described as follows: Ad(x,y)(X,Y) is that element of $\underline{V}$ which corresponds to $V(x,y)\underline{V}(X,Y)V(x,y)^{-1}$, i.e.

$$\mathrm{Ad}(x,y)(X,Y) = (X,\ Y - 4\mathrm{Im}(xX^*)) .$$

The (real)dual $\underline{V}^*$ of $\underline{V}$ is just $\mathrm{Hom}_R(\underline{V}, R)$; we identify this with $F^{n-1}\times\mathrm{Im}(F)$ in the following manner. By (α,λ) we denote the linear functional:

$$\langle(\alpha,\ \lambda)|(X,\ Y)\rangle = \mathrm{Re}(\alpha X^* + \lambda Y^*) \qquad (X,Y) \in \underline{V}$$

The coadjoint representation Cd of V on $\underline{V}^*$ is the contragredient of the adjoint representation:

$$\langle \mathrm{Cd}(x,y)(\alpha,\lambda)|(X,Y)\rangle = \langle(\alpha,\lambda)|\mathrm{Ad}(x,y)^{-1}(X,Y)\rangle \quad ,$$

whence, by straightforward calculation,

$$\mathrm{Cd}(x,y)(\alpha,\lambda) = (\alpha - 4\lambda x,\lambda) \quad .$$

There are two distinct cases of this action: $\lambda = 0$, and $\lambda \neq 0$. In the former case, then the V-orbit is a point, and otherwise it is a plane O_λ:

$$\mathrm{Cd}(V)(\alpha,0) = \{(\alpha,0)\}$$

$$O_\lambda = \mathrm{Cd}(V)(\alpha,\lambda) = (\mathcal{F}^{n-1},\lambda) \qquad\qquad (\text{if } \lambda \neq 0) \ .$$

The starting point of Kirillov's theory is that we can define distributions $T'_{(\alpha,\lambda)}$ on V by the formula

$$T_{(\alpha,\lambda)}(f) = \int_{\underline{V}} dXdY\ f\circ\exp(X,Y)\ e(\langle(\alpha,\lambda)|(X,Y)\rangle) \ ,$$

where e(r) denotes exp(ir). We observe that

$$T_{(\alpha,\lambda)}(f) = (f\circ\exp)\hat{}(\alpha,\lambda) \quad ,$$

where $\hat{g}$ denotes the Euclidean Fourier transform of a function g on a Euclidean space, in this case $\underline{V}$. When $\lambda = 0$, these are characters of V, (i.e. homomorphisms of V into $\mathcal{T}$), but otherwise they are not: some "V-invariance" is needed. The necessary V-invariance can be obtained by integrating over the orbit O_λ. The appropriate measure is obtained by transferring Lebesgue measure on $\mathcal{F}^{n-1}$ (i.e. V/ker(Cd)). We define the distribution T_λ on V by the formula

$$T_\lambda(f) = \int_{\mathcal{F}^{n-1}} dx\ T_{\mathrm{Cd}(x,0)(0,\lambda)}(f) \quad .$$

Then $T_\lambda(f)$ is the integral of $(f\circ\exp)\hat{}$ over O_λ.

We shall now study T_λ, and show that it is a "central" distribution; this result could be obtained by applying Lemma 3.2 (below) which describes T_λ explicitly.

Suppose that f be a function on V, that $(x,y) \in V$, and that f' be defined by the formula:

$$f'((x',y')) = f((x,y)(x',y')(x,y)^{-1}) \qquad (x',y') \in V \ .$$

Then we say that f is central if f = f'; we say that a distribution T is central if T(f) = T(f') (for all choices of (x,y) in V). The distributions T_λ and $T_{(\alpha,0)}$ are central: we shall check this only for the former.

LEMMA 3.1. The distributions T_λ defined above are central.

Proof. First we express T_λ in terms of the Fourier transform of f, and then we compare the Fourier transforms of f and f', defined as above. We have the formula for T_λ:

$$(3.1)\qquad T_\lambda(f) = \int_{F^{n-1}} dx\, T_{Cd(x,0)(0,\lambda)}(f) = \int_{F^{n-1}} dx\, (f\circ\exp)^\wedge(Cd(x,0)(0,\lambda)) \quad .$$

Furthermore,

$$\begin{aligned}
&(f'\circ\exp)^\wedge(\alpha,\lambda)\\
&= \int_V dXdY\, f'(\exp(X,Y))\, e(\langle(\alpha,\lambda)|(X,Y)\rangle)\\
&= \int_V dXdY\, f((x,y)\exp(X,Y)(x,y)^{-1})\, e(\langle(\alpha,\lambda)|(X,Y)\rangle)\\
&= \int_V dXdY\, f\circ\exp(Ad(x,y)(X,Y))\, e(\langle(\alpha,\lambda)|(X,Y)\rangle)\\
&= \int_V dXdY\, f\circ\exp((X,Y))\, e(\langle(\alpha,\lambda)|Ad(x,y)^{-1}(X,Y)\rangle)\\
&= \int_V dXdY\, f\circ\exp((X,Y))\, e(\langle Cd(x,y)(\alpha,\lambda)|(X,Y)\rangle) \quad .
\end{aligned}$$

This last formula is combined with the preceding equality (3.1) to conclude:

$$\begin{aligned}
&T_\lambda(f')\\
&= \int_{F^{n-1}} dx'\, (f'\circ\exp)^\wedge(Cd(x',0)(0,\lambda))\\
&= \int_{F^{n-1}} dx'\, (f\circ\exp)^\wedge(Cd(x,y)Cd(x',0)(0,\lambda))\\
&= \int_{F^{n-1}} dx'\, (f\circ\exp)^\wedge(Cd(x+x',\, y-2\mathrm{Im}(xx'^*))(0,\lambda))\\
&= \int_{F^{n-1}} dx'\, (f\circ\exp)^\wedge(Cd(x+x',0)(0,\lambda))\\
&= \int_{F^{n-1}} dx'\, (f\circ\exp)^\wedge(Cd(x',0)(0,\lambda))\\
&= T_\lambda(f) \quad . \qquad \square
\end{aligned}$$

In fact, it is possible to describe T_λ more explicitly.

LEMMA 3.2. The distribution T_λ is given by the following formula:

$$T_\lambda(f) = (\pi/2|\lambda|)^p \int_{\mathrm{Im}(F)} dy\, f((0,y))\, e(\mathrm{Re}(\lambda y^*))$$

for all f in $C_c^\infty(V)$, where $p = \dim(F^{n-1})$.

Proof. We omit this proof, which uses the ideas already ventilated together with standard Fourier analytic techniques.

The next step of the Kirillov theory is to associate a representation to each orbit O_λ in $\underline{V}^*$. The philosophy behind this is that since the orbits give rise to central distributions, and since representations give rise to central distributions (by taking the character), there might be some connection between the orbits and the representations. We shall describe the general procedure very briefly.

To the element θ of $\underline{V}^*$, we associate a bilinear form B_θ on $\underline{V}$, by defining

$$(3.2) \qquad B_\theta(W,Z) = \theta([W,Z]) \qquad W,Z \in \underline{V}$$

(where $[W,Z]$ is the Lie product of W and Z, i.e. their commutator as matrices). A "polarisation" of $\underline{V}$ is a maximal subalgebra $\underline{H}$ (i.e. a maximal [,]-closed subspace of $\underline{V}$) such that

$$B_\theta(\underline{H},\underline{H}) = \{0\} \quad .$$

Associated to the subalgebra $\underline{H}$ of $\underline{V}$, there is a subgroup H of V; associated to the functional θ of $\underline{V}^*$, there is a character χ_θ of H (i.e. χ_θ is a homomorphism of H into T, where T is the group of complex numbers of modulus one under multiplication), given by the formula

$$\chi_\theta(\exp(W)) = e(\theta(W)) \qquad W \in \underline{H} \quad .$$

One "induces" this character to a representation of V. One of the main results of Kirillov's theory is that we obtain (essentially) the same representation as we vary the polarisation.

Let us now apply this general method to the case in hand. We take λ in $\mathrm{Im}(F)\backslash\{0\}$, and consider the bilinear form B_λ (cf. (3.2)) given by the rule:

$$\begin{aligned} B_\lambda((X,Y),(X',Y')) &= \mathrm{Re}(\lambda[-2\mathrm{Im}(XX'^*)]^*) \\ &= 2\mathrm{Re}(\lambda XX'^*) \quad . \end{aligned}$$

We now meet one of the unsatisfactory aspects of the theory, viz, there is no canonical choice of polarisation. We suppose that $F \neq R$, for otherwise this discussion would be futile. If $F = C$, then it is possible to take $\underline{H}$ as follows

$$(3.3) \qquad \underline{H} = \underline{iR}^{n-1} + \underline{V}_2 \subset \underline{V}_1 + \underline{V}_2 ,$$

while if $F = Q$, we take

$$\underline{H} = (aR^{n-1} + bR^{n-1}) + \underline{V}_2 \subset \underline{V}_1 + \underline{V}_2 \quad ,$$

where a and b are two perpendicular vectors in the (real) space F which also satisfy a λ-dependent condition. The point is that a and b must vary with λ, but there is no particularly good way to choose them.

Once $\underline{H}$ has been chosen, we employ the induction procedure. We obtain first the character χ_λ of H, and then consider the space of (measurable) functions on V which transform by χ_λ on the left, i.e. such that

$$f(hv) = \chi_\lambda(h)\ f(v) \qquad h \in H,\ v \in V \quad ,$$

and which satisfy the integrability condition that $\|f\|$ be finite, where

$$\|f\| = (\int_{H\backslash V} d\dot{v}\ |f(\dot{v})|^2)^{1/2} \quad .$$

We shall illustrate matters for the case where $F = C$.

If $F = C$, more explicit parametrisation is possible. We write $\lambda = \underline{i}\nu$, with ν in R, and $x = u + \underline{i}v$, with u,v in R^{n-1}, and also $y = \underline{i}w$, with w in R. We take $\underline{H}$ to be the subalgebra indicated in the formula (3.3) above and functions f such that

$$f((\underline{i}v',\underline{i}w')(u+\underline{i}v,\underline{i}w)) = e(\nu w')f((u+\underline{i}v,\underline{i}w)) \quad .$$

This transformation condition may be rephrased as follows:

$$f((u + \underline{i}v,\ \underline{i}w)) = e(\nu[w + 2vu^*])\ f((u,0)) \quad ,$$

and f is determined by its restriction to $R^{n-1}\times\{0\}$, which is a "section" for $H\backslash V$. We use the norm:

$$\|f\| = [\int_{R^{n-1}} dr\ |f(r,0)|^2]^{1/2} \quad .$$

Now we let V act on the space of such functions, on the right, thus:

$$(\pi_\lambda(v)f)(v') = f(v'v) \qquad v,v' \in V \quad .$$

With the notation $(x,y) = (u + \underline{i}v,\ \underline{i}w)$, we find that (for the complex case)

$$\begin{aligned}
&[\pi_\lambda(u + \underline{i}v,\ \underline{i}w)f]((r,\ 0)) \\
&= f((r,\ 0)(u + \underline{i}v,\ \underline{i}w)) \\
&= f((r + u + \underline{i}v,\ \underline{i}w + 2\underline{i}rv^*)) \\
&= e(\nu[w + 2rv^* + 2v(r + u)^*])\ f((r + u,\ 0)) \\
&= e(\nu[w + 4rv^* + 2vu^*])\ f((r + u,\ 0)) \quad .
\end{aligned}$$

If we write F(r) instead of f((r, 0)), we have that

$$[\pi_\lambda(u + \underline{i}v,\ \underline{i}w)F](r) = \exp(i\nu[w + 4rv^* + 2vu^*])\ F(r + u) \quad .$$

Thus $\pi_\lambda((u,\ 0))$ acts by translations, $\pi_\lambda((\underline{i}v,\ 0))$ acts by multiplication by characters, and $\pi_\lambda((0,\ \underline{i}w))$ by scalars.

Let us consider some Fourier transforms. It is natural to define $\pi_\lambda(X)$ by the rule that

$$\pi_\lambda(X)f(r) = [d/dt\ \pi_\lambda((tX,\ 0))f(r)]_{t=0} \, ,$$

and so on. If $X_1,\ldots,X_{n-1}$ are the vectors $(1,0,\ldots,0),\ldots,\ (0,0,\ldots,1)$ in $\underline{V}_{-1}$, then

$$\pi_\lambda(X_j)f(r) = (\partial/\partial r_j)f(r)$$

and

$$\pi_\lambda(\underline{i}X_j)f(r) = 4i\nu r_j\, f(r) \quad .$$

It follows that $-\{\pi_\lambda(X_j)^2 + \pi_\lambda(\underline{i}X_j)^2\}$ is the operator $-\partial^2/\partial r_j^{\,2} + 16\nu^2 r_j^{\,2}$; this operator becomes the Hermite operator $(-d^2/dt^2 + t^2)$ when we effect the change of variables $s = 2|\lambda|^{1/2} r$. In fact

$$-\{\pi_\lambda(X_j)^2 + \pi_\lambda(\underline{i}X_j)^2\}\, f(s) = 4|\lambda|\, \{-\partial^2/\partial s_j^{\,2} + s_j^{\,2}\}\, f(s) \quad .$$

The Hermite operator on R is known to have a discrete spectrum, with eigenvalues 1, 3, 5, ..., and 1-dimensional eigenspaces generated by the Hermite functions D_0, D_1, D_2, ... It follows that, if we form functions of the form

$$D_{m_1} \otimes D_{m_2} \otimes \ldots \otimes D_{m_{n-1}}$$

on R^{n-1}, then these are eigenfunctions for $\pi_\lambda(\Delta)$, and

$$\pi_\lambda(\Delta)[D_{m_1} \otimes \ldots \otimes D_{m_{n-1}}] = 4|\lambda|[\textstyle\sum_{j=1}^{n-1} (2m_j + 1)]D_{m_1} \otimes \ldots \otimes D_{m_{n-1}} \quad ;$$

the expansion we obtain in this way is a complete eigenfunction expansion for $\pi_\lambda(\Delta)$. The eigenvalues are $4|\lambda|(n - 1)$, $4|\lambda|(n + 1)$, $4|\lambda|(n + 3)$, and so on. It may be worth pointing out that the action of M on V gives rise to an action on each of the eigenspaces, which permutes the order of the D_{m_j} and takes weighted sums of these.

It is appropriate to view $\pi_\lambda(\Delta)$ as a diagonal operator of the form

$$2|\lambda| \begin{vmatrix} p & 0 & 0 & \\ 0 & p+4 & 0 & \ldots \\ 0 & 0 & p+8 & \\ & \ldots & & \ldots \end{vmatrix}$$

where the blocks correspond to the different eigenspaces. Thus $\pi_\lambda(\Delta^2)$ corresponds to

$$4|\lambda|^2 \begin{vmatrix} p^2 & 0 & 0 & \\ 0 & (p+4)^2 & 0 & \ldots \\ 0 & 0 & (p+8)^2 & \\ & \ldots & \ldots & \ldots \end{vmatrix}$$

The same holds when $F = Q$: p is $\dim_R(F^{n-1})$, as before.

Analogously, $\pi_\lambda(\square)$ corresponds to $|\lambda|^2 I$, where I is the identity operator.

The reason for this analysis is the following lemma, which is proved by the methods of §1.

LEMMA 3.3. The following equality holds:

$$(\Delta^2 + B(\zeta)\square)\, N^{\zeta-r} = C(\zeta)\, N^{\zeta-r-4}$$

where $B(\zeta) = -4(\zeta-2)^2$ and $C(\zeta) = (\zeta-r)(\zeta-2)(\zeta-p-2)(\zeta-4)$.

Proof. Omitted.

On the Fourier transform side, we find that there are recurrence relations which connect the Fourier transforms of the functions $N^{\zeta-r}$. These Fourier transforms are also diagonal, and are made up of blocks of the same size as the blocks which make up the Fourier transform of Δ. We obtain, for the kth block, the equality

$$C(\zeta)\, \pi_\lambda(N^{\zeta-r-4})_k = 4|\lambda|^2\, [(p + 4k - 4)^2 - (\zeta - 2)^2]\, \pi_\lambda(N^{\zeta-r})_k$$

($k = 1, 2, 3, \ldots$). It is clear that, if $N^{\zeta-r-4}$ is positive (or negative) definite, then $N^{\zeta-r}$ is also positive (or negative) definite if and only if

$$(3.4) \qquad (p + 4k - 4)^2 - (\zeta - 2)^2 > 0$$

for all k (this expression could not possibly be negative for all k). This observation will be of use later in our understanding of the "complementary series". Now we note that, if $F = C$, then the condition (3.4) is satisfied if $0 < \zeta < p + 2 = p + 2q$, while if $F = Q$, then this same condition holds only if $\zeta < p + 2 < p + 2q$. This will correspond to the fact that the groups Sp(n,1) ($F = Q$) have "property T", while the groups SU(n,1) ($F = C$) do not.

4. SOME SIMPLE GROUPS

We shall consider the vector space F^{n+1} over F, where scalars act on the left, with basis $(e_1, e_2, \ldots, e_n)$ over F, and the sesquilinear form q given by the formula

$$q(z, \zeta) = \sum_{j=1}^{n} z_j \bar{\zeta}_j - z_0 \bar{\zeta}_0 \qquad z, \zeta \in F^{n+1} .$$

We denote by O(q) the group of linear transformations of F^{n+1} which preserve this form (these transformations act on the right). We shall consider harmonic analysis on O(q). In what follows, we must assume that F is either C or Q, since in the real case there are some (silly) complications due to the non-connectness of O(q) (which has four components); the results stated below are sometimes true for the connected component of O(q) only and sometimes true for the whole group only. We leave to the reader the task of deciding for which of these groups the theorems hold. To tackle harmonic analysis on O(q), we shall need harmonic analysis on some other groups.

In order to describe some of the other groups that we shall meet, we shall consider a second basis of F^{n+1}: let $f^0 = (e_n - e_0)/\sqrt{2}$, $f^n = (e_n + e_0)/\sqrt{2}$, and and $f^j = e_j$ if $1 \leq j \leq n-1$. Coordinates in the new system will be written z^j. Then the following formulae hold:

$$(z^j) = (z_j)\, U \qquad (z_j) = (z^j)\, U \qquad z \in F^{n+1} ,$$

where

$$U = \begin{vmatrix} -c & 0 & c \\ 0 & 1 & 0 \\ c & 0 & c \end{vmatrix} ,$$

the matrix U being made up of blocks of side 1 or n-1, and the central block being (n-1)×(n-1); the number c is $1/\sqrt{2}$. In the new system, the form q is represented as follows:

$$q(z, \zeta) = \sum_{j=1}^{n-1} z^j \bar{\zeta}^j + z^0 \bar{\zeta}^n + z^n \bar{\zeta}^0 .$$

One of the groups of interest is the following:

$$K = O(q) \cap O(|.|) ,$$

where $O(|.|)$ is the compact group of all linear transformations of F^{n+1} which preserve the length of vectors, defined in the usual way. The group K is made up of elements k(v, w), which, in the first coordinate system, are described by matrices of the following form:

$$k(v, w) = \begin{vmatrix} v & 0 \\ 0 & w \end{vmatrix}$$

where v and w are 1×1 and n×n norm-preserving matrices respectively. Every element of K is the exponential of an element of its Lie algebra $\underline{K}$, which is the set of all matrices of the same form, but which satisfy the conditions that $v^* = -v$ and $w^* = -w$. (The adjoint w^* of w is the conjugate transpose.)

Other groups which will attract our attention are best viewed in the other coordinate system: $A = \{a(t)\colon t \in R\}$, where

$$(4.1) \qquad a(t) = \begin{vmatrix} e^{-t} & 0 & 0 \\ 0 & 1 & 0 \\ 0 & 0 & e^{t} \end{vmatrix} \qquad t \in R \quad ,$$

$V = \{v(x, y)\colon x \in F^{n-1}, y \in \mathrm{Im}(F)\}$, where v(x, y) is the element V(x, y) of §3; $N = \{n(x, y)\colon x \in F^{n-1}, y \in \mathrm{Im}(F)\}$, where

$$(4.2) \qquad n(x, y) = \exp\begin{vmatrix} 0 & 0 & 0 \\ -x & 0 & 0 \\ y/2 & x & 0 \end{vmatrix} = \begin{vmatrix} 1 & 0 & 0 \\ -x & 1 & 0 \\ (y-xx^*)/2 & x & 1 \end{vmatrix} \quad ;$$

$M = \{m(v, u)\colon m(v, u) \in K\}$, where

$$(4.3) \qquad m(v, u) = \begin{vmatrix} v & 0 & 0 \\ 0 & u & 0 \\ 0 & 0 & v \end{vmatrix} \quad .$$

It turns out that v satisfies the condition $|v| = 1$, while u is an isometric $(n-1)\times(n-1)$ matrix. The group V is exactly the group which we have discussed for the first three sections.

It is easy to check that M centralises A (in fact M is the centraliser of A in K), and both M and A normalise V and N. Thus MA, MN, AN, MAN, MV, AV, and MAV are all subgroups of G. The action of M and A on V of §1 is conjugation.

It will be helpful to study certain natural geometric actions of the group G. Since G preserves the form q, it preserves the following cone Γ:

$$\Gamma = \{z \in F^{n+1}\colon q(z, z) < 0\} \quad .$$

The unit disc D in F^n ($D = \{z \in F^n\colon |z| < 1\}$) can be identified with the projective space $F^*\backslash\Gamma$ as follows:

$$z \leftrightarrow (1, z) \leftrightarrow \{(w, wz)\colon w \in F^*\}$$

where $F^* = F\backslash\{0\}$. Then, since the G-action commutes with the F-action, we have a G-action on D by fractional linear transformations:

$$z\circ g = (a + zc)^{-1}(b + zd) \qquad z \in F^n ,$$

where, in the first coordinate system,

$$g = \begin{vmatrix} a & b \\ c & d \end{vmatrix} ,$$

a being a 1×1 matrix and d being n×n. We find immediately that

$$0\circ g = 0$$

if and only if g ∈ K, and that K acts on D by rotations. Hence 0 is the unique K-fixed point of D. The action of A on D may be calculated as follows: in the first coordinate system,

$$a(t) = U \begin{vmatrix} e^{-t} & 0 & 0 \\ 0 & 1 & 0 \\ 0 & 0 & e^{t} \end{vmatrix} U = \begin{vmatrix} c(t) & 0 & s(t) \\ 0 & 1 & 0 \\ s(t) & 0 & c(t) \end{vmatrix} ,$$

where c(t) and s(t) are the hyperbolic cosine and sine of t respectively. Therefore

$$0\circ a(t) = \tanh(t)\, e_n \qquad t \in R .$$

We thus obtain a "polar decomposition" of D:

$$D = 0\circ AK$$

and hence the "Cartan decomposition" of G:

$$G = KAK .$$

We now consider the same G-action in the second system of coordinates. Since $(z\bar{w})^- = w\bar{z}$ $(w, z \in F)$,

$$q(z, z) = \sum_{j=1}^{n-1} |z^j|^2 + 2\mathrm{Re}(z^0\bar{z}^n) \qquad z \in F^{n+1} .$$

If $q(z, z) < 0$, then $\mathrm{Re}(z^0\bar{z}^n) < 0$ and $z^0 \neq 0$. Thus the projective cone may be identified with the "Siegal domain" S:

$$S = \{z \in F^n : h(z) < 0\} ,$$

where h is the height function given by the formula

$$h(z) = \mathrm{Re}(z^n) - (1/2)\sum_{j=1}^{n-1} |z^j|^2 \qquad z \in F^n .$$

The identification proceeds by the two correspondences:

$$z \leftrightarrow (-1, z) \leftrightarrow \{(-w, wz): w \in F^*\} \qquad z \in F^n .$$

Before we study the action of G on S, we state formally the relation between D and S.

LEMMA 4.1. The spaces D and S are isomorphic. In particular, let ϕ and ψ be the maps given by the formulae:

$$\phi(z)^j = (1 - z_j)^{-1}\, 2\, z_j \qquad (1 \leqslant j \leqslant n - 1),$$
$$\phi(z)^n = (1 - z_n)^{-1}\, (1 + z_n) \quad ,$$
$$\psi(z)_j = (z^n + 1)^{-1}\, 2\, z^j \qquad (1 \leqslant j \leqslant n - 1),$$
$$\psi(z)_n = (z^n + 1)^{-1}\, (z^n - 1) \quad .$$

Then ϕ is a bijection of D onto S and of ∂D onto $\partial S \cup \{\infty\}$, which intertwines the G-actions, and ψ is its inverse. Moreover, $\phi(0, 0) = (0, 1)$, $\phi(-1, 0) = (0, 0)$, and

$$h(\phi(z)) = |1 - z_n|^{-2}\, (1 - |z|^2) \qquad z \in D \quad ,$$

whilst
$$|\psi(z)| = 1 - |z^n + 1|^{-2} 4h(z) \qquad z \in S \quad .$$

Proof. Obvious from the considerations above.

We now consider the action of G on S, and on ∂S. First, if $x \in F^{n-1}$, and $y \in \mathrm{Im}(F)$, then for any z in F^{n-1} and w in F,

$$(-1, z, w)\, v(x, y) = (-1, z - x, w - zx^* + (xx^* - y)/2) \quad ,$$

whence
$$(z, w)\circ v(x, y) = (z - x, w - zx^* + (xx^* - y)/2) \quad .$$

We note that, since $\mathrm{Re}(y) = 0$,

$$\begin{aligned} h((z, w)\circ v(x, y)) &= \mathrm{Re}(w - zx^* + (xx^* - y)/2) - (1/2)|z - x|^2 \\ &= \mathrm{Re}(w - zx^*) + (1/2)|x|^2 - (1/2)|z - x|^2 \\ &= \mathrm{Re}(w) - (1/2)|z|^2 \\ &= h((z, w)) \quad , \end{aligned}$$

i.e., V acts on S and on ∂S, preserving the height. Since, if $h \in R$,

$$(4.4) \qquad (0, h)\circ v(x, y) = (-x, h + (xx^* - y)/2) \quad ,$$

then an arbitrary element of S of height h is the image of (0, h) by an appropriate element of V, and further only the identity element e of V fixes (0, h), i.e. V acts simply transitively on the leaves of the foliation induced by h.

Next, we consider the action of M. We have that

$$\begin{aligned} (-1, z, w)\, m(v, u) &= (-v, zu, wv) \\ &= v\, (-1, v^{-1} zu, v^{-1} wv) \quad , \end{aligned}$$

whence in S,

$$(4.5) \quad (-1, z\, , w)\, m(v, u) = (v^{-1} zu, v^{-1} wv) \quad .$$

In particular, M stabilises the elements (0, h) ($h > 0$) of S and (0, 0) in ∂S.

To proceed, we consider the action of A. It is clear that

$$(-1,\ z,\ w)\ a(t) = (-e^{-t},\ z,\ we^{t}) = e^{-t}(-1,\ e^{t}z,\ e^{2t}w)$$

whence in S

(4.6) $$(z,\ w)\circ a(t) = (e^{t}z,\ e^{2t}w) \quad .$$

In particular, $(0,\ 1)\circ a(t) = (0,\ e^{2t})$.

We conclude our study of the action of G on S by considering one special element of K. Let w be that element of G given by the matrices

$$\begin{vmatrix} -1 & 0 \\ 0 & 1 \end{vmatrix} \quad \text{and} \quad \begin{vmatrix} 0 & 0 & 1 \\ 0 & 1 & 0 \\ 1 & 0 & 0 \end{vmatrix}$$

in the first and second coordinate systems respectively (the first matrix has an n×n identity submatrix, while the second contains an (n−1)×(n−1) submatrix in the central position). It is clear that w acts on D by multiplication by −1. We notice also that $w = w^{-1}$, and that

$$w\ a(t)\ w = a(-t) \qquad t \in R \quad .$$

It is easy to check that w commutes with each m in M, and that

$$w\ v(x,\ y)\ w = n(x,\ y) \quad .$$

We find that, in the second coordinate system,

$$(-1,\ z,\ x)\ w = (x,\ z,\ -1) = -x\ (-1,\ -x^{-1}z,\ x^{-1}) \quad ,$$

whence the action of w on S is given by the formula

(4.7) $$(z,\ x)\circ w = (-x^{-1}z,\ x^{-1}) \quad .$$

The same formula applies for (z, x) in ∂S. We now prove a lemma.

LEMMA 4.2. Suppose that v(x, y) be in V, and that $v(x,\ y) \neq e$. Then, for the action of G on ∂S already described, we have the formula

$$(0,\ 0)\circ v(x,\ y)\circ w = (0,\ 0)\circ v(x^{\dagger},\ y^{\dagger}) \quad ,$$

where $$x^{\dagger} = -(xx^{*} - y)^{-1}\ 2x$$

and $$y^{\dagger} = -|xx^{*} - y|^{-2}\ 4y \quad .$$

Proof. The formulae above ((4.4) and (4.7)) imply that

$$(0,\ 0)\circ v(x,\ y) = (-x,\ (xx^{*} - y)/2)$$

and then $$(0,\ 0)\circ v(x,\ y)\circ w = (-2(xx^{*} - y)^{-1}(-x),\ 2(xx^{*} - y)^{-1}) = ((xx^{*} - y)^{-1}2x,\ (xx^{*} - y)^{-1}2) \quad ,$$

while $$(0,\ 0)\circ v(x^{\dagger},\ y^{\dagger}) = ((xx^{*} - y)^{-1}2x,\ (\,|(xx^{*} - y)^{-1}2x|^{2} - y^{\dagger})/2)$$

$$= ((xx^* - y)^{-1}2x,\ |xx^* - y|^{-2}2(xx^* + y))$$
$$= ((xx^* - y)^{-1}2x,\ (xx^* - y)^{-1}2)\ ,$$

as required. □

It follows from our study of the actions of G on D and on S that

$$G = KAV$$

(because (0,1) is the K-fixed point in S, and there is an obvious bijection of AV onto S, given by the formula $av \mapsto (0, 1)\circ av$),

$$G = KAN$$

(obtained from the previous decomposition by conjugating by w), and

$$G = NAK$$

(obtained from the preceding decomposition by taking inverses). Finally,

$$G = ANK$$

(since AN = NA); this is what we shall call the Iwasawa decomposition of G. Further, consideration of the action of G on ∂S leads to the so-called Bruhat decomposition:

$$G = MANV \cup MANw\ .$$

This is a disjoint union, and MANw is of a lower dimension than MANV. For the sake of completeness, we offer a brief discussion of this decomposition.

LEMMA 4.3. Every element g of G has either a unique decomposition of the form g = manv or a unique decomposition of the form g = manw.

Proof. We recall that M centralises A, and that M and A both normalise N, so that MAN is a group. Moreover, it is clear from the definitions ((4.1), (4.2), (4.3)) that

$$M \cap A = M \cap N = N \cap A = \{e\}\ ,$$

so that each element of MAN has a unique expression of the form man.

We now claim that if $g \in G$, and $(0, 0)\circ g = (0, 0)$ (with the action of G on S and ∂S already described), then $g \in MAN$. To show this, we write g (in the second coordinate system) as the following matrix:

$$g = \begin{vmatrix} a & b & c \\ d & e & f \\ g & h & i \end{vmatrix}\ .$$

Since (1, 0, 0) g = (a, b, c), it must be that $(0, 0)\circ g = (0, 0)$ if and only if

$b = 0$ and $c = 0$. Moreover, if g is of this form, then

$$(y, x, 0)\circ g = (ya + xd, xe, xf) \qquad x \in F^{n-1}, y \in F .$$

Since G preserves the form q, the following equality holds for any x in F^{n-1} and y in F:

$$2\mathrm{Re}(y\bar{0}) + |x|^2 = 2\mathrm{Re}((ya + xd)(xf)^-) + |xe|^2 .$$

By letting y vary, we conclude that $xf = 0$; by letting x vary, we deduce that $f = 0$ and that $|x|^2 = |xe|^2$ for all x in F^{n-1}. It now follows that g lies in MAN, as claimed.

If, on the other hand, g lies in MAN then g is of the form just described. It is clear that $(0, 0)\circ g = (0, 0)$. We have thus proved that MAN is the isotropy group of (0, 0) in ∂S.

Now suppose that $g \in G$, and consider $(0, 0)\circ g$. There are two possibilities: either $(0, 0)\circ g \in \partial S$ or $(0, 0)\circ g = \infty$. We shall consider only the first of these possibilities, since the second is treated in a similar manner. There exists a unique v in V such that

$$(0, 0)\circ g = (0, 0)\circ v,$$

and for this v,

$$(0, 0)\circ g\circ v^{-1} = (0, 0) ,$$

i.e., $gv^{-1} \in$ MAN. By now expressing gv^{-1} uniquely in the form man, it follows that g = manv. In the other case, we find analogously that g = manw: it is enough to remark that $(0, 0)\circ w = \infty$. □

More general information about the geometric side of things can be found in S. Helgason's comprehensive work ([Hel]) and the references there cited.

We now need an analytic consequence of the algebraic and geometric considerations, viz, an expression for the invariant measure on G. The first step is to find the commutator subgroup of G, H say. It is easy to see that the commutator subgroup of AN is N and that, with the action of G on the disc D, $0\circ N$ is a connected subset of D which has a limit point on the boundary ∂D. It follows that the closed subgroup KH of G has the property that $0\circ KH = D$. We conclude that KH = G, and that H is co-compact in G.

We may now argue that G must be unimodular, i.e. that a left Haar measure on G is automatically a right Haar measure. For the kernel of the modular function, a homomorphism of G into R^+, contains the commutator subgroup, and is a fortiori co-compact; the modular function must be trivial, and so G is indeed unimodular.

Since G = ANK, it must be possible to express the Haar measure on G by means of a formula of the following type:

$$\int_G dg\ f(g) = \int_A da \int_N dn \int_K dk\ \phi(a, n, k)\ f(ank) \quad ,$$

where dn, da, and dk are the Haar measures on N, A, and K. This formula implies that

$$\begin{aligned}\int_G dg\ f(gk') &= \int_A da \int_N dn \int_K dk\ \phi(a, n, k)\ f(ankk') \\ &= \int_A da \int_N dn \int_K dk\ \phi(a, n, kk'^{-1})\ f(ank)\end{aligned}$$

(by a change of variables on K), and so

$$\phi(a, n, k) = \phi(a, n, e) \quad .$$

An analogous argument shows that ϕ does not depend on the first variable (a). Finally, since

$$\int_G dg\ f(n'g) = \int_A da \int_N dn \int_K dk\ \phi(a, n, k)\ f(a\tilde{n}k) \quad ,$$

where $\tilde{n} = a^{-1}na$, and since $\tilde{n} \in N$, the same reasoning permits us to conclude that

$$\phi(a, n, k) = \phi(e, e, e) \quad .$$

Since we may normalise Haar measure as we like, we choose ϕ to be 1. Thus

$$\int_G dg\ f(g) = \int_A da \int_N dn \int_K dk\ f(ank) \ .$$

It is possible to establish by exactly the same method that

$$(4.8) \qquad \int_G dg\ f(g) = C_G \int_M dm \int_A da \int_N dn \int_V dv\ f(manv) \quad ,$$

where C_G depends only on G.

We conclude this introductory section with some very brief remarks about the adjoint representation, the coadjoint representation, and the orbit structure. We consider only the case where $F = R$ and n = 2, since in general, things get rather messy. For the case in consideration, the Lie algebra consists of all matrices of the form X(a, b, c), where, in the first basis,

$$X(a, b, c) = \begin{vmatrix} 0 & a & b \\ a & 0 & c \\ b & -c & 0 \end{vmatrix} \qquad a, b, c \in R \quad .$$

The adjoint representation, which acts by conjugation, obviously preserves the bilinear form $\tilde{q}$ — $\tilde{q}(X, Y) = tr(XY)/2$ — and since

$$q^{\sim}(X(a, b, c), X(a, b, c)) = a^2 + b^2 - c^2 \quad ,$$

there is an obvious homomorphism of G into $O(q^{\sim})$, whose kernel is $\{\pm I\}$. The coadjoint representation is essentially equivalent to the adjoint representation, and the orbit structure is as follows: there are two-sheeted hyperboloids lying inside the cone $\Gamma^{\sim}$, there are one-sheeted hyperboloids wrapping around the cone, there are the two halves of the cone, and there is the origin.

Attempts to copy Kirillov's theory in this situation meet with several difficulties. One of these is that not all orbits in the coadjoint representation orbit space can give rise to representations: one of the steps in the Kirillov method is the "exponentiation" of a bilinear form on a subalgebra to a character of the corresponding subgroup. Now, just as the linear functional $\theta \longmapsto \lambda\theta$ gives rise to a character of the torus $e^{i\theta} \longmapsto e^{i\lambda\theta}$ only if λ is integral, so in our case only a discrete set of the elliptic orbits (i.e. the two-sheeted hyperboloids) orbits give rise to representations, the so-called discrete series. There is no such obstruction to to extending functionals associated to the hyperbolic (i.e. one-sheeted hyperboloid) orbits, and these give rise to a continuous series of representations, called the principal series, and realised by induction. A second problem with the orbit method is that the construction of the discrete series is not a straightforward use of the method of induction, and is in general quite difficult.

Nevertheless, the orbit picture offers some useful intuitions. For instance, the nilpotent orbits (the half-cones) are limits of both elliptic and hyperbolic orbits; the latter must be cut in half, as it were, to match the two half-cones. These in fact correspond to representations linked with both the principal and the discrete series.

We conclude this discussion by noting that w, the special element of K, maps the orbits into themselves, and that, restricted to the plane perpendicular to the axis of the cone, takes elements to their negatives.

5. THE PRINCIPAL SERIES AND THE INTERTWINING OPERATORS.

From the orbit theory or for other reasons, one suspects that the following series of representations will be of interest. One takes an irreducible unitary representation μ of the group M on a finite-dimensional Hilbert space H_μ, and a character α of A, and forms the representation $\mu \otimes \alpha$ of MAN given by the formula

$$\mu \otimes \alpha : man \longmapsto \mu(m)\,\alpha(a) \quad .$$

This is a representation because N is normal in MAN. We then induce this representation to G. This could be done by considering H_μ-valued functions f on G which transform according to the rule

$$f(mang) = \mu \otimes \alpha(man)\, f(g)$$

(where, obviously, $m \in M$, $a \in A$, etc.) and then letting G act on the right on this space, thus obtaining a representation $\pi^\sim_{\mu,\alpha}$:

$$(\pi^\sim_{\mu,\alpha}(g)f)(g') = f(g'g) \qquad g,g' \in G \quad .$$

In order to have a Hilbert space norm, we should integrate over V or K (because f is essentially determined by its values on V or on K): we define the norms $\|.\|_2^{(K)}$ and $\|.\|_2^{(V)}$ by the formulae:

$$\|f\|_2^{(K)} = (\int_K dk\ \|f(k)\|^2)^{1/2}$$
$$\|f\|_2^{(V)} = (\int_V dv\ \|f(v)\|^2)^{1/2} \quad .$$

We now encounter the problem that G does not act isometrically on these spaces. For instance,

$$\|\pi^\sim_{\mu,\alpha}(a)f\|_2^{(V)} = (\int_V dv\ \|f(aa^{-1}va)\|^2)^{1/2}$$
$$= (\int_V dv\ \|f(a^{-1}va)\|^2)^{1/2}$$
$$\neq (\int_V dv\ \|f(v)\|^2)^{1/2} \quad ,$$

unless a is trivial. The dilations $D^s : v \longmapsto a_{-t}va_t$ $(s = e^t)$ are not measure-preserving; a twist with the Radon-Nikodym derivative is needed to put things right.

We define $\rho: A \longrightarrow \mathbb{R}^+$ by the rule

$$\rho(a_t) = \exp(rt/2) \quad ,$$

where, as usual,

$$r = \dim_{\mathbb{R}}(F^{n-1}) + 2\dim_{\mathbb{R}}(\mathrm{Im}(F)) \quad .$$

Then the following integral formulae hold:

$$\text{(5.1)} \qquad \int_V dv\, f(a^{-1}va) = \rho(a)^{-2} \int_V dv\, f(v)$$

$$\text{(5.2)} \qquad \int_N dn\, f(a^{-1}na) = \rho(a)^{2} \int_N dn\, f(n) \quad .$$

We may now define the induced representations in such a way that they are unitary.

LEMMA 5.1. Let μ be an irreducible unitary representation of M on the space H_μ and let α: A $\longrightarrow$ T be a character of A. Let f be a measurable H_μ-valued function such that

$$f(\mathrm{mang}) = \mu(m)\ (\rho\alpha)(a)\ f(g) \quad .$$

Then the following equality holds:

$$\int_K dk\, \|f(k)\|^2 = C_G \int_V dv\, \|f(v)\|^2 \quad ,$$

where C_G is the constant of the formula (4.8), and consequently the group G acts unitarily on $H_{\mu,\alpha}^{(V)}$, the Hilbert space of functions f for which the above integral converges, equipped with the norm $\|.\|_2^{(V)}$.

Proof. We take any function ϕ in $C_c(AN)$ such that

$$\int_{AN} dadn\, \phi(an) = 1 \quad ,$$

and $\phi(an) = \phi(manm^{-1})$, and define the function ψ on G by the formula

$$\psi(ank) = \phi(an)\ \rho(a)^{-2} \quad .$$

Then, by changing variables, we find that

$$\begin{aligned}
&\int_K dk\, \|f(k)\|^2 \\
&= \int_{AN} dadn \int_K dk\, \phi(an)\ \|f(k)\|^2 \\
&= \int_{AN} dadn \int_K dk\, [\psi(ank)\ \rho(a)^2]\ |(\rho\alpha)(a)^{-1}\|f(ank)\||^2 \\
&= \int_G dg\, \psi(g)\ \|f(g)\|^2 \\
&= C_G \int_{MAN} dmdadn \int_V dv\, \psi(manv)\ \|f(manv)\|^2 \\
&= C_G \int_{AN} dadn \int_V dv\, \psi(anv)\ \rho(a)^2\ \|f(v)\|^2 \quad .
\end{aligned}$$

We write $v = A^{\sim}(v)N^{\sim}(v)K^{\sim}(v)$ to express the Iwasawa decomposition of v in V, and let $n^{\sim}$ be $A^{\sim}(v)^{-1}nA^{\sim}(v)$; then

$$\int_K dk\, \|f(k)\|^2$$

$$= C_G \int_{AN} da\,dn \int_V dv\, \psi(aA^\sim(v)n^\sim N^\sim(v)K^\sim(v))\, \rho(a)^2\, \|f(v)\|^2$$

$$= C_G \int_{AN} da\,dn \int_V dv\, \psi(aA^\sim(v)n^\sim)\, \rho(a)^2\, \|f(v)\|^2$$

$$= C_G \int_{AN} da\,dn \int_V dv\, \psi(aA^\sim(v)n)\, \rho(a)^2\, \rho(A^\sim(v))^2\, \|f(v)\|^2$$

$$= C_G \int_{AN} da\,dn \int_V dv\, \phi(an)\, \|f(v)\|^2$$

$$= C_G \int_V dv\, \|f(v)\|^2 \quad ;$$

the critical step is the change of variables on N, justified by the formula (5.2). Now

$$\int_K dk\, \|f(k)\|^2 = C_G \int_V dv\, \|f(v)\|^2 \quad .$$

To show that G acts unitarily on $H^{(V)}_{\mu,\alpha}$ is now trivial. Both K and V act unitarily, since right translations preserve the $L^2(K)$- and $L^2(V)$-norms, and K and V generate the whole group G. □

It will be convenient to reparametrise these representations: by α_ζ we denote the exponential given by the formula

$$\alpha_\zeta : a(t) \longmapsto \exp(\zeta t/2) \quad ,$$

and we write $\pi_{\mu,\zeta}$ for the representation of G on $H^{(V)}_{\mu,\zeta}$, the Hilbert space of Lemma 5.1. Of course, in the above ζ is purely imaginary; if ζ has a real component then the corresponding exponential is unbounded (cf. ρ). It is not hard to modify the proof of Lemma 5.1 to obtain the following result.

LEMMA 5.2. Suppose that ζ is in the tube T, given by the formula

$$T = \{\zeta \in \mathbb{C} : \mathrm{Re}(\zeta) \in [-r, r]\} \, ,$$

and let p be given by the formula

$$1/p = \mathrm{Re}(\zeta)/2r + 1/2 \quad .$$

Then, if f is a measurable function on G such that

$$f(mang) = \mu(m)\, (\rho\alpha_\zeta)(a)\, f(g) \quad ,$$

and if C_G is defined as in the formula (4.8), then

$$\int_K dk\, \|f(k)\|^p = C_G \int_V dv\, \|f(v)\|^p \quad ,$$

and G acts isometrically on $H^{(V,p)}_{\mu,\zeta}$, the space of such functions f for which $\|f\|^{(V)}_p$ is finite, equipped with the same norm. If $p = \infty$, the integral is replaced by the essential supremum.

Proof. Suffice it to remark that, as in the proof of Lemma 5.1 we use that fact that

$$|\rho\alpha(a)|^2 = \rho(a)^2 ,$$

in the present situation we have the formula

$$|\rho\alpha_\zeta(a)|^p = \rho(a)^2 ;$$

the factor $\rho(a)^2$ is the Jacobian which makes the argument work. □

In studying these representations, it is especially interesting to consider the "K-finite vectors", i.e., those f which transform under K by finite dimensional representations. For this study, it is convenient to use the so-called compact picture of the representations $\pi_{\mu,\zeta}$: one restricts attention to K. Thus

$$\begin{aligned}[\pi_{\mu,\zeta}(g)f](k) &= f(kg) \\ &= f(A^{\sim}(kg)N^{\sim}(kg)K^{\sim}(kg)) \\ &= (\rho\alpha_\zeta)(A^{\sim}(kg))\ f(K^{\sim}(kg)) \quad .\end{aligned}$$

If we consider the H_μ-norms, then

$$\|\pi_{\mu,\zeta}(g)f(k)\| \leqslant |\rho\alpha_\zeta(A^{\sim}(kg))|\ \|f(K^{\sim}(kg))\| \quad .$$

For fixed g in G, $A^{\sim}(kg)$ and $K^{\sim}(kg)$ range over compact sets as k varies in K, and the representation always acts by bounded operators on all the spaces $H_{\mu,\zeta}^{(K,p)}$ for any ζ in $\mathcal{C}$. The representation obviously depends holomorphically on ζ. As is customary (at least at the present time), we shall consider only those representations for which the K-finite vectors form a dense subspace (in some sensible topology) of the representation space.

We now ask how the K-finite vectors or the C^∞-vectors (i.e. those whose restriction to K is C^∞) look in the "noncompact picture", obtained when we restrict our attention to V. If $f \in H_{\mu,\zeta}^K$ or $H_{\mu,\zeta}^\infty$ (the spaces of K-finite and C^∞ vectors respectively), then

$$\begin{aligned} f(v) &= f(A^{\sim}(v)N^{\sim}(v)K^{\sim}(v)) \\ &= (\rho\alpha_\zeta)(A^{\sim}(v))\ f(K^{\sim}(v)) \quad ,\end{aligned}$$

and since $\|f(K^{\sim}(v))\|$ is bounded,

(5.3) $$\|f(v)\| \leqslant C\ \rho\alpha_\zeta(A^{\sim}(v)) \quad .$$

It is therefore interesting to know what $A^{\sim}(v)$ looks like.

LEMMA 5.3. With the notation established above, the following formulae are valid:

$$\rho(A^\sim(v))^2 = |1 + (|x|^2 + y)/2|^{-r} ,$$

$N^\sim(v) = n(x^\sim, y^\sim)$, where

$$x^\sim = -[1 + (|x|^2 - y)/2]\,|1 + (|x|^2 + y)/2|^{-1}\, x$$

$$y^\sim = -y,$$

and $K^\sim(v) = N^\sim(v)^{-1} A^\sim(v)^{-1} v$.

Proof. To calculate $A^\sim(v)$ and $N^\sim(v)$, there is a simple trick. If $v = ank$, then $v^{-1} = k^{-1} n^{-1} a^{-1}$, and $v^{-1} w = k^{-1} w(wn^{-1} w)a$. Thus, with the usual G-action on S, we have that

$$(0, 1)\circ v^{-1} w = (0, 1)\circ k^{-1} w(wn^{-1} w)a = (0, 1)\circ(wn^{-1} w)a.$$

Since V preserves height, while

$$h((x, y)\circ a(t)) = e^{2t}\, h((x, y)) ,$$

we calculate the number $A^\sim(v)$ by finding the height of $(0, 1)\circ v^{-1}w$, which is easy. We find $N^\sim(v)$ similarly. □

The formula for $(A^\sim(v))$ we obtained permits the calculation of the constant C_G in the integral formula for Haar measure: it must be true that

$$C_G \int_V dxdy\, |1 + (|x|^2 + y)/2|^{-r} = 1 ,$$

whence

$$C_G = 2^{p/2}\, \omega(p + q + 1)^{-1} ,$$

where $\omega(p)$ is the "area" of the unit sphere in R^p.

For later use, we establish another integral formula. We write a generic element g of G in the form

$$g = M(g)A(g)N(g)V(g)$$

to express its "Bruhat decomposition".

LEMMA 5.4. The Jacobian of the map $v \mapsto V(vg)$ is $\rho(A(vg))^2$.

Proof. Suppose that ψ be an integrable function on V. By ψ° we denote the function on G given by the formula

$$\psi^\circ(manv) = \rho(a)^2\, \psi(v) .$$

Then ψ° lies in $H_{1,\rho}^{(V,1)}$, on which G acts isometrically, whence

$$\int_V dv\, \rho(A(vg))^2\, \psi(V(vg)) = \int_V dv\, \psi^\circ(vg) = \int_V dv\, \psi(vg) ,$$

and the result follows. □

We now introduce the intertwining operators. There is reason to suspect that the representations $\pi_{\mu,\zeta}$ and $\pi_{\mu,-\zeta}$ might be "equivalent", and the intertwining operators will express this equivalence. We might also suspect that the Weyl group element w of G might play a special role in all this. It is so.

It is evident that, if

$$f(mang) = \mu(m)\ (\rho\alpha_\zeta)(a)\ f(g) \quad ,$$

then, if we define ${}_wf$ by setting ${}_wf(g)$ equal to f(wg), we have that

$$\begin{aligned} {}_wf(mang) &= f(wmang) \\ &= f(ma^{-1}wng) \\ &= \mu(m)\ (\rho\alpha_\zeta)(a)^{-1}\ {}_wf(ng) \\ &= \mu(m)\ \rho(a)^{-2}\ (\rho\alpha_{-\zeta})(a)\ {}_wf(ng) \quad ; \end{aligned}$$

we are near to effecting a transformation from $H_{\mu,\zeta}$ to $H_{\mu,-\zeta}$, but we have an undesirable factor of $\rho(a)^{-2}$ to eliminate, and further we cannot "pull out the n". An integration over N will fix both these problems.

We define the intertwining operator A(w, μ, α), also written as A(w, μ, ζ) if α is α_ζ, by the formula

$$A(w,\ \mu,\ \alpha)f(g) = \int_N dn\ {}_wf(ng) \quad .$$

At least formally, A(w, μ, α) has the desired property:

$$\begin{aligned} A(w,\ \mu,\ \alpha)f(\tilde{n}g) &= \int_N dn\ {}_wf(n\tilde{n}g) \\ &= \int_N dn\ {}_wf(ng) \quad , \end{aligned}$$

by the translation-invariance of Haar measure on N. By similar change of variable arguments we see that

$$\{A(w,\ \mu,\ \alpha)f\}(mg) = \mu(m)\ A(w,\ \mu,\ \alpha)f(g) \quad ,$$

and

$$\begin{aligned} \{A(w,\ \mu,\ \alpha)f\}(ag) &= \int_N dn\ f(wnag) \\ &= \int_N dn\ f(a^{-1}wa^{-1}nag) \\ &= \int_N dn\ \rho(a)^2\ f(a^{-1}wng) \\ &= \int_N dn\ \rho(a)^2\ (\rho\alpha)(a^{-1})\ f(wg) \\ &= (\rho\alpha^{-1})(a)\ A(w,\ \mu,\ \alpha)f(g) \quad . \end{aligned}$$

The question is, of course, the existence of the integral.

There are two different methods to attack the problem of the existence of the intertwining operator: one may work on the group K or on the group V. We shall use both these approaches. Still working formally, we have that

$$A(w,\ \mu,\ \alpha)f(g) = \int_N dn\ f(wng)$$
$$= \int_V dv\ f(vwg) \quad .$$

Thus, on K, we have that

$$A(w,\ \mu,\ \alpha)f(k) = \int_V dv\ f(vwk)$$
$$= \int_V dv\ f(A^{\sim}(v)N^{\sim}(v)K^{\sim}(v)wk)$$
$$= \int_V dv\ (\rho\alpha)(A^{\sim}(v))\ f(K^{\sim}(v)wk) \quad .$$

If $\zeta = \xi + i\eta$, with ξ positive, and if $f \in H^{\infty}_{\mu,\zeta}$, then, from Lemma 5.3,

$$\int_V dv\ \|(\rho\alpha_\zeta)(A^{\sim}(v))\ f(K^{\sim}(v)wk)\|$$
$$\leq C \int_V dv\ (\rho\alpha_\xi)(A^{\sim}(v))$$
$$= C \int_V dxdy\ |1 + (|x|^2 - y)/2|^{-r(1+\xi)/2}$$
$$< \infty \quad .$$

The integral therefore converges for smooth functions f.

The alternative approach to the existence problem is to write the operator as a convolution on V. We have that

$$A(w,\ \mu,\ \alpha)f(v^*) = \int_V dv\ f(vwv^*) \quad .$$

Now we write vw in the form M(vw)A(vw)N(vw)V(vw); note that, if $v^\dagger = V(vw)$, then

$$v^\dagger w = [M(vw)A(vw)N(vw)]^{-1} v$$

whence
$$v^\dagger w = M(v^\dagger w)A(v^\dagger w)N(v^\dagger w)v,$$

where
$$M(v^\dagger w) = M(vw)^{-1} \quad , \qquad A(v^\dagger w) = A(vw)^{-1} \quad .$$

It follows that

$$A(w,\ \mu,\ \alpha)f(v^*) = \int_V dv\ f(M(vw)A(vw)N(vw)v^\dagger v^*)$$
$$= \int_V dv^\dagger\ f(M(v^\dagger w)A(v^\dagger w)N(v^\dagger w)vv^*)$$
$$= \int_V dv\ (dv^\dagger/dv)\ \mu(M(vw))^{-1}\ (\rho\alpha)(A(vw))^{-1}\ f(vv^*) \quad .$$

But since Lemma 5.4 tells us that $(dv^\dagger/dv) = \rho(A(vw))^2$,

$$A(w,\ \mu,\ \alpha)f(v^*) = \int_V dv\ (\rho\alpha^{-1})(A(vw))\ \mu(M(vw))^{-1}\ f(vv^*)$$
$$= K_{\mu,\alpha} * f(v^*) \quad ,$$

where
$$K_{\mu,\alpha}(v) = (\rho\alpha^{-1})(A(v^{-1}w))\ \mu(M(v^{-1}w))^{-1} \quad .$$

It is clearly worth our while to investigate more fully the Bruhat decomposition.

LEMMA 5.5. Suppose that v is v(x, y) in V. Then, in the second coordinate system

$$M(vw) = \begin{vmatrix} c & 0 & 0 \\ 0 & d & 0 \\ 0 & 0 & c \end{vmatrix} ,$$

where $$c = (y - |x|^2)/|y - |x|^2|$$

and $$d = 1 - x^*(|x|^2 - y)^{-1} 2x ,$$

and so $$M((D^s v)w) = M(vw) ,$$

while $$A(vw) = a(t) ,$$

where $$e^t = |(y - |x|^2)/2|^{-1} = 2N(v)^{-2} ,$$

and therefore $$\rho(A((D^s v)w)) = s^{-r} \rho(A(vw)) .$$

Proof. In Lemma 4.2, we showed that, for the action of G on ∂S, we have the formula

$$(0, 0)\circ v(x, y)w = (0, 0)\circ v(x^+, y^+)$$

where $$x^+ = -(|x|^2 - y)^{-1} 2x$$

Since $(0, 0)\circ man = (0, 0)$, the element $v(x^+, y^+)$ described in Lemma 4.2 is exactly the element $v^+(x, y)$, i.e. $V(v(x, y)w)$. We now make an explicit calculation: if $vw = m^+a^+n^+v^+$, then $m^+a^+n^+ = vwv^{+-1}$, whence, with our coordinates,

$$M(vw)A(vw)N(vw)$$

$$= \begin{vmatrix} 1 & x & (y-|x|^2)/2 \\ 0 & 1 & -x^* \\ 0 & 0 & 1 \end{vmatrix} \begin{vmatrix} 0 & 0 & 1 \\ 0 & 1 & 0 \\ 1 & 0 & 0 \end{vmatrix} \begin{vmatrix} 1 & -x^+ & (-y^+-|x^+|^2)/2 \\ 0 & 1 & -x^{+*} \\ 0 & 0 & 1 \end{vmatrix}$$

$$= \begin{vmatrix} (y-|x|^2)/2 & * & * \\ * & 1 + x^*x^+ & * \\ * & * & * \end{vmatrix} .$$

Since the matrix represents an element of MAN, the upper right hand entries must be zero, and the lemma follows. □

It now follows that, on V, the kernel of the intertwining operator is of the form

$$K_{\mu,\zeta} = 2^{r/2 - \zeta/2} \Omega_\mu N^{\zeta - r} ,$$

where $\Omega_\mu(v) = \mu(M(v^{-1}w)^{-1})$, and $M(vw)$ is given by the preceding lemma. We note that Ω_μ is even, in the sense that $\Omega_\mu(v(x, y)) = \Omega_\mu(v(-x, y))$. Such kernels have been analysed (this kernel is $\mathrm{Hom}(H_\mu)$-valued, but H_μ is finite-dimensional and there are no problems in checking that the earlier theorems hold in this case).

It is now plain sailing to show that the integral defining $A(w, \mu, \zeta)$ converges if $\mathrm{Re}(\zeta) > 0$ (the possible problems at infinity can be controlled by the decrease of $H_{\mu,\zeta}^{\infty}$-functions on V (see formula (5.3) and Lemma 5.3)). Further, the analytic map $\zeta \mapsto A(w, \mu, \zeta)$ defined for ζ such that $\mathrm{Re}(\zeta) > 0$ extends meromorphically to the whole complex plane, with possible poles at 0, -2, -4, ...

It may be worthy of note that, using the compact picture, we may also establish the meromorphic continuation of the operator, but it is harder, and certain fine points, such as the conclusion that -1, -3, -5, etc. are not poles, are not at all obvious.

The seventh section will be devoted to, amongst other things, some applications of the earlier analysis on V to the representation theory of G via the intertwining operators. The rest of this section will be used to describe the asymptotic behaviour of matrix coefficients.

Before we discuss the asymptotics, let us note that, if $f \in H_{\mu,\zeta}$ and $f' \in H_{\mu,-\bar{\zeta}}$, then we may form the inner product $\langle f, f'\rangle$ in H_{μ}, and we thus obtain a function on G which satisfies the relation:

$$\begin{aligned}\langle f, f'\rangle(mang) &= \langle f(mang), f'(mang)\rangle \\ &= \langle \mu(m)\,(\rho\alpha_{\zeta})(a)\, f(g),\ \mu(m)\,(\rho\alpha_{-\bar{\zeta}})(a)\, f'(g)\rangle \\ &= \rho(a)^2 \langle f(g), f'(g)\rangle \quad .\end{aligned}$$

We may therefore form the inner products $\langle f, f'\rangle^{(K)}$ and $\langle f, f'\rangle^{(V)}$, defined by the formulae

$$\begin{aligned}\langle f, f'\rangle^{(K)} &= \int_K dk\, \langle f(k), f'(k)\rangle \\ \langle f, f'\rangle^{(V)} &= \int_V dv\, \langle f(v), f'(v)\rangle \quad ,\end{aligned}$$

and, as in Lemma 5.1, we find that

$$\langle f, f'\rangle^{(K)} = C_G \langle f, f'\rangle^{(V)} \quad .$$

We deduce that the inner products above are G-invariant, and $H_{\mu,\zeta}$ and $H_{\mu,-\bar{\zeta}}$ are naturally dual. In particular, it follows that

$$\langle \pi_{\mu,\zeta}(g)f, f'\rangle^{(V)} = \langle f, \pi_{\mu,-\bar{\zeta}}(g^{-1})f'\rangle^{(V)} \quad .$$

It is interesting that, if $1/p = \xi/2r + 1/2$ and if $1/p' = -\xi/2r + 1/2$, then $1/p + 1/p' = 1$, and the Lebesgue spaces associated to the dual spaces $H_{\mu,\zeta}$ and $H_{\mu,-\bar{\zeta}}$ are dual in the functional analytic sense.

We conclude this section by describing the asymptotic behaviour of the matrix coefficients of the representations $\pi_{\mu,\zeta}$.

LEMMA 5.6. Let f_1 and f_2 be in $H_{\mu,-\bar{\zeta}}^{\infty}$ and $H_{\mu,\zeta}^{\infty}$ with ξ positive. Then there exists a positive ξ-dependent number ε such that, if $t \in R^+$,

$$
\begin{aligned}
&|(\rho\alpha_{-\zeta})(a(t))\langle \pi_{\mu,\zeta}(k_1 a(t) k_2) f_2, f_1\rangle \\
&\qquad - \langle A(w, \mu, \zeta) f_2(wk_2), f_1(k_1^{-1})\rangle| \\
&\leqslant C(\xi, \varepsilon)\, C(f_1|_K, f_2|_K)\, (1 + |\eta|) \exp(-\varepsilon t) \quad .
\end{aligned}
$$

Proof. Omitted.

LEMMA 5.7. Let f_1 and f_2 be in $H_{\mu,-\bar{\zeta}}^{\infty}$ and $H_{\mu,\zeta}^{\infty}$ with ξ equal to 0. Then there exists a positive number ε such that, if $t \in R^+$,

$$
\begin{aligned}
&|\rho(a(t))\langle \pi_{\mu,\zeta}(k_1 a(t) k_2) f_2, f_1\rangle \\
&\qquad - \alpha_{\zeta}(a(t))\langle A(w, \mu, \zeta) f_2(wk_2), f_1(k_1^{-1})\rangle \\
&\qquad - \alpha_{-\zeta}(a(t))\langle f_2(wk_2), A(w, \mu, -\bar{\zeta}) f_1(k_1)\rangle| \\
&\leqslant C(\eta, \varepsilon)\, C(f_1|_K, f_2|_K) \exp(-\varepsilon t) \quad .
\end{aligned}
$$

Proof. Omitted.

6. L^p-HARMONIC ANALYSIS ON SOME SIMPLE GROUPS

In this section, we investigate the role played by the analytic families of representations $\pi_{\mu,\zeta}$ in the harmonic analysis of G. First we consider the convolution algebra $L^1(G)$, and later we discuss L^p-analysis on G.

There has been some interest in the spectral properties of the algebra $L^1(G)$. L. Ehrenpreis and F.I. Mautner [EM1], [EM2], [EM3] studied the algebra for certain particular groups G. They showed that it has two strange properties: first, it is not symmetric, which means that there exist functions f in $L^1(G)$ such that $f = f^\sim$ (here $^\sim$ is the involution of the convolution algebra $L^1(G)$: $f^\sim(g) = \bar{f}(g^{-1})$ for unimodular groups) but whose spectrum is not real, and second, that it contains "non-Tauberian ideals", i.e. proper closed ideals which are annihilated by no irreducible (Banach) representation of the group. M. Duflo, in a letter to H. Leptin, gave a quick proof that the algebra $L^1(G)$ is not symmetric for any noncompact semisimple Lie group G (see [Lep]). At least for the real-rank-one groups (essentially those which we are considering) the result on the existence of non-Tauberian ideals was apparently obtained by R. Krier in an apparently unpublished thesis. The most recent work on this argument is presumably that of A. Sitaram [Sit], in which partial results for general semisimple Lie groups are obtained. It may be supposed that the recent work of Y. Weit [Wei] will stimulate some further development in the study of non-Tauberian ideals.

It seems that L^p-analysis requires noncommutative techniques. We shall discuss a characterisation of the "Fourier transform" of certain subspaces of $L^p(G)$ due to P.C. Trombi and V.S. Varadarajan [TrV], whose proof has been elegantly simplified by J.-L. Clerc [Cle]. We shall also consider the Fourier transforms of general L^p-functions, the "Kunze-Stein phenomenon", discovered by R.A. Kunze and E.M. Stein [KS1] for $SL(2,\mathbb{R})$ and (after various generalisations based on uniformly bounded representations (v.i.)) established in general by M. Cowling [Co1]. It may be worthy of note that D. Poguntke, following up a suggestion of M. Duflo, has recently found analogous phenomena for some solvable groups, and that M. Picardello independently proved the same results. In the solvable case, it is essential to use analytic continuations which act isometrically on L^p-spaces rather than uniformly bounded representations, for uniformly bounded representations of solvable groups are equivalent to unitary representations.

Let us consider the convolution algebra $L^1(G)$, armed with the involution $\tilde{}$, defined by the formula

$$f^{\sim}(g) = \bar{f}(g^{-1}) \qquad\qquad f \in L^1(G) .$$

We shall denote by $L^1(K\backslash G/K)$ the subalgebra of $L^1(G)$ of K-biinvariant functions, i.e. those functions f such that

$$f(g) = f(kgk') \qquad\qquad g \in G,\ k,\ k' \in K .$$

In this section, we shall denote by μ the measure on G, supported on K, given by integration against the normalised Haar measure of K. The map P, defined by the formula

$$Pf = \mu * f * \mu \qquad\qquad f \in L^1(G) ,$$

is a non-norm-increasing projection of $L^1(G)$ onto $L^1(K\backslash G/K)$, whose restriction to the subspace $L^1(K\backslash G/K)$ is the identity map.

It is quite easy to produce pathological examples in $L^1(K\backslash G/K)$, because this is a commutative Banach algebra with a well-defined spectrum. Then one lifts these examples to the whole group, using the projection P. This is the essential point behind the $L^1(G)$ results which we discuss here. On the other hand, this approach is not fine enough to yield the Kunze-Stein theorem.

PROPOSITION 6.1. The convolution algebra $L^1(K\backslash G/K)$ is commutative.

Proof. To show that $L^1(K\backslash G/K)$ is an algebra is easy: one notes that f in $L^1(G)$ lies in $L^1(K\backslash G/K)$ if and only if f = Pf. Then, if f, f' $\in L^1(K\backslash G/K)$

$$\begin{aligned} \mu * f * f' * \mu &= \mu * \mu * f * \mu * \mu * f' * \mu * \mu \\ &= \mu * f * \mu * \mu * f' * \mu \\ &= f * f' , \end{aligned}$$

because μ is an idempotent measure.

To show that $L^1(K\backslash G/K)$ is commutative is no more difficult. We recall that any g in G may be written in the form kak', with k, k' in K and a in A.

Thus $$KgK = KaK = KwawK = Ka^{-1}K = Kg^{-1}K \ .$$

This implies that if $f \in L^1(K\backslash G/K)$, then f is equal to its reflection $f^\vee$. Now

$$(f * f')^\vee = f'^\vee * f^\vee \qquad\qquad f, f' \in L^1(G)$$

(compare with Lemma 1.8), and so, for f and f' in $L^1(K\backslash G/K)$,

$$f * f' = (f * f')^\vee = f'^\vee * f^\vee = f' * f \ . \qquad \square$$

Both the algebra of radial functions on V and the algebra $L^1(K\backslash G/K)$ can be treated in a unified manner by using the theory of Gelfand pairs. A. Korányi first noticed this fact, which explains the similarities between the proofs of Lemma 1.8 and Proposition 6.1.

When one deals with a commutative Banach algebra, one looks for its spectrum. In this case, we are lead to the theory of spherical functions. We describe these as follows: in the spaces $H_{1,\zeta}$ and $H_{1,-\bar{\zeta}}$ of the representations $\pi_{1,\zeta}$ and $\pi_{1,-\bar{\zeta}}$ there are unique functions f_ζ and $f_{-\bar{\zeta}}$ whose restrictions to K are identically 1. The spherical function ϕ_ζ is given by the formula

$$\phi_\zeta(g) = \langle \pi_{1,\zeta}(g) f_\zeta, f_{-\bar{\zeta}} \rangle^{(K)} \ .$$

It is not hard to show that, if we define the spherical transform $\hat{f}$ of f in $C_c(K\backslash G/K)$ by the rule

$$\hat{f}(\zeta) = \int_G dg\ \phi_\zeta(g)\ f(g) \quad ,$$

then $$(f * f')^\wedge(\zeta) = \hat{f}(\zeta)\ \hat{f}'(\zeta) \ .$$

This follows from the analogous multiplicative formula for the "Fourier transforms" $\pi_{1,\zeta}(f)$:

$$\pi_{1,\zeta}(f) = \int_G dg\ f(g)\ \pi_{1,\zeta}(g) \qquad\qquad f \in C_c(G) \ .$$

The representations $\pi_{1,\zeta}$ act isometrically with the Banach norms $\|.\|_p^{(K)}$, when $1/p = \xi/2r + 1/2$, and so if $\xi \in T$, the tube where $\xi \in [-r, r]$, then

$$\begin{aligned} |\phi_\zeta(g)| &\leq \|\pi_{1,\zeta}(g)\ f_\zeta\|_p^{(K)}\ \|f_{-\bar{\zeta}}\|_{p'}^{(K)} \\ &= \|f_\zeta\|_p^{(K)} \\ &= 1 \ . \end{aligned}$$

Therefore the spherical transform extends to a multiplicative linear functional on $L^1(K\backslash G/K)$, as long as $\xi \in T$. We note that the duality between $\pi_{1,\zeta}$ and $\pi_{1,-\bar{\zeta}}$ implies that

$$\phi_\zeta = \phi_{-\zeta} \ ;$$

moreover the asymptotic behaviour of the spherical functions, described in §5, implies that there are no other equivalences of the ϕ_ζ, and that no other ϕ_ζ are bounded. It may be shown that the Gelfand spectrum of $L^1(K\backslash G/K)$ is exhausted by the integrations against the bounded spherical functions, as defined above, but we shall not do that here.

We need only the obvious fact that the map $\zeta \mapsto \hat{f}(\zeta)$ is analytic in order to prove the first result on $L^1(K\backslash G/K)$, and hence on $L^1(G)$.

THEOREM 6.2. The algebra $L^1(G)$, equipped with the involution $\tilde{}$, is not symmetric.

Proof. Let f' be a function in $L^1(K\backslash G/K)$ which does not annihilate all ϕ_ζ $(\zeta \in T)$, and let f be $f' * f'^{\sim}$. Since $(f^{\sim})^{\wedge}(\zeta) = [\hat{f}(\zeta)]^{-}$ if $\zeta \in i\mathbb{R}$ (because the associated representations are unitary), it is clear that f is nonzero. Choose a point ω where $\hat{f}$ is not real; such a point exists because $\hat{f}$ is analytic and not constant (in fact, by the appropriate version of the Riemann-Lebesgue lemma, $\hat{f}$ vanishes at infinity in T). We claim that $\hat{f}(\omega)$ is in the spectrum of f. This is clear if we consider only $L^1(K\backslash G/K)$, and if there were F in $\mathbb{C}\delta \oplus L^1(G)$ such that $F * (f - \hat{f}(\omega)\delta) = \delta$, then we should also have that $PF * (f - \hat{f}(\delta)\mu) = \mu$, a contradiction. □

The next result is less trivial.

THEOREM 6.3. There exist proper closed ideals in $L^1(G)$ which are not contained in the kernel of any irreducible bounded representation of G.

Sketch of the proof. It should suffice to prove the analogous result for $L^1(K\backslash G/K)$: For if I is such an ideal in $L^1(K\backslash G/K)$, then define J by the formula

$$J = \{f \in L^1(G): P(f_1 * f * f_2) \in I, \quad f_1, f_2 \in L^1(G)\} \quad ;$$

J should be a non-Tauberian ideal in $L^1(G)$.

Let $a: T \longrightarrow \mathbb{C}$ be the map given by the formula

$$a(\zeta) = ([\exp(i\pi\zeta/2r) - 1]/[\exp(i\pi\zeta/2r) + 1])^2 \qquad \zeta \in T .$$

With the aid of this map and the spherical transform $f \mapsto \hat{f}$, we may realise $L^1(K\backslash G/K)$ as a subalgebra of the algebra A of all continuous functions on the closed unit disc $\bar{D}$ which vanish at 1 and are holomorphic in the interior.

One then seeks non-Tauberian ideals in A. These can be found with the aid of work of A. Beurling [Beu], who, by using factorisation into "inner" and "outer" functions, showed that the space of all functions in A which vanish as fast as the function $z \mapsto \exp(2/[z-1])$ is a non-Tauberian ideal.

The proof finishes by "pulling back" this ideal to $L^1(K\backslash G/K)$, and showing that the "pull-back" is nontrivial. For this, it is necessary to have some sort of characterisation of the space of spherical transforms of $L^1(K\backslash G/K)$-functions.

□

The problem of characterising the spherical transforms of all $L^1(K\backslash G/K)$-functions seems to be very difficult. One substitute for this characterisation is the result of P. Trombi and V.S. Varadarajan [TrV] already mentioned. One considers the subspace of $L^1(K\backslash G/K)$ of those functions which, together with all their derivatives, vanish so rapidly at infinity that, even when multiplied by a polynomial (which here means an expression of the form $ka(t)k' \mapsto t^n$), they still belong to $L^1(G)$. This space has been named $I^1(G)$. The space $I^p(G)$ is defined analogously.

THEOREM 6.4. The image of $I^p(G)$ under the spherical transform is exactly the space of all even functions in the strip T_p —

$$T_p = \{\zeta \in C: |\xi/2r| \leq 1/p - 1/2\} —$$

which are analytic in the interior of T_p and continuous on the whole of T_p, and whose restrictions to the vertical lines in the strip belong to the usual Schwartz space on these lines, with uniform estimates holding on all the lines contained in T_p.

Proof. The idea of J.-L. Clerc's proof [Cle] of this theorem is as follows. The result must first be proved for $I^2(G)$. Next, certain spherical functions, ϕ_ν say, are coefficients of finite-dimensional representations of G, and for these L. Vretare [Vre] has proved a multiplication formula:

$$\phi_\zeta \phi_\nu = \sum_{\zeta'} c(\nu, \zeta, \zeta')\, \phi_{\zeta'} \quad .$$

(This sum extends over a finite set whose size is bounded independent of ζ and ν.) The product of an I^p-function f with the spherical function ϕ_ν lies in a different space $I^q(G)$, say, and one has that

$$(\phi_\nu f)\hat{}(\zeta) = \sum_{\zeta'} c(\nu, \zeta, \zeta')\, \hat{f}(\zeta') \quad .$$

$$(\phi_\nu f)^\frown(\zeta) = \sum_{\zeta'} c(\nu, \zeta, \zeta') \hat{f}(\zeta') \quad .$$

Clerc first characterises the spherical transforms of the I^p-functions for certain values of p, viz, those for which the products $\phi_\nu f$, with f in $I^p(G)$, lie in $I^2(G)$, and then applies complex interpolation to fill in the gaps. But the essential idea behind the proof is really to study the tensor products of the representations $\pi_{1,\zeta}$ with certain finite-dimensional representations. This same idea has made its appearance in some other problems. □

We shall conclude by considering the Fourier transform of more general functions. For the sake of simplicity, we shall restrict our attention to the representations $\pi_{1,\zeta}$, although the other families could be treated similarly. We should like to describe an analytic Fourier transform $\pi_{1,\zeta}(f)$, where f lies in $L^1(G)$, just as we have considered the analytic spherical transforms $\hat{f}(.)$ for f in $L^1(K\backslash G/K)$.

One problem arises immediately: the spaces on which the representations $\pi_{1,\zeta}$ act vary with ζ. There are two possibilities to consider. We might fix two functions ϕ and ψ on K, say, and then extend these to functions ϕ_ζ and $\psi_{-\zeta}$ in $H_{1,\zeta}$ and $H_{1,-\zeta}$ respectively. We could then deal with the functions

$$g \longmapsto \langle \pi_{1,\zeta}(g)\, \phi_\zeta, \psi_{-\zeta} \rangle^{(K)}$$

and define our Fourier transform by integrating against these. We should have to impose the restriction that ϕ and ψ belong to $L^\infty(K)$, in order to ensure that ϕ_ζ and ψ_ζ belong to $H_{1,\zeta}^{(p)}$ and $H_{1,-\zeta}^{(p')}$, when $1/p = \xi/2r + 1/2$ and p' is the dual index. Our Fourier transform would be an expression of the form

$$\langle \pi_{1,\zeta}(f)\, \phi_\zeta, \psi_{-\zeta} \rangle^{(K)} \quad ,$$

obtained by integrating f against the function above, or rather a collection of such, for ϕ and ψ may vary.

This way of doing things is not fine enough to obtain some analytic results, because restrictions on ϕ and ψ are necessary, while one may wish to consider general ϕ and ψ, for instance in $L^2(K)$. Further, one may ask why the functions ϕ and ψ should be fixed on K rather than on V, or on some other section for MAN in G. If we fix ϕ and ψ on K, then ϕ_ζ and $\psi_{-\zeta}$ vary holomorphically and antiholomorphically on V. It turns out to be more canonical, at least in many situations, to allow ϕ_ζ and $\psi_{-\zeta}$ to vary in this way everywhere.

It is possible and natural to allow ϕ_ζ and $\psi_{-\zeta}$ to vary with ζ. The best way to do this is standard in complex interpolation theory, where, given a function θ on a measure space X, one considers functions of the form $\theta\,|\theta|^z$ as the parameter z varies. We apply this idea. If ϕ lies in $H_{1,0}^{(2)}$, we define the analytic family of functions ϕ_ζ by the formula

$$\phi_\zeta = \phi\ |\phi|^{\zeta/r} .$$

We note first that

$$\begin{aligned}\phi_\zeta(\text{mang}) &= \phi(\text{mang})\ |\phi(\text{mang})|^{\zeta/r}\\ &= \rho(a)\ \phi(g)\ |\rho(a)\ \phi(g)|^{\zeta/r}\\ &= \rho\alpha_\zeta(a)\ \phi_\zeta(g) \quad ,\end{aligned}$$

i.e. ϕ_ζ belongs to $H_{1,\zeta}$. (If we make the same definition for ϕ in $H_{\mu,0}$, then ϕ_ζ will lie in $H_{\mu,\zeta}$.) Moreover, if $1/p = \xi/2r + 1/2$, then

$$\begin{aligned}\int_V dv\ |\phi_\zeta(v)|^p &= \int_V dv\ |\phi(v)|^{(1+\xi/r)p}\\ &= \int_V dv\ |\phi(v)|^2 \quad ,\end{aligned}$$

and ϕ_ζ belongs to the appropriate L^p-space. In fact every function in $H_{1,\zeta}^{(p)}$ arises in this way.

There is a useful generalisation of this construction: if we deal with vector-valued functions ϕ such that, on K, $\phi(k) = \gamma(k)\phi(e)$ for some representation γ of K, then the ϕ_ζ we produce are multiples of ϕ on K. Such functions are in fact considered in the theories of generalised spherical functions and of the Eisenstein integral. Of course, in more general situations one may construct analytically varying sections of analytically varying vector bundles.

We finish off this discussion by giving an application of the above construction. If $f \in L^1(G)$, and if ϕ and ψ are in $H_{1,0}^{(2)}$, with $\|\phi\|_2^{(V)} = \|\psi\|_2^{(V)} = 1$, then

$$\zeta \longmapsto \langle \pi_{1,\zeta}(f)\ \phi_\zeta,\ \psi_\zeta\rangle$$

is a continuous bounded function of ζ in the strip T, which is analytic in the interior of T. Further, one has the estimate

$$\begin{aligned}&|\langle \pi_{1,\zeta}(f)\ \phi_\zeta,\ \psi_{-\zeta}\rangle^{(V)}|\\ &\le \|\pi_{1,\zeta}(f)\ \phi_\zeta\|_p^{(V)}\ \|\psi_{-\zeta}\|_{p'}^{(V)}\\ &\le \|f\|_1\ \|\phi_\zeta\|_p^{(V)}\end{aligned}$$

$$= \|f\|_1 \quad .$$

(This estimate is much simpler than the estimate one obtains if one seeks to use the functions ϕ_ζ and $\psi_{-\zeta}$ defined earlier which were constant on K.)

A different sort of estimate is available from the Plancherel formula for $L^2(G)$ (or for $L^2(K\backslash G/K)$). One has that (omitting the superscript V)

$$\left[\int_R d\eta\ \mu(\eta)\ |\langle \pi_{1,i\eta}(f)\ \phi_{i\eta},\ \psi_{i\eta}\rangle|^2\right]^{1/2}$$
$$\leq \left[\int_R d\eta\ \mu(\eta)\ \|\pi_{1,i\eta}(f)\|_{op}^2\right]^{1/2}$$
$$\leq \left[\int_R d\eta\ \mu(\eta)\ \|\pi_{1,i\eta}(f)\|_{HS}^2\right]^{1/2}$$

where $\|.\|_{op}$ and $\|.\|_{HS}$ denote the operator and Hilbert-Schmidt norms. If μ be the Plancherel measure of G associated to the representations $\pi_{1,i\eta}$, then

$$\left[\int_R d\eta\ \mu(\eta)\ .|\langle \pi_{1,i\eta}(f)\ \phi_{i\eta},\ \psi_{i\eta}\rangle|^2\right]^{1/2}$$
$$\leq \|f\|_2 \quad .$$

It follows by a messy but elementary interpolation argument that, if $f \in L^p(G)$ and $1 \leq p < 2$, then the function

$$\zeta \mapsto \langle \pi_{1,\zeta}(f)\ \phi_\zeta,\ \psi_{-\zeta}\rangle$$

is analytic on the interior of the strip T_p —

$$T_p = \{\zeta \in C\colon\ |\zeta| \leq 2r(1/p - 1/2)\} \quad \text{—}$$

and on the edges of this strip, non-tangential limits exist almost everywhere, and moreover

$$\left[\int_R d\eta\ |\langle \pi_{1,\theta+i\eta}(f)\ \phi_{\theta+i\eta},\ \psi_{-\theta+i\eta}\rangle|^{p'}\right]^{1/p'}$$
$$\leq C(G,\theta)\ \|f\|_p$$

if $|\theta| = 2r(1/p - 1/2)$. This is a partial characterisation of the Fourier transform of a general L^p-function, akin to the Hausdorff-Young inequality.

An easy corollary of this result is obtained by means of Cauchy's theorem. One integrates around the perimeter of the strip T_p and obtains that

$$|\langle \pi_{1,0}(f)\ \phi,\ \psi\rangle| = (2\pi)^{-1}\ \left|\int_{\partial T_p} \frac{d\zeta}{\zeta} \langle \pi_{1,}\ (f)\ \phi_\zeta, \psi_{\ \zeta}\rangle\right|$$
$$\leq C(G,p)\ \|f\|_p \quad .$$

This is the essential step in the proof of the Kunze-Stein convolution theorem, which we state here, without proof.

THEOREM 6.5. If $1 \leq p < 2$, then $L^p(G) * L^2(G) \subseteq L^2(G)$.

7. UNIFORMLY BOUNDED REPRESENTATIONS AND COMPLEMENTARY SERIES.

We recall that $H_{\mu,\zeta}^{\infty}$ is the space of smooth H_μ-valued functions f on G satisfying the condition

$$f(\mathrm{mang}) = \mu(m)\ (\rho\alpha_\zeta)(a)\ f(g)\ ,$$

and that $\pi_{\mu,\zeta}$ is the right translation representation of G on this space. We have seen that, if we define $\|.\|_p^{(V)}$ by the formula

$$\|f\|_p^{(V)} = (\int_V dv\ \|f(v)\|^p)^{1/p},$$

and if $1/p = \xi/2r + 1/2$, then $\pi_{\mu,\zeta}$ acts isometrically. So we have a strip of isometric Banach representations $\{\pi_{\mu,\zeta} : \xi \in [-r, r]\}$.

In the preceding section, we saw that this analytic family of representations plays an important role in the L^p-harmonic analysis of G. It would therefore be desirable to develop a calculus for representations which would permit us to analyse and synthesize bounded Banach representations of G. Since Hilbert space theory is an essential feature of the known techniques for such calculi, it is natural to ask if we can realise these representations on a Hilbert space. We shall devote our attention to this question.

First, since the L^p-spaces on which the representations act isometrically are not Hilbert spaces, we shall have to modify them somewhat. The situation is rather suggestive: in R^n the potential spaces H^ξ —

$$H^\xi = \{f: \Delta^{-\xi/4} * f \in L^2(R^n)\} \text{ —}$$

are very similar to the L^p-spaces, where $1/p = 1/2 + \xi/2n$. They behave in the same way with respect to rotations and dilations, and if $\xi \geqslant 0$, then $L^p \supseteq H^\xi$, while if $\xi \leqslant 0$, then $H^\xi \subseteq L^p$. These are certainly good candidates for Hilbert spaces on which the representations $\pi_{\mu,\zeta}$ might act. In fact R.A. Kunze and E.M. Stein [KS1] showed that $SL(2,R)$ has uniformly bounded representations on these spaces, that is, there exists a uniform bound for the operator norms $\|\pi(g)\|_{op}$, as g runs over the group being represented. Various persons have subsequently considered uniformly bounded representations: we mention here the work of P. Sally [Sal], R.L. Lipsman [Li1], [Li2], [Li3], [Li4], E.N. Wilson

[Wil], N. Lohoué [Loh], and L. Bamazi [Bam], all of whom attempted to develop the work of Kunze and Stein [KS1], [KS2], [KS3]; some general comments on these representations may be found in [Co2].

We shall show that the representations $\pi_{\mu,\zeta}$ of the groups O(q) act uniformly boundedly on the potential spaces H^ξ defined in terms of the sub-Laplacian Δ on V, provided that $\xi \in (-r, r)$. This result is best possible, in the sense that no other nontrivial representations could be uniformly bounded, for the matrix coefficients must go to zero at infinity, and this excludes the other representations.

After studying the uniformly bounded representations, we shall turn our attention to the possibility that some of the representations $\pi_{\mu,\zeta}$ might be unitarisable, i.e., with an appropriate choice of inner product, they might become unitary. This question has already been considered by other persons, and in fact all the unitarisable representations of the groups SO(1, n) and SU(1, n), corresponding to the cases where F is either R or C, are known, and it seems that the case of Sp(1, n), corresponding to the case where $F = Q$, has just been solved. We shall therefore only touch on the philosophy of our approach to these representations, after our analysis of the uniformly bounded representations.

Before we state our theorem, let us give a provisional definition of the space $H_{\mu,\zeta}^\xi$. This will be superceded later. Temporarily, we denote by $H_{\mu,\zeta}^\xi$, or just H^ξ, the closure of the subspace of $H_{\mu,\zeta}^\infty$ of those functions which, when restricted to V, have compact support, in the norm $\| \ \|^\xi$, given by the formula

$$\|f\|^\xi = \|\Delta^{-\xi/4} * f\|_2^{(V)} .$$

It follows from the proof of the theorem that H^ξ is actually the closure of $H_{\mu,\zeta}^\infty$ in the same norm. Of this, more anon.

THEOREM 7.1. The representation $\pi_{\mu,\zeta}$ acts uniformly boundedly on the space $H_{\mu,\zeta}^\xi$ provided that $\xi \in (-r, r)$.

Proof. We first consider the actions of V, M, and A on H^ξ. Since $\|.\|^\xi$ is defined in terms of generalised differential operators acting on the left and V acts on the right by translations, V acts unitarily. Next,

$$\Delta^{-\xi/4} \pi_{\mu,\zeta}(m)f = \Delta^{-\xi/4} \mu(m)(f\circ R^m) = \mu(m)([\Delta^{-\xi/4} f]\circ R^m) ;$$

it follows that M acts unitarily. The action of A is treated similarly, but is a little less easy:

$$\pi_{\mu,\zeta}(a(t))f(v) = f(va(t)) = (\rho\alpha_{\zeta})(a(t))\ f(D^s v) \quad ,$$

where $s = e^t$. The necessary fact is that

$$\Delta(f\circ D^s) = s^2 (\Delta f)\circ D^s \quad ,$$

whence, from the spectral calculus,

$$\Delta^{-\xi/4}(f\circ D^s) = s^{-\xi/2} (\Delta^{-\xi/4} f)\circ D^s \ .$$

It follows that

$$\begin{aligned} \|\pi_{\mu,\zeta}(a(t))f\|^{\xi} &= (\rho\alpha_{\xi})(a(t))\ s^{-\xi/2}\ \|(\Delta^{-\xi/4} * f)\circ D^s\|_2^{(V)} \\ &= \rho(a(t))\ \|(\Delta^{-\xi/4} * f)\circ D^s\|_2^{(V)} \\ &= \|\Delta^{-\xi/4} * f\|_2^{(V)} \quad , \end{aligned}$$

and A acts unitarily.

We claim that it suffices to show that $\pi_{\mu,\zeta}(w)$ acts boundedly. For if this is so, then the operators of the form $\pi_{\mu,\zeta}(g)$, where g is either mawvwv' or mawv, act boundedly, and

$$\|\pi_{\mu,\zeta}(\text{mawvwv}')\|_{op} \leqslant \|\pi_{\mu,\zeta}(w)\|_{op}^2$$

$$\|\pi_{\mu,\zeta}(\text{mawv})\|_{op} \leqslant \|\pi_{\mu,\zeta}(w)\|_{op} \quad ,$$

but every element of G is of the form manv or manw (see Lemma 4.3), whence the claim.

To complete the demonstration, we use an elementary but intricate argument. First, if $\xi = 0$, then we are done. For H^0 "is" just $L^2(V)$, and $\pi_{\mu,i\eta}$ acts unitarily on $L^2(V)$ by Lemma 5.1.

Next, we consider the case where $\xi = -2$. In order to show that $\pi_{\mu,\zeta}(w)$ acts boundedly on H^{-2}, it suffices to show that

$$\|X_j \pi_{\mu,\zeta}(w)f\|_2 \leqslant C(\zeta)\ \|f\|^{-2} \qquad 1 \leqslant j \leqslant p$$

where $\{X_j : 1 \leqslant j \leqslant p\}$ is a basis of the subspace $\underline{V}_{-1}$ of the Lie algebra $\underline{V}$. For this, we write everything out explicitly:

$$\pi_{\mu,\zeta}(w)f(v) = \mu(M(vw))\ (\rho\alpha_{\zeta})(A(vw))\ f(V(vw)) \quad ,$$

where M(vw) and A(vw) are described in Lemma 5.5, and V(vw) is given in Lemma 4.2. We recall that M(vw), as a function of v, is homogeneous of degree 0 and further is C^∞ on $V\setminus\{e\}$. On the other hand,

$$(\rho\alpha_\zeta)(A(vw)) = N(v)^2/2^{-(r+\zeta)/2} ,$$

which is homogeneous of degree $-(r + \zeta)$. Thus we may write

$$\pi_{\mu,\zeta}(w)f^{\dagger}(v) = m(v)\ f(v^{\dagger}) = mf^{\dagger}(v),$$

where $v^{\dagger} = V(vw)$, $f^{\dagger}(v) = f(v^{\dagger})$, and $m \in M^{-(r+\zeta)}(V)$. Clearly

$$X_j(\pi_{\mu,\zeta}(w)f) = (X_j m)f^{\dagger} + m(X_j f^{\dagger}) .$$

We consider the two pieces separately.

Since differentiation with X_j lowers the degree of homogeneity by 1, $X_j m$ is homogeneous of degree $-(r + \zeta + 1)$. We may estimate the L^2-norm of the first piece as follows:

$$\|(X_j m)f^{\dagger}\|_2 = (\int_V dv\ \|(X_j m)(v)\ f^{\dagger}(v)\|^2)^{1/2}$$
$$= (\int_V dv\ \ (dv/dv^{\dagger})\ \|(X_j m)(v)\ f(v^{\dagger})\|^2)^{1/2} .$$

We recall that $(dv/dv^{\dagger})$ is $\rho(A(v^{\dagger}w))^2$, which is homogeneous of degree $-2r$ in v, and in $C^{\infty}(V\setminus\{e\})$. Next, since the map $v^{\dagger} \mapsto v$ is homogeneous of degree -1, in the obvious sense, the function $v^{\dagger} \mapsto (X_j m)(v)$ is homogeneous of degree $(r + (-2 + i\eta) + 1)$, and is of course in $C^{\infty}(V\setminus\{e\})$. We must therefore estimate

$$(\int_V dv^{\dagger}\ \|m^{\dagger}(v^{\dagger})\ f(v^{\dagger})\|^2)^{1/2} ,$$

where $m^{\dagger}$ is homogeneous of degree $(-1 + i\eta)$. But we showed in Section 2 that multiplication by functions of negative homogeneity takes one space H^{ξ} into another (Proposition 2.16), and we are in exactly the right situation to be able to apply this result. We conclude that

$$(\int_V dv^{\dagger}\ \|m^{\dagger}(v^{\dagger})\ f(v^{\dagger})\|^2)^{1/2} \leq C\ \|f\|_{-2} .$$

It remains to estimate $mX_j(f^{\dagger})$. We note that the map $^{\dagger}: v \mapsto v^{\dagger}$ is a diffeomorphism when restricted to $V\setminus\{e\}$. It follows that $X_j(f^{\dagger})$ can be written as $(E_j f)^{\dagger}$, where E_j is a smooth vector field on $V\setminus\{e\}$. Since

$$(f\circ D^s)^{\dagger} = (f^{\dagger})\circ D^{1/s} ,$$

$$X_j((f\circ D^s)^{\dagger}) = (1/s)\ (X_j(f^{\dagger}))\circ D^{1/s} ;$$

written differently, we have that

$$(E_j(f\circ D^s))^{\dagger} = (1/s)\ ((E_j f)^{\dagger})\circ D^{1/s}$$

whence

$$E_j(f\circ D^s) = (1/s)\ (E_j f)\circ D^s .$$

(This is a generalisation of the fact that $d/dx\ (f(x^{-1})) = -x^{-2}\ (df/dx)(x^{-1})$.)

The rest of the proof for the case where $\xi = -2$ is now easy: it is clear that

$$\begin{aligned}\|m\, X_j(f^{\dagger})\|_2^{(V)} &= \left(\int_V dv\, \|m(v)\, X_j(f^{\dagger})(v)\|^2\right)^{1/2} \\ &= \left(\int_V dv\, \|m(v)\, (E_j f)^{\dagger}(v)\|^2\right)^{1/2} \\ &= \left(\int_V dv^{\dagger}\, (dv/dv^{\dagger})\, \|m(v)\, (E_j f)(v^{\dagger})\|^2\right)^{1/2},\end{aligned}$$

and the usual homogeneity arguments will complete the proof.

The proof for general values of ξ contains no new ingredients; one applies complex interpolation and duality arguments to first prove the result for the case where $-2 \leqslant \xi \leqslant 2$, and then applies the above ideas to treat the general case. □

We now return to the question of the definition of $H^{\xi}_{\mu,\zeta}$. It is clear that, if $f \in H^{\infty}_{\mu,\zeta}$, then we may write f as the sum of two $H^{\infty}_{\mu,\zeta}$-functions, one of which vanishes near w in K and the other which vanishes near e in K. The function which vanishes near w is, when restricted to V, of compact support, and so belongs to $H_{\mu,\zeta}$. Further, $\pi_{\mu,\zeta}(w)$ applied to the latter function is of the same form as the former, and so belongs to $H^{\xi}_{\mu,\zeta}$. The theorem now implies that f is in $H^{\xi}_{\mu,\zeta}$, and the definition which we announced would be satisfactory is just that.

This completes our discussion of uniformly bounded representations, and we now turn to the question of complementary series.

We attempt to describe the philosophy of A.W. Knapp and E.M. Stein [KnS]. If it were possible to find an inner product B(,) relative to which $\pi_{\mu,\zeta}$ acts unitarily, then we should be able to define a G-invariant map b from $H_{\mu,\zeta}$ to $H_{\mu,-\zeta}$ by setting

$$\langle b(f), f' \rangle = B(f, f') \qquad f, f' \in H^{\infty}_{\mu,\zeta} .$$

This map is G-invariant in the sense that

$$b\, \pi_{\mu,\zeta}(g)\, f = \pi_{\mu,-\zeta}(g)\, b\, f .$$

Thus we should have some sort of equivalence between the representations $\pi_{\mu,\zeta}$ and $\pi_{\mu,-\zeta}$. This would imply either that ζ was imaginary, and b would then be trivial, or that ζ was real, in which case b would have to be a multiple of the intertwining operator $A(w, \mu, \zeta)$. In this latter case, a multiple of $A(w, \mu, \zeta)$ would be positive definite or semidefinite. Conversely, if $A(w, \mu, \zeta)$ (or a multiple of it) were positive definite or semidefinite, then we could reverse the above argument and define a G-invariant inner product. In short, we find "new" unitary representations exactly when the intertwining operator is positive definite or semidefinite.

One important "simplification" of the problem of finding "all" the unitary representations of G is due to R.P. Langlands [Lan]. We shall describe briefly the idea.

The intertwining operators $A(w, \mu, \zeta)$ have the property that the operator

$$A(w, \mu, -\zeta)\ A(w, \mu, \zeta)$$

takes $H_{\mu,\zeta}$ into itself, and commutes with $\pi_{\mu,\zeta}$. It may be shown that, for generic ζ in C, $\pi_{\mu,\zeta}$ acts irreducibly. Therefore there exists a meromorphic function $d(w, \mu, \zeta)$ such that

$$A(w, \mu, -\zeta)\ A(w, \mu, \zeta) = d(w, \mu, \zeta)\ I \quad ,$$

from Schur's lemma. For most ζ in R, $A(w, \mu, \zeta)$ is invertible. In this case, $A(w, \mu, -\zeta)$ is positive definite if and only if $A(w, \mu, \zeta)$ is, and we might as well restrict our attention to the case where ζ lies in R^+.

The case where $A(w, \mu, \zeta)$ is not invertible is more interesting. If we suppose that $A(w, \mu, \zeta)$ be "normalised", i.e. multiplied by functions of the form $(\zeta - \gamma)$ to eliminate the poles where one would like to study the operator, then it will be true that

$$A(w, \mu, -\zeta)\ A(w, \mu, \zeta) = 0$$

and

$$A(w, \mu, \zeta)\ A(w, \mu, -\zeta) = 0 \quad .$$

Suppose that $\zeta > 0$. Then unitary representations could arise from the representations $\pi_{\mu,\zeta}$ or $\pi_{\mu,-\zeta}$ if $A(w, \mu, \zeta)$ or $A(w, \mu, -\zeta)$ were positive semidefinite, by the procedure outlined above. The matrix coefficients associated to the unitary representation arising if $A(w, \mu, -\zeta)$ is positive semidefinite are of the form

$$g \mapsto \langle A(w, \mu, -\zeta)\ \pi_{\mu,-\zeta}(g)\ f_2, f_1\rangle \qquad f_1, f_2 \in H_{\mu,-\zeta} \quad ,$$

i.e.

$$g \mapsto \langle \pi_{\mu,\zeta}(g)\ A(w, \mu, -\zeta)\ f_2, f_1\rangle \qquad f_1, f_2 \in H_{\mu,-\zeta} \quad .$$

The asymptotic formula for the matrix coefficients (cf. Lemma 5.6) is

$$\begin{aligned} \big| (\rho\alpha_{-\zeta})(a(t))\ &\langle \pi_{\mu,\zeta}(k\ a(t)k\ )\ A(w, \mu, -\zeta)f_2, f_1\rangle \ . \\ &- \langle A(w, \mu, \zeta)\ A(w, \mu, -\zeta)f_2(wk_2), f_1(k_1^{-1})\rangle \big| \\ &= O(\exp(-\varepsilon t)) \quad . \end{aligned}$$

This means that the matrix coefficients of the representation decay at infinity faster than $(\rho^{-1}\alpha_\zeta)(a(t))$.

Langlands' contribution consists in showing that any irreducible unitary representation π (in fact the hypothesis of unitarity can be greatly relaxed) is either tempered, which means that the exponential controlling the vanishing at infinity is like that which controls the "unitary principal series" or the rate of decay at infinity is even faster (in which case we have "discrete series"), or the representation may be embedded in a principal series representation $\pi_{\mu,\zeta}$ whose rate of vanishing is the same of that of π. This means that the representations coming from $\pi_{\mu,-\zeta}$ can be safely ignored, since either they are associated to the regular representation (whose decomposition into irreducibles is known) or they occur in some other $\pi_{\mu,\zeta}$.

At this point, we make a conjecture.

CONJECTURE 7.2. Suppose that $0 < \zeta < r$. Then the intertwining operator $A(w, \mu, \zeta): H_{\mu,\zeta} \longrightarrow H_{\mu,-\zeta}$ is positive definite or semidefinite if and only if the convolution operator with kernel $\operatorname{tr}(K_{\mu,\zeta})$ on V is positive definite or semidefinite.

REMARK 7.3. It is easy to show that if $A(w, \mu, \zeta)$ is positive definite or semidefinite, then so is convolution with $\operatorname{tr}(K_{\mu,\zeta})$. Further, if $\zeta = 0$, this condition is necessary and sufficient.

We now illustrate the problem of describing the complementary series with a couple of examples.

EXAMPLE 7.4. The class-one complementary series.

The intertwining operator $A(w, 1, \zeta)$ is given by convolution with a scalar kernel on V, which is nothing but a multiple of the norm function $N^{\zeta-r}$. The Fourier transform of this is given by the formula

$$\pi_\lambda(N^{\zeta-r})_k = C\,|\lambda|^{-\zeta/2}\,\frac{\Gamma((p+4k-2-\zeta)/4)}{\Gamma((p+4k-2+\zeta)/4)}\cdot\frac{\Gamma(\zeta/2)}{\Gamma((r-\zeta)/4)\,\Gamma((p+2-\zeta)/4)},$$

where C depends only on G. This formula is to be interpreted as follows: the "Fourier transform" of a cylindrical function, of which the functions $\operatorname{tr}(K_{\mu,\zeta})$ are examples, is diagonal and constant on blocks, which correspond to the eigenvalues of the Hermite operator; $\pi_\lambda(N^{\zeta-r})_k$ is the Fourier transform of $N^{\zeta-r}$ at the parameter λ on the kth block. This formula will be proved in the next and final section.

If $0 < \zeta < r$, then we have a multiple of

$$\Gamma((p+4k-2-\zeta)/4)/\Gamma((p+2-\zeta)/4)$$

(the other terms are always positive); if $k = 1$, then this factor is just equal to one, but if $k > 1$, then it is positive when $\zeta < p + 2$ and negative when $\zeta > p + 2$.

It follows that the class-one complementary series of the groups SU(1, n) (which correspond to the case where $F = C$) "goes all the way" to r, where the identity representation occurs. But for the groups Sp(1, n), which correspond to the case where $F = Q$, $p + 2 < r$, and the complementary series stops before one reaches the trivial representation. This is a result of B. Kostant [Kos].

Before we take up our next example, we remark that, even though the groups Sp(1, n) have "property T", i.e. the identity representation is isolated in the unitary dual, the identity representation may be approached by uniformly bounded representations. This is not true for many other groups, SL(3, R) for instance. One should presumably discuss "property TUB", by which we mean that the identity representation is isolated in the uniformly bounded dual. See the author's paper [Co2] for a further discussion of this phenomenon, which is not without interest in harmonic analysis.

EXAMPLE 7.5. (Communicated by W. Baldoni Silva).

For the groups Sp(1, n), there is a natural representation of M, which is just $Sp(n - 1)\times Sp(1)$, which is trivial on the "small" factor and which acts on $C^{2(n-1)}$ in the "large" factor. For this example, if $n \geqslant 3$, there is a complementary series in the "critical strip" $(0, 4n - 6)$, and then an isolated unitary representation when the parameter ζ is equal to $4n - 2$. In this latter case, the representation $\pi_{\mu,\zeta}$ is reducible.

8. AN EXTRAPOLATION PRINCIPLE.

We conclude with some applications of a theorem proved in the thesis of F. Carlson (1914). This theorem states that if f is an analytic function in the half-plane $\{\zeta \in C: \mathrm{Re}(\zeta) > 0\}$ and $f(\zeta)$ is $O(\exp(k|\zeta|))$, where $k < \pi$, and if moreover $f(\zeta) = 0$ when $\zeta = 1, 2, 3, \ldots$, then f is identically zero.

The first application we offer is to the calculation of Fourier transforms on V. We shall treat the kernels $N^{\zeta-r}$ explicitly, in a reasonably painless manner. This analysis is of interest not only in the study of the intertwining operators, but also in the finer aspects of nilpotent harmonic analysis.

THEOREM 8.1. Let the function $a_{\lambda,k}$ be defined by the formula

$$a_{\lambda,k}(\zeta) = 2^{1-p/2}\,\pi^{(p+q+1)/2}\,|\lambda|^{-\zeta/2}\,b_{\lambda,k}(\zeta) \quad ,$$

where

$$b_{\lambda,k}(\zeta) = \frac{\Gamma((p+4k-2-\zeta)/4)}{\Gamma((p+4k-2+\zeta)/4)}\,\frac{\Gamma(\zeta/2)}{\Gamma((r-\zeta)/4)\,\Gamma((p+2-\zeta)/4)} \quad .$$

Then $a_{\lambda,k}(\zeta)$ is the value of $\pi_\lambda(N^{\zeta-r})$ on the kth block, in the sense of Section 3.

Proof. We shall denote by $c_{\lambda,k}(\zeta)$, or by $\pi_\lambda(N^{\zeta-r})_k$, the value of $\pi_\lambda(N^{\zeta-r})$ on the kth block.

The proof can be divided into several stages. First, by evaluating two integrals, we determine $\pi_\lambda(N^{\zeta-r})$ for two particular values of ζ, namely 0 and 2. Next, we use the recurrence relations of Lemma 3.3 to show that $a_{\lambda,k}(\zeta)$ and $c_{\lambda,k}(\zeta)$ coincide for many values of ζ. Third, we show that the difference between the two functions considered does not grow too fast at infinity. Last, we apply Carlson's theorem. Here are the details of the proof.

We consider $N^{\zeta-r}$ as ζ approaches 0. We first claim that

$$\lim_{\varepsilon\to 0+} N^{\varepsilon-r} = [\,\omega(p)\;\omega(q)\;B(p/4,\; q/2)/2\,]\;\delta \tag{8.1}$$

and

$$\lim_{\varepsilon\to 0+} \Delta * N^{\varepsilon+2-r} = [\,\omega(p)\;\omega(q)\;B(p/4+1/2,q/2)(r-2)/2\,]\;\delta\ , \tag{8.2}$$

where $\omega(p)$ is the area of the unit sphere in R^p. To prove the equality (8.1), we recall from the proof of Lemma 1.1 that, if $f \in C_c^\infty(V)$, then

$$\lim_{\varepsilon\to 0+} \int_V dxdy\ N^{\varepsilon-r}(x, y)\ f(x, y)$$
$$= \lim_{\varepsilon\to 0+} \int_0^a ds\ s^{\varepsilon-1} \int_\Sigma d\sigma\ f(0, 0)$$
$$= \int_\Sigma d\sigma\ f(0, 0) \quad .$$

It therefore suffices to evaluate the area of Σ. Now

$$\int_\Sigma d\sigma$$
$$= r \int_0^1 ds\ s^{r-1} \int_\Sigma d\sigma$$
$$= r \int_{|x|^4 + |y|^2 \leqslant 1} dxdy$$
$$= r\ \omega(p)\ \omega(q) \iint_{u^4 + w^2 \leqslant 1} dudw\ u^{p-1}\ w^{q-1} \quad ,$$

where $w = |y|$ and $u = |x|$. We now put v equal to u^2 and obtain that

$$\int_\Sigma d\sigma$$
$$= r\ \omega(p)\ \omega(q)/2 \iint_{v^2 + w^2 \leqslant 1} dvdw\ v^{p/2-1}\ w^{q-1}$$
$$= \omega(p)\ \omega(q) \int_0^{\pi/2} d\theta\ \cos^{p/2-1}(\theta)\ \sin^{q-1}(\theta)$$
$$= \omega(p)\ \omega(q)\ B(p/4, q/2)/2 \quad ,$$

proving the equality (8.1).

In order to prove the equality (8.2), we recall (from Proposition 1.9 and Remark 1.10) that

$$(\Delta * N^{\varepsilon+2-r})(x, y) = \varepsilon(r-2-\varepsilon)\ |x|^2\ N(x, y)^{\varepsilon-2-r} \quad ,$$

whence $\Delta * N^{2-r} = C\ \delta$, with

$$C = (r-2) \int_\Sigma d\sigma\ |x|^2 \quad .$$

The calculation of C proceeds much as above, and we leave to the reader the task of verifying our claim.

The equality (8.1) is that

$$\mathrm{Res}(N^{\zeta-r};\ \zeta = 0) = [\omega(p)\ \omega(q)\ B(p/4, q/2)/2]\ \delta \ ,$$

which implies that

$$\begin{aligned}
\mathrm{Res}(c_{\lambda,k}(\zeta);\ \zeta = 0) &= \omega(p)\ \omega(q)\ B(p/4, q/2)/2 \\
&= 2\ \pi^{(p+q)/2}\ \Gamma(p/2)^{-1}\ \Gamma(q/2)^{-1}\ B(p/4, q/2) \\
&= 2\ \pi^{(p+q)/2}\ \Gamma(p/2)^{-1}\ \Gamma(p/4)\ \Gamma(p/4 + q/2)^{-1} \\
&= 2^{2-p/2}\ \pi^{(p+q+1)/2}\ \Gamma(p/2 + 1/2)^{-1}\ \Gamma(r/4)^{-1} \ ,
\end{aligned}$$

from the Legendre duplication formula (see, e.g., E.C. Titchmarsh [Tit], 1.86) and the fact that $r = p + 2q$. This is clearly equal to $\mathrm{Res}(a_{\lambda,k}(\zeta);\ \zeta = 0)$.

Analogously, from the equality (8.2), we have that

$$N^{2-r} = [(r-2)\ \omega(p)\ \omega(q)\ B(p/4 + 1/2,\ q/2)/2]\ \Delta^{-1} \ ,$$

whence it follows that

$$\begin{aligned} c_{\lambda,k}(2) &= \pi_\lambda(N^{2-r})_k \\ &= [(r-2)\ \omega(p)\ \omega(q)\ B(p/4 + 1/2,\ q/2)/2]\ \pi_\lambda(\Delta)_k^{-1} \\ &= [(r-2)\ \omega(p)\ \omega(q)\ B(p/4 + 1/2,\ q/2)/2][2\,|\lambda|\,(p+4k-4)]^{-1} \ . \end{aligned}$$

Similar calculations now lead to the conclusion that $c_{\lambda,k}(2) = a_{\lambda,k}(2)$.

The second step of the proof involves an examination of recurrence relations. We recall (Lemma 3.3) that

$$(\Delta^2 + B(\zeta)\square) * N^{\zeta-r} = C(\zeta)\ N^{\zeta-r-4} \ ,$$

where $B(\zeta) = -4(\zeta - 2)^2$ and $C(\zeta) = (\zeta-r)(\zeta-2)(\zeta-p-2)(\zeta-4)$. By passing to the Fourier transform, we obtain that

$$(\pi_\lambda(\Delta)^2 + B(\zeta)\pi_\lambda(\square))\ \pi_\lambda(N^{\zeta-r}) = C(\zeta)\ \pi_\lambda(N^{\zeta-r-4}) \ ,$$

whence, for k equal to 1, 2, 3, ...,

$$(4|\lambda|^2(p+4k-4)^2 + B(\zeta)\ |\lambda|^2)\ c_{\lambda,k}(\zeta) = C(\zeta)\ c_{\lambda,k}(\zeta-4) \ ,$$

i.e.
$$c_{\lambda,k}(\zeta-4)/c_{\lambda,k}(\zeta) = 4|\lambda|^2\ ((p+4k-4)^2 - (\zeta-2)^2)\ C(\zeta)^{-1} \ .$$

On the other hand, it is immediate from the definitions of $a_{\lambda,k}$ and $b_{\lambda,k}$ that

$$a_{\lambda,k}(\zeta-4)/a_{\lambda,k}(\zeta) = |\lambda|^2\ b_{\lambda,k}(\zeta-4)/b_{\lambda,k}(\zeta)$$

and the quotient on the right hand side is an expression involving ten gamma-functions. By using the recursion relation for the gamma-function —

$$\Gamma(z+1) = z\ \Gamma(z) \quad —$$

it follows very easily that

$$\begin{aligned} (8.3) \qquad a_{\lambda,k}(\zeta-4)/a_{\lambda,k}(\zeta) &= c_{\lambda,k}(\zeta-4)/c_{\lambda,k}(\zeta) \\ &= 4\,|\lambda|^2\ ((p+4k-4)^2 - (\zeta-2)^2)\ C(\zeta)^{-1} \ . \end{aligned}$$

The third stage of the proof is the study of the growth of $a_{\lambda,k}$ and $c_{\lambda,k}$ as ζ tends to infinity in the left half plane. Since $N^{\zeta-r}$ is homogeneous, so is its Fourier transform; $a_{\lambda,k}$ is homogeneous in λ by definition. In what follows, then we may assume that $|\lambda| = 1$. We shall also fix k, arbitrarily.

It is clear from the formula (8.3) that both the quotients involved are bounded by 1 when $\mathrm{Re}(\zeta) < 0$ and $|\zeta|$ is large enough. If we can show that $a_{\lambda,k}$ and $c_{\lambda,k}$ are bounded in the set S_0, given by the formula

$$S_0 = \{\zeta \in \mathbb{C}:\ \mathrm{Re}(\zeta) \in [-5, -1]\ ,\ |\mathrm{Im}(\zeta)| \geq 1\} \ ,$$

then it follows from the recurrence relation that $a_{\lambda,k}$ and $c_{\lambda,k}$ are bounded in the set S, given by the formula

$$S = \{\zeta \in C: \mathrm{Re}(\zeta) \leq -1, \ |\mathrm{Im}(\zeta)| \geq 1\} .$$

We notice also that a continuous function which satisfies the recurrence relation is automatically bounded in the set

$$\{\zeta \in C: \mathrm{Re}(\zeta) \leq -1, \ |\mathrm{Im}(\zeta)| \leq 1\} ,$$

by the same argument.

On the one hand, since when $|b|$ tends to infinity,

$$|\Gamma(a + ib)| \sim (2\pi)^{1/2} \ |b|^{a - 1/2} \ \exp(-\pi |b| /2) ,$$

uniformly for a in a compact interval (see, e.g., E.C. Titchmarsh [Tit], 4.42), it is true that $a_{\lambda,k}$ is bounded in S_0 and hence in S.

On the other hand, if f_k is a unit vector in the Hilbert space of π_λ, in the kth block, then the function $\langle \pi_\lambda f_k, f_k \rangle$ is both infinitely differentiable and bounded. If $\mathrm{Re}(\zeta) \in [-5, -1]$, then the distribution $N^{\zeta-r}$ is the sum of a compactly supported distribution of order at most 6 and a uniformly (in ζ) integrable piece away from (0, 0) (see Lemma 1.1 and Remark 1.3). It follows from the analysis of the proof of Lemma 1.1 that

$$\begin{aligned} |c_{\lambda,k}(\zeta)| &= |\langle \pi_\lambda (N^{\zeta-r}) f_k, f_k \rangle| \\ &= |\textstyle\int_V dv \ N^{\zeta-r}(v) \ \langle \pi_\lambda (v) f_k, f_k \rangle | \\ &\leq C, \end{aligned}$$

if $\zeta \in S_0$. Therefore $c_{\lambda,k}$ is also bounded in S.

To finish off the proof, we apply Carlson's theorem. Both $a_{\lambda,k}$ and $c_{\lambda,k}$ are meromorphic functions whose only possible poles in the left half-plane are simple poles at 0, -2, -4, -6, etc.; their difference $d_{\lambda,k}$ —

$$d_{\lambda,k} = a_{\lambda,k} - c_{\lambda,k} \quad —$$

is another function of the same kind. Moreover, we have the following information about $d_{\lambda,k}$. First, $d_{\lambda,k}(2) = 0$ and $\mathrm{Res}(d_{\lambda,k}(\zeta); \ \zeta = 0) = 0$, i.e. $d_{\lambda,k}$ has no pole at 0. Next, $d_{\lambda,k}$ satisfies the same recurrence relation as $a_{\lambda,k}$ and $c_{\lambda,k}$ ((8.3)). Finally, $d_{\lambda,k}$ is bounded in the set S.

We recall the recurrence relation:

$$(\zeta-2)(\zeta-4)(\zeta-r)(\zeta-p-2) \ d_{\lambda,k}(\zeta-4) = 4((p+4k-4)^2 - (\zeta-2)^2) \ d_{\lambda,k}(\zeta) \quad .$$

By letting ζ approach 2, we find that

$$\mathrm{Res}(d_{\lambda,k}(\zeta);\ \zeta = -2) = 0 ,$$

i.e. there is no pole at -2. Now if $\mathrm{Re}(\zeta) < 2$, the factor $(\zeta-2)(\zeta-4)\ldots$ which multiplies $d_{\lambda,k}(\zeta-4)$ is nonzero; inductively, we find that there are no poles at -4, -6, -8, ... But the recurrence relation actually gives us more than this. By putting ζ equal to $6 - p - 4k$, we find that $d_{\lambda,k}(2 - p - 4k) = 0$. Another inductive application of the recurrence formula implies that

$$(8.4) \qquad d_{\lambda,k}(2 - p - 4k - 4n) = 0 \qquad\qquad n \in N ,$$

where N denotes the set of natural numbers.

We may now conclude the $d_{\lambda,k}$ is an analytic function in the left half-plane, with many zeroes ((8.4)), and which is bounded in the set S and thence in the whole half-plane $\{\zeta \in C\colon \mathrm{Re}(\zeta) \leqslant -1\}$. Carlson's theorem (see E.C. Titchmarsh [Tit], 5.81) now implies that $d_{\lambda,k}$ is zero everywhere, as required to complete the proof. □

The second application of Carlson's theorem is in the decomposition of products of spherical functions. This application is due to L. Vretare [Vre], and is an essential part of J.-L. Clerc's treatment [Cle] of the spherical transform, already touched on in Section 6. We shall merely outline Vretare's result.

If one considers the space $H_{\mu,\zeta}$ of all functions on G such that

$$f(mang) = \mu(m)\ (\rho\alpha_{\zeta})(a)\ f(g) ,$$

one finds that for certain values of the parameters μ and ζ there is a finite-dimensional subspace of $H_{\mu,\zeta}$ invariant under the action of G. When this happens, we have a finite-dimensional representation of G embedded in the principal series. In N. Wallach's book [Wal], one may find the following results about these subrepresentations, which we summarise here as a theorem.

THEOREM 8.2. If $H_{\mu,\zeta}$ contains a nontrivial finite-dimensional G-invariant subspace, then this is unique. Consequently we may denote this subspace unambiguously by $V_{\mu,\zeta}$, and by $\sigma_{\mu,\zeta}$ the restriction of $\pi_{\mu,\zeta}$ to $V_{\mu,\zeta}$. If there is no such subspace, then we put $V_{\mu,\zeta}$ equal to {0}.

Every finite-dimensional representation of G occurs in this way.

The space $V_{1,-3r/2}$ contains a (K-fixed)vector $f_{-3r/2}$ whose restriction to K is identically 1.

If $H_{\mu,\zeta}$ contains a nontrivial subspace $V_{\mu,\zeta}$, then $H_{\mu,\zeta-r}$ contains the nontrivial finite-dimensional G-invariant subspace $V_{\mu,\zeta}\,V_{1,-3r/2}$. The map $f \mapsto f\, f_{-3r/2}$ is a K-module isomorphism of the K-modules $V_{\mu,\zeta}$ and $V_{\mu,\zeta}\, f_{-3r/2}$.

The representations $\pi_{\mu,\zeta}$ invade $H_{\mu,\zeta}$, in the sense that if we fix a representation κ of K occurring in $\pi_{\mu,\zeta}|_K$, then there exists ζ_κ in C such that, if $\zeta = \zeta_\kappa - nr$, with n in N, then

$$m(\kappa,\ (\sigma_{\mu,\zeta})|_K) = m(\kappa,\ (\pi_{\mu,\zeta})|_K)\quad ,$$

where $m(\kappa, \rho)$ is the multiplicity of the irreducible representation κ of K in the representation ρ of K.

Proof. Omitted.

We consider the so-called spherical principal series,i.e. the series of representations $\pi_{1,\zeta}$, with ζ in C. We denote by f_ζ the function in $H_{1,\zeta}$ whose restriction to K is identically 1. According to the above theorem, f_ζ lies in $V_{1,\zeta}$ if ζ is of the form $-(2n+1)r/2$, with n in N. For these values of ζ, the spherical functions ϕ_ζ —

$$\phi_\zeta(g) = \langle \pi_{1,\zeta}(g) f_\zeta,\ f_{-\bar{\zeta}} \rangle^{(K)} \quad —$$

may be viewed as matrix coefficients of the finite-dimensional representations $\sigma_{1,\zeta}$. Finite-dimensional representation theory gives us a product formula for $\phi_\zeta(g)\ \phi_\nu(g)$ when both ζ and ν lie in the set $-(2N + 1)r/2$. Analytic extrapolation allows us to extend to the case where ζ is arbitrary.

In order to understand better the finite-dimensional representations, we shall need a little more notation. We shall consider only the cases where F is C or Q, for simplicity. Let H and B be the subgroups of G which are given by the formulae

$$H = \{m(v,\ u) : u \in D\}$$

(cf. (4.3)), where D is the group of all $(n-1)\times(n-1)$ diagonal matrices whose diagonal entries are of the form $e^{\underline{i}\psi}$ (where $\underline{i} \in F$ and $\psi \in R$, by which we mean $\cos(\psi) + \underline{i}\sin(\psi)$, of course),

$$B = \{b_\theta : \theta \in R\}\quad ,$$

where b_θ is described in the second coordinate system as a matrix thus:

$$b_\theta = \begin{vmatrix} \cos(\theta) & 0 & \underline{i}\sin(\theta) \\ 0 & 1 & 0 \\ \underline{i}\sin(\theta) & 0 & \cos(\theta) \end{vmatrix} ,$$

the central submatrix being an $(n-1)\times(n-1)$ identity. The reader may wish to consult Section 4 for the definitions of A and M.

It is clear that both HA and HB are abelian subgroups of G. They are in fact so-called Cartan subgroups of G (maximal abelian subgroups of semisimple elements, in our case). We notice also that H is a Cartan subgroup of M and that HB is a maximal abelian subgroup of K.

The matrix coefficients of finite-dimensional representations of G, when restricted to the groups HA and HB, may be written as finite sums of exponentials. When HB is considered, these exponentials are all characters of the compact group HB. However, when one deals with HA, the exponentials are of a mixed nature: real exponentials appear on the group A. Thus the coefficients of the representation $\sigma_{\mu,-(2n+1)r/2}$, when restricted to HA, are of the form

$$f(h(\psi)a(t)) = \sum_{p,q} c(p,\ q)\ e^{ip\psi}\ e^{qt} ,$$

where ψ is a "multiparameter" and p is a multiindex, while q is an integer between nr and -nr.

In particular, for the spherical functions $\phi_{-(2n+1)r/2}$, where $n \in N$, we have that

(8.5) $$\phi_{-(2n+1)r/2}(h(\psi)a(t)) = \sum_q c(q)\ e^{qt} ,$$

where c(q) depends also on n. By using finite-dimensional representation theory, one could determine c(q) explicitly.

If we consider the product $\phi_\sigma\phi_\nu$ of such spherical functions, then it must be a sum of matrix coefficients of finite-dimensional representations. In fact we may determine exactly which finite-dimensional representations intervene in the tensor product of $\sigma_{1,\zeta}$ and $\sigma_{1,\nu}$; in particular, the number of representations involved is at most the order of the Weyl group of the complexification of G, a finite number. More precisely, the only ζ' which occur are of the form $\zeta + \gamma$, where γ runs over a finite set Γ which depends on ν but not on ζ.

Since the product of two K-biinvariant functions is again K-biinvariant, it is possible to find a formula

(8.6) $$\phi_\zeta\phi_\nu = \sum_\gamma C(\zeta,\ \nu;\ \gamma)\ \phi_{\zeta+\gamma} .$$

This formula holds when ζ and ν are in $(2N + 1)r/2$. It may be worth commenting that the coefficients in the previous formula (8.5) determine the coefficients $C(\zeta, \nu; \gamma)$ in the formula (8.6). In particular, if we consider that γ for which the absolute value of $\zeta + \gamma$ is largest, then the associated C-function is the product of two of the c-functions in the formula (8.5). This may be interpreted as the statement that asymptotically, as t tends to $+\infty$, $\phi_{(2n+1)r/2}(a(t))$ behaves like $c(nr)\, e^{nrt}$, and the asymptotic behaviour of the product is the product of the asymptotic behaviours.

It is possible to find rational functions, denoted by $C(\zeta, \nu; \gamma)$, of ζ which interpolate the function $C(\zeta, \nu; \gamma)$ determined by the formula (8.6). Let us now fix g in G, and consider the meromorphic function

$$\zeta \mapsto \phi_\zeta(g)\, \phi_\nu(g) \; - \; \textstyle\sum_{\gamma\in\Gamma} C(\zeta, \nu; \gamma)\, \phi_{\zeta+\gamma}(g) \quad .$$

One multiplies by the denominator of $C(\zeta, \nu; \gamma)$ to obtain an analytic function which is zero if $\zeta \in -(2N + 1)r/2$. Simple growth estimates show that Carlson's theorem is applicable; it follows that the formula (8.6) holds for all ζ in C.

It is an amusing exercise to show that the asymptotic behaviour of the spherical functions could also be obtained from Carlson's theorem and finite-dimensional representation theory. It is a natural question, at this point, to ask to what extent harmonic analysis on the groups considered may be derived in this elementary manner.

We should like to finish off by suggesting that the existence of discrete series may be established by this "extrapolation principle".

We consider the finite-dimensional representations $\sigma_{\mu,\zeta}$ of G. Restricted to K, these break up into a finite sum of irreducible representations of K. Certain representations of K occur with multiplicity one in $\sigma_{\mu,\zeta}|_K$; these are the "highest weight representations", by which we mean the following: the representation $\sigma_{\mu,\zeta}$ restricted to HB breaks up into a sum of one-dimensional representations, i.e. characters of HB. A certain number ($|W_C|$, the order of the Weyl group of the complexification of G) of these are extreme points of the set of all characters which occur in this way. There are certain representations of K for which at least one of these extreme points is again an extreme point; these representations we call the highest weight representations. There are $|W_C|/|W_K|$ of these, where W_K is the Weyl group of K.

We look for certain functions f^ζ in $V_{\mu,\zeta}$ of $L^2(K)$-norm one, which transform under HB by one of the extremal characters just described, and let ψ^ζ be the associated matrix coefficients:

$$\psi^\zeta : g \longmapsto \langle \pi_{\mu,\zeta}(g) f^\zeta , f^{-\zeta} \rangle^{(K)} .$$

Restricted to HB, this function is the extremal character of HB.

It would seem that product formulae of the form

$$\psi^\zeta \cdot \psi^\nu = \psi^{\zeta+\gamma}$$

hold, and that these formulae allow us to infer the existence of ψ^ζ which vanish rapidly at infinity, and hence the existence of discrete series. More precisely, such formulae hold when ζ and ν are negative integers, in which case γ is a ν-dependent negative integer. By "analytic extrapolation" such formulae continue to hold when ζ is a positive integer, and in this case as ζ increases, ψ^ζ vanishes more rapidly at infinity; when ζ is large enough we obtain square integrable ψ^ζ, which correspond to discrete series.

It seems necessary to consider other groups along the way, à la M. Flensted-Jensen [FJ1], [FJ2], in order to work in an environment where HB is noncompact, so that the (not necessarily unitary) characters of HB are parametrised by C rather than Z.

REFERENCES

[Bam] L. Bamazi, *Représentations sphériques uniformement bornées des groupes de Lorentz; Analyse Harmonique sur les Groupes de Lie II.* Lecture Notes in Math. 739. Springer-Verlag, Berlin, Heidelberg, New York, 1979.

[Beu] A. Beurling, *On two problems concerning linear transformations in Hilbert spaces,* Acta Math. 81 (1948), 239-255.

[Cal] A.P. Calderón, *Intermediate spaces and interpolation, the complex method,* Studia Math. XXIV (1964), 113-190.

[CaZ] A.P. Calderón and A. Zygmund, *On singular integrals,* Amer. J. Math. 78 (1956), 289-309.

[Cle] J.-L. Clerc, *Transformation de Fourier sphérique des espaces de Schwartz,* preprint, Université de Nancy I.

[CoW] R.R. Coifman and G. Weiss, *Extensions of Hardy spaces and their use in analysis,* Bull. Amer. Math. Soc. 83 (1977), 569-645.

[Co1] M.G. Cowling, *The Kunze-Stein phenomenon,* Annals of Math. 107 (1978), 209-234.

[Co2] M.G. Cowling, *Sur les coefficients des représentations unitaires des groupes de Lie simples; Analyse Harmonique sur les Groupes de Lie II.* Lecture Notes in Math. 739. Springer-Verlag, Berlin, Heidelberg, New York, 1979.

[CoM] M.G. Cowling and A.M. Mantero, *Intertwining operators and representations of semisimple Lie groups,* in preparation.

[EM1;2;3] L. Ehrenpreis and F.I. Mautner, *Some properties of the Fourier transform on semisimple Lie groups. I,* Annals of Math. 61 (1955), 406-439; *II,* Trans. Amer. Math. Soc. 84 (1957), 1-55; *III,* Trans. Amer. Math. Soc. 90 (1959), 431-484.

[FJ1] M. Flensted-Jensen, *Spherical functions on a real semisimple Lie group. A method of reduction to the complex case,* J. Funct. Anal. 30 (1978), 106-146.

[FJ2] M. Flensted-Jensen, *Discrete series for semisimple symmetric spaces,* Annals of Math. 111 (1980), 253-311.

[Fo1] G.B. Folland, *A fundamental solution for a subelliptic operator*, Bull. Amer. Math. Soc. 79 (1973), 373-376.

[Fo2] G.B. Folland, *Subelliptic estimates and function spaces on nilpotent groups*, Arkiv för Mat. 13 (1975), 161-207.

[FoS] G.B. Folland and E.M. Stein, *Estimates for the* $\bar{\partial}_b$ *complex and analysis on the Heisenberg group*, Comm. Pure Appl. Math. 27 (1974), 429-522.

[Hel] S. Helgason, *Differential Geometry, Lie Groups, and Symmetric Spaces*. Academic Press, New York, San Francisco, London, 1978.

[Her] C.S. Herz, *Harmonic synthesis for subgroups*, Ann. Inst. Fourier (Grenoble) 23 (1973), 91-123.

[Hun] R.A. Hunt, *On L(p,q) spaces*, L'Enseignement Math. XII (1956), 249-276.

[Kap] A. Kaplan, *Fundamental solutions for a class of hypoelliptic pde generated by composition of quadratic forms*, Trans. Amer. Math. Soc. 258 (1980), 147-153.

[Ki1] A.A. Kirillov, *Unitary representations of nilpotent Lie groups*, Russian Math. Surveys 17 (1962), 53-104.

[Ki2] A.A. Kirillov, *Eléments de la Théorie des Représentations*. MIR, Moscou, 1974.

[KnS] A.W. Knapp and E.M. Stein, *Intertwining operators for semisimple groups*, Annals of Math. 93 (1971), 489-578.

[KoV] A. Korányi and S. Vági, *Singular integrals on homogeneous spaces and some problems of classical analysis*, Ann. Scuola Norm. Sup. Pisa 25 (1971), 575-648.

[Kos] B. Kostant, *On the existence and irreducibility of a certain series of representations*, Bull. Amer. Math. Soc. 75 (1969), 627-642.

[KS1] R.A. Kunze and E.M. Stein, *Uniformly bounded representations and harmonic analysis on the 2×2 real unimodular group*, Amer. J. Math. 82 (1960), 1-62.

[KS2] R.A. Kunze and E.M. Stein, *Uniformly bounded representations II: Analytic continuation of the principal series of representations of the n×n complex unimodular group*, Amer. J. Math. 83 (1961), 723-786.

[KS3] R.A. Kunze and E.M. Stein, *Uniformly bounded representations III:*

Intertwining operators for the principal series on semisimple groups, Amer. J. Math. 89 (1967), 385-442.

[Lan] R.P. Langlands, *On the classification of irreducible representations of real algebraic groups*, preprint, I.A.S., Princeton.

[Lep] H. Leptin, *Ideal theory in group algebras of locally compact groups*, Invent. Math. 31 (1976), 259-278.

[Li1] R.L. Lipsman, *Uniformly bounded representations of SL(2,C)*, Amer. J. Math. 91 (1969), 47-66.

[Li2] R.L. Lipsman, *Harmonic analysis on SL(n,C)*, J. Funct. Anal. 3 (1969), 126-155.

[Li3] R.L. Lipsman, *Uniformly bounded representations of the Lorentz group*, Amer. J. Math. 91 (1969), 938-962.

[Li4] R.L. Lipsman, *An explicit realisation of Kostant's principal series with applications to uniformly bounded representations*, preprint, University of Maryland.

[Loh] N. Lohoué, *Sur les représentations uniformement bornées et le théorème de convolution de Kunze-Stein*, preprint, Université de Paris XI.

[Puk] L. Pukanszky, *Leçons sur les Représentations des groupes*. Dunod, Paris, 1967.

[Sae] S. Saeki, *Translation invariant operators on groups*, Tôhoku Math. J. 22 (1970), 409-419.

[Sal] P.J. Sally, Jr, *Analytic continuation of the irreducible unitary representations of the universal covering group of SL(2,R)*, Mem. Amer. Math. Soc. 69 (1967).

[Sit] A. Sitaram, *An analogue of the Wiener tauberian theorem for spherical transforms on semisimple Lie groups*, to appear, Pacific J. Math.

[Ste] E.M. Stein, *Interpolation of linear operators*, Trans. Amer. Math. Soc. 83 (1956), 482-492.

[SW1] E.M. Stein and S. Wainger, *The estimation of an integral arising in multiplier transformations*, Studia Math. XXXV (1970), 101-104.

[SW2] E.M. Stein and S. Wainger, *Problems in harmonic analysis related to curvature*, Bull. Amer. Math. Soc. 84 (1978), 1239-1295.

[StW] E.M. Stein and G. Weiss, *Introduction to Fourier Analysis on Euclidean Spaces*. Princeton University Press, Princeton, 1975.

[Str] R.S. Strichartz, *Multipliers on fractional Sobolev spaces*, J. Math. Mech. 16 (1967), 1031-1060.

[Tit] E.C. Titchmarsh, *The Theory of Functions*. Oxford University Press, Oxford, etc., 1978.

[TrV] P.C. Trombi and V.S. Varadarajan, *Spherical transforms on semisimple Lie groups*, Annals of Math. 94 (1971), 246-303.

[Vre] L. Vretare, *Elementary spherical functions of symmetric spaces*, Math. Scand. 39 (1976), 343-358.

[Wal] N. Wallach, *Harmonic Analysis on Homogeneous Spaces*. Marcel Dekker, Inc., New York, 1973.

[War] G. Warner, *Harmonic Analysis on Semi-Simple Lie Groups. Vols I and II.* Springer-Verlag, Berlin, Heidelberg, New York, 1972.

[Wei] Y. Weit, *On the one-sided Wiener's theorem for the motion group*, Annals of Math. 111 (1980), 415-422.

[Wil] E.N. Wilson, *Uniformly bounded representations for the Lorentz groups*, Trans. Amer. Math. Soc. 166 (1972), 431-438.

[Zyg] A. Zygmund, *Trigonometric Series. Vol. I, English translation*, 2nd ed.. Cambridge University Press, London and New York, 1978.

CENTRO INTERNAZIONALE MATEMATICO ESTIVO

(C.I.M.E.)

CONSTRUCTION DE REPRESENTATIONS UNITAIRES D'UN GROUPE DE LIE

MICHEL DUFLO
Université Paris 7

RESUME :

Soit G un groupe de Lie d'algèbre de Lie $\underline{g}$. La "méthode des orbites" veut associer à certaines formes linéaires g sur $\underline{g}$ une famille de représentations unitaires de G. Je construis de telles représentations lorsque g est admissible et bien polarisable. Elles sont paramétrées par certaines représentations projectives du groupe des composantes connexes du stabilisateur de g dans G. Lorsque G est algébrique, l'ensemble des représentations unitaires irréductibles de G obtenues par ce procédé suffit à décomposer $L^2(G)$.

Summary :

Let G be a Lie group with Lie algebra $\underline{g}$. The "orbit method" should associate to some linear forms g on $\underline{g}$ a set of unitary representations of G. I construct such representations when g is admissible and has a good polarization. They are parametrized by some projective representations of the group of connected components of the stabilizer of g in G. When G is algebraic, the set of the irreducible unitary representations of G obtained this way is large enough to decompose $L^2(G)$.

TABLE DES MATIERES :

Appendice :

Parametrization of the set of regular orbits of the co-adjoint representation of a Lie group. (Un texte préparé pour une conférence à l'Université de Maryland, décembre 1978).

INTRODUCTION :

Soit G un groupe de Lie d'algèbre de Lie $\underline{g}$. Notre but est de construire un ensemble de représentations unitaires irréductibles de G, assez gros dans un sens que nous préciserons plus bas. Il est bien connu que cela peut se faire dans le cadre de la "méthode des orbites" lorsque G est résoluble connexe (cf. Kirillov [17] , Pukanszky [24]), ou lorsque G est réductif connexe (il est facile d'interpréter les résultats d'Harish-Chandra [12] dans ce cadre). Nous le faisons ici dans le cas général.

* * * *

Le groupe G opère dans son algèbre de Lie $\underline{g}$ par la représentation adjointe, et dans le dual $\underline{g}^*$ de $\underline{g}$ par la représentation co-adjointe. Le stabilisateur d'un élément $g \in \underline{g}^*$ est noté $G(g)$, et son algèbre de Lie $\underline{g}(g)$. J'ai introduit dans [8] un certain revêtement d'ordre 2 de $G(g)$. Notons le $G(g)^{\underline{g}}$. C'est l'ensemble des couples (x,m) où $x \in G(g)$, m est dans le groupe métaplectique associé à l'espace symplectique $\underline{g}/\underline{g}(g)$, tels que x et m aient même image dans le groupe symplectique de $\underline{g}/\underline{g}(g)$. Notons $(1,-1)$ l'élément non trivial du noyau de la projection de $G(g)^{\underline{g}}$ sur $G(g)$. Nous noterons $X(g)$ l'ensemble des classes de représentations unitaires τ de $G(g)^{\underline{g}}$ dont la différentielle est la restriction de $ig\,\mathrm{Id}$ à $\underline{g}(g)$, et telle que $\tau(1,-1) = -\mathrm{Id}$.

Une forme linéaire g est dite <u>admissible</u> si $X(g)$ est non vide. Nous aurons besoin aussi de la notion de forme $g \in \underline{g}^*$ <u>bien polarisable</u> . La définition sera donnée au chapitre II. Disons simplement que si $\underline{g}$ est résoluble, toutes les formes sont bien polarisables, et que si $\underline{g}$ est semi-simple, g est bien polarisable si et seulement si $\underline{g}(g)$ est une sous-algèbre de Cartan de $\underline{g}$.

Dans ces notes, nous construisons, si g est admissible et bien polarisable, et si $\tau \in X(g)$, une classe de représentations unitaires de G, notée $T_{g,\tau}$. Nous démontrons les propriétés suivantes.

(i) Le commutant de $T_{g,\tau}$ est isomorphe à celui de τ. En particulier, $T_{g,\tau}$ est irréductible (resp. factorielle, factorielle semi-finie) si et seulement s'il en est de même de τ.

(ii) Si a est un automorphisme de G, on a : ${}^{a}T_{g,\tau} = T_{ag,{}^{a}\tau}$ (où l'on a posé ${}^{a}\tau = \tau \circ a^{-1}$, etc...).

(iii) Si g' est admissible et bien polarisable, si $\tau' \in X(g')$ et si $g' \notin Gg$, alors les représentations $T_{g,\tau}$ et $T_{g',\tau'}$ sont disjointes.

(iv) Soient τ, τ' dans $X(g)$. L'espace des opérateurs d'entrelacement entre $T_{g,\tau}$ et $T_{g,\tau'}$ est isomorphe à l'espace des opérateurs d'entrelacement entre τ et τ'.

(v) Supposons τ factorielle. Soit γ un élément du centre de G. Soit $(\gamma,1)$ l'élément correspondant du centre de $G(g)^{\underline{g}}$. Alors $T_{g,\tau}(\gamma)$ et $\tau(\gamma,1)$ sont la multiplication par un même scalaire.

* * * *

Il est naturel de demander si l'on obtient par ce procédé beaucoup de représentations de G (ou même, plus précisément, quelles sont les représentations de G ainsi obtenues). J'espère revenir plus tard sur cette question. Je me contenterai de donner par des exemples un début de réponse.

Lorsque G est connexe et localement algébrique, les représentations $T_{g,\tau}$ avec τ irréductible, suffisent pour décomposer $L^2(G)$.

Ceci est encore vrai lorsque G est connexe de type I. Plus généralement, si G est connexe (mais pas nécessairement de type I), et si $\alpha \in L^1(G)$ est une fonction telle que $T_{g,\tau}(\alpha) = 0$ pour toutes les représentations $T_{g,\tau}$ avec τ irréductible, alors $\alpha = 0$.

Je ne démontrerai ces résultats que lorsque G est connexe et localement algébrique. En effet, ce cas fournit déjà (à mon avis) une bonne motivation de travail entrepris ici. D'autre part, il doit être établi avant le cas général. Le cas général demande la comparaison de notre construction avec celles faites dans l'article fondamental de Pukanszky [25] , ce qui déborde le cadre de ces notes.

Je donnerai cependant quelques applications aux groupes de Lie moyennables.

* * * *

L'idée de base pour définir les représentations $T_{g,\tau}$ est très simple. En appliquant, autant de fois qu'il le faut, la théorie de Mackey aux sous-groupes invariants nilpotents fermés connexes, on se ramène au cas où $\underline{g}$ est réductive. En appliquant une nouvelle fois la théorie de Mackey, à la composante neutre de G cette fois, on se ramène au cas où G est réductif et connexe. Comme g est bien polarisable, G(g) est un sous-groupe de Cartan de G et l'on est dans une situation où l'on peut appliquer des résultats d'Harish-Chandra.

Evidemment, cette idée n'est pas nouvelle et a déjà été appliquée avec succès dans de nombreux cas. Citons les groupes résolubles connexes (Auslander - Kostant [2]), les groupes connexes à radical co-compact (Pukanszky [26]), les groupes compacts non connexes (Kostant [19]).

Ce qui est important ici est d'avoir une formulation des résultats qui survive à l'application de la théorie de Mackey, et donc qui incorpore les obstructions cohomologiques qu'elle comporte. C'est précisément le rôle que joue le revêtement $G(g)^{\underline{g}}$ de G(g).

* * * *

Le groupe métaplectique et sa représentation métaplectique interviennent de manière essentielle dans la formulation et dans la démonstration de nos résultats. (Il est peut-être curieux de remarquer que la représentation métaplectique elle-même n'est pas associée à une forme linéaire bien polarisable sur l'algèbre de Lie du groupe symplectique, et n'est donc pas une de nos représentations $T_{g,\tau}$). J'ai rassemblé dans le chapitre I un certain nombre de résultats plus ou moins bien connus sur la représentation métaplectique, pour faciliter la lecture de ces notes.

* * * *

Une de mes motivations en entreprenant ce travail était de donner un peu de consistance à des conjectures relatives à la formule de Plancherel des groupes algébriques unimodulaires, énoncées dans une conférence que j'ai faite à l'Université de Maryland. Pensant que cela pourrait être une motivation aussi pour le lecteur, j'ai adjoint en appendice le texte que j'ai écrit à cette occasion - en harmonisant les notations avec celles employées ici.

Principales conventions :

Toutes les constructions dépendent d'un choix d'un caractère unitaire non trivial de $\underline{\underline{R}}$. Nous noterons i une racine carrée de -1 fixée une fois pour toutes, et nous choisissons le caractère $x \to e^{ix}$.

Ainsi si V est un espace vectoriel réel de dimension finie et si V^* est son dual, alors V^* est canoniquement en bijection avec le dual unitaire de V par l'application qui à $v \in V^*$ associe le caractère $T_f : x \to e^{v(x)}$ de V .

* * * *

Tous les groupes de Lie considérés ici sont dénombrables à l'infini. Lorsque T est une représentation d'un groupe de Lie dans un espace de Banach H , on suppose toujours H séparable et T fortement continue.

Principales notations :

1. Si V est un espace vectoriel, on note V^* son dual.

2. Si V est un espace vectoriel réel, on note $V_{\underline{\underline{C}}}$ son complexifié, et $v \to \bar{v}$ la conjugaison complexe.

3. Si $\underline{g}$ est une algèbre de Lie sur un corps commutatif, on note $U(\underline{g})$ son algèbre enveloppante, $Z(\underline{g})$ son centre.

4. Si G est un groupe de Lie, on note G_0 sa composante neutre. Si G opère dans un espace vectoriel V de dimension finie par une représentation et si $v \in V$, on note $G(v)$ le stabilisateur de v dans V. Si $\underline{g}$ est l'algèbre de Lie de G, on note $\underline{g}(v)$ l'algèbre de Lie de $G(v)$.

5. Soient $V \supset V_1 \supset V_2$ des espaces vectoriels, et $x \in GL(V)$ laissant stable V_1 et V_2. On note x_{V_1/V_2} l'élément de $GL(V_1/V_2)$ qui s'en déduit.

Si g est une forme linéaire sur V, on notera $g|V_1$ sa restriction à V_1.

6. Soit H un sous-groupe fermé de G d'algèbre de Lie $\underline{h}$. Si R est une représentation unitaire de H dans un espace de Hilbert $\mathcal{H}$, on note $\mathrm{Ind}_H^G(R)$ la représentation unitaire de G induite par R. Rappelons que l'espace de cette représentation est l'espace de Hilbert formé des fonctions α sur G, à valeurs dans $\mathcal{H}$, qui sont mesurables, et vérifient :

$$(1) \qquad \alpha(yx) = |\det \mathrm{Ad}\, x_{\underline{g}/\underline{h}}|^{\frac{1}{2}} R(x)^{-1} \alpha(y)$$

pour tout $y \in G$, $x \in H$.

$$(2) \qquad \int_{G/H} |\alpha|^2 \, dx < \infty.$$

Le groupe G opère par translations à gauche dans cet espace (voir [3] chapitre V par exemple).

7. Si $u \in U(\underline{g}_{\underline{\underline{C}}})$, on note par la même lettre l'opérateur différentiel à gauche sur G qui lui correspond. En particulier, si $X \in \underline{g}$, $\alpha \in C^\infty(G)$,

$x \in G$, on a :

$$(X\alpha)(x) = \frac{d}{dt}\,\alpha(x \exp tX)\,\Big|_{t=0}\,.$$

CHAPITRE I - LE GROUPE METAPLECTIQUE ET LA REPRESENTATION METAPLECTIQUE

Soit V un espace vectoriel réel symplectique. Nous donnons (par. 1-5 .) une description du groupe métaplectique et de sa représentation métaplectique, sous la forme où nous l'utiliserons. Cette présentation est due à Souriau [31] et Lion [20] . Pour les démonstrations et l'historique, nous renvoyons au livre [22] . Pour tout sous-espace Lagrangien L de $V_{\mathbb{C}}$, nous définissons un caractère ρ_L du sous-groupe du groupe métaplectique stabilisant L et donnons plusieurs propriétés amusantes de ces caractères (par. 6-9.).

I. 1. Indice de Maslov :

Dans tout ce chapitre, V est un espace vectoriel réel symplectique i.e. un espace vectoriel réel muni d'une forme bilinéaire alternée non dégénérée que nous noterons B. La dimension de V est donc paire. Nous la noterons $2d$.
Si E est un sous-espace de V, on note $E^{\perp}$ le sous-espace de V orthogonal.
Un sous-espace E est dit isotrope si $E^{\perp}$ contient E, et lagrangien si $E^{\perp} = E$. On note $\Lambda(V)$ l'ensemble des sous-espaces lagrangiens de V .
Si $L \in \Lambda(V)$, et si E est un sous-espace isotrope de V , on pose

$$L^E = (L \cap E^{\perp}) + E \text{ . On a } L^E \in \Lambda(V).$$

Soient L_1 , L_2 , L_3 trois sous-espaces lagrangiens. Sur l'espace $L_1 \oplus L_2 \oplus L_3$ on considère la forme quadratique

$$Q(x_1,x_2,x_3) = B(x_1,x_2) + B(x_2,x_3) + B(x_3,x_1) .$$

On pose

$$t(L_1,L_2,L_3) = p - q$$

où (p,q) est la signature de Q .

C'est l'indice de Maslov, comme défini par M. Kashiwara. L'indice de Maslov est donc une fonction à valeurs dans $\underline{\underline{Z}}$ définie sur $\Lambda(V) \times \Lambda(V) \times \Lambda(V)$.

Nous noterons $Sp(V)$ le groupe symplectique de V, i.e. le sous-groupe de $GL(V)$ conservant B.

Soient L_0, L_1, L_2, L_3 des sous-espaces lagrangiens de V, E un sous-espace isotrope de V contenu dans $L_1 \cap L_2 + L_2 \cap L_3 + L_3 \cap L_1$, et $x \in Sp(V)$. On a les formules :

(1) $t(L_1,L_2,L_3) = t(L_2,L_3,L_1) = -t(L_2,L_1,L_3)$

(2) $t(L_1,L_2,L_3) - t(L_0,L_2,L_3) + t(L_0,L_1,L_3) - t(L_0,L_1,L_2) = 0$

(3) $t(xL_1,xL_2,xL_3) = t(L_1,L_2,L_3)$

(4) $t(L_1^E,L_2^E,L_3^E) = t(L_1,L_2,L_3)$.

On posera

(5) $\gamma(L_1,L_2,L_3) = \exp(i\,\frac{\pi}{4}\,t(L_1,L_2,L_3))$.

I. 2. Indice de deux sous-espaces lagrangiens orientés :

Un sous-espace lagrangien orienté est un couple (L,e) où e est une composante connexe de $\Lambda^d L$. (Rappelons que tous les sous-espaces lagrangiens sont de dimension d). Notons $\Lambda(V)$ l'ensemble des sous-espaces lagrangiens orientés de V.

Soient $L = (L,e)$ et $L' = (L',e')$ dans $\Lambda(V)$. Nous allons définir un nombre $\varepsilon(L,L')$ qui vaut ± 1. Supposons d'abord que l'on a : $L \cap L' = \{0\}$. On peut choisir une base $e_1,\dots,e_d$ de L et une base $f_1,\dots,f_d$ de L' telles que $e_1 \wedge \dots \wedge e_d \in e$, et $B(e_j,f_k) = \delta_{jk}$ pour $1 \le f,k \le d$. Alors $\varepsilon(L,L') = +1$ si et seulement si $f_1 \wedge \dots \wedge f_d \in e'$. Dans le cas général, posons $E = L \cap L'$, et choisissons une orientation de E. Par passage au quotient, on en déduit des orientations e_E et e'_E de L/E et L'/E.

L'espace $E^{\perp}/E$ est canoniquement muni d'une structure symplectique, et l'on obtient ainsi des éléments L/E et L'/E de $\Lambda(E^{\perp}/E)$. On pose alors $\varepsilon(L,L') = \varepsilon(L/E,L'/E)$. (On vérifie immédiatement que cette définition ne dépend pas du choix de l'orientation sur E). Enfin on pose

(6) $\quad s(L,L') = i^{d-\dim L\cap L'}\,\varepsilon(L,L')$.

Soient L_1,L_2,L_3 des éléments de $\Lambda(V)$, et $x\in Sp(V)$. L'indice s a les propriétés suivantes :

(7) $\quad s(L_1,L_2) = s(L_2,L_1)^{-1}$

(8) $\quad s(xL_1,xL_2) = s(L_1,L_2)$

(9) $\quad s(L_1,L_2)s(L_2,L_3)s(L_3,L_1) = \gamma(L_1,L_2,L_3)^2$.

I. 3. Le groupe métaplectique :

Soit $x\in Sp(V)$. On définit une fonction s_x sur $\Lambda(V)$ à valeurs dans les racines quatrièmes de 1, en posant :

(10) $\quad s_x(L) = s(L,xL)$

où $L\in\Lambda(V)$, et L est une orientation de L. La définition ne dépend pas du choix de l'orientation.

Il existe deux fonctions ϕ sur $\Lambda(V)$ vérifiant les relations

(11) $\quad \phi^2 = s_x^{-1}$

(12) $\quad \phi(L') = \phi(L)\gamma(L',L,xL)\gamma(L',L,x^{-1}L')^{-1}$

pour tout L, $L'\in\Lambda(V)$. Si $L\in\Lambda(V)$, une telle fonction est complètement déterminée par le choix d'une racine carrée de $s_x(L)^{-1}$.

On note $Mp(V)$ l'ensemble des couples (x,ϕ), où $x\in Sp(V)$, et où ϕ est une fonction sur $\Lambda(V)$ vérifiant (11) et (12).

On vérifie que l'on munit Mp(V) d'une loi de groupe en posant :

(13) $(x,\phi)(x,\phi') = (xx',\phi'')$

où ϕ'' est la fonction sur $\Lambda(V)$ définie par la formule.

(14) $\phi''(L) = \phi(L)\phi'(L)\gamma(L,xL,xx'L)$.

L'application $(x,\phi) \to x$ est un homomorphisme surjectif de noyau $\{(1,1),\ (1,-1)\}$. On vérifie qu'il y a sur Mp(V) une unique topologie qui en fait un groupe localement isomorphe à Sp(V). Si $V \neq \{0\}$, on démontre que Mp(V) est connexe. Si $V = \{0\}$, $Mp(V) = \mathbb{Z}/2\mathbb{Z}$. Le groupe Mp(V) s'appelle le groupe métaplectique (pour tout ceci, voir [22]).

I. 4. Le groupe d'Heisenberg :

On note $\underline{n}$ l'algèbre de Lie dont l'espace vectoriel sous-jacent est $V \oplus \mathbb{R}e$ muni du crochet $[v + te, v' + t'e] = B(v,v')e$. On note N le groupe de Lie simplement connexe d'algèbre de Lie $\underline{n}$. C'est le groupe d'Heisenberg. Il est de dimension $2d + 1$.

Le théorème de Stone-Von Neumann affirme qu'il y a une, et une seule, classe d'équivalence de représentations unitaires irréductibles de N dont la restriction au sous-groupe $\exp(\mathbb{R}e)$ est multiple du caractère $\exp(te) \to e^{it}$. Nous noterons T une telle représentation et $\mathcal{H}$ l'espace de Hilbert dans lequel elle opère.

Soit $L \in \Lambda(V)$. A L est associé un modèle concret de T. On note $\underline{b}_L$ l'espace $L + \mathbb{R}e$. C'est une sous-algèbre abélienne de $\underline{n}$. On note B_L le sous-groupe analytique correspondant, et χ_L le caractère unitaire de B_L tel que $\chi_L(\exp(v+te)) = e^{it}$ pour $v + te \in \underline{b}_L$. On pose $T_L = \mathrm{Ind}_{B_L}^{N}(\chi_L)$ et on note $\mathcal{H}_L$ l'espace de Hilbert dans lequel T_L opère (la définition des représentations induites est rappelée dans les "principales notations"). Les représentations T et T_L sont équivalentes.

Soient $L, L' \in \Lambda(V)$. Comme les représentations T_L et $T_{L'}$ sont équivalentes, il existe un opérateur d'entrelacement unitaire de $\mathcal{H}_L$ dans $\mathcal{H}_{L'}$. Nous allons en choisir un canoniquement.

Soit dX une mesure de Haar sur $L \cap L'$. Soit $\alpha \in \mathcal{H}_L$ une élément représenté par une fonction continue sur N à support compact modulo B_L. On définit une fonction $F_{L'L}(\alpha)$ sur N en posant :

$$(15) \qquad F_{L'L}(\alpha)(x) = \int_{L'/L\cap L'} \alpha(x \exp X)dX$$

pour tout $x \in N$. Alors $F_{L'L}(\alpha)$ est dans $\mathcal{H}_{L'}$. On démontre que l'on peut choisir dX de telle sorte que $F_{L'L}$ se prolonge par continuité en un opérateur unitaire de $\mathcal{H}_L$ dans $\mathcal{H}_{L'}$ qui entrelace T_L et $T_{L'}$.

Soient $L_1, L_2, L_3 \in \Lambda(V)$. On a (d'après Souriau [31] et Lion [20]).

$$(16) \qquad F_{L_1L_2} F_{L_2L_3} F_{L_3L_1} = \gamma(L_1,L_2,L_3)\mathrm{Id}_{\mathcal{H}_{L_1}}$$

$$(17) \qquad F_{L_1L_2} F_{L_2L_1} = \mathrm{Id}_{\mathcal{H}_{L_1}}$$

I. 5. La représentation métaplectique :

Le groupe $Sp(V)$ opère comme groupe d'automorphismesde $\underline{n}$ par la formule $x(v + te) = x(v) + te$ $(x \in Sp(V), t \in \mathbb{R})$. Il opère donc dans N.

Pour tout $x \in Sp(V)$, la représentation ${}^xT = T \circ x^{-1}$ est équivalente à T d'après le théorème de Stone-Von-Neumann. Il existe donc un opérateur unitaire $S'(x)$ dans $\mathcal{H}$ tel que $S'(x)T(n)S'(x)^{-1} = T(x(n))$ pout tout $n \in N$, et S' est une représentation projective de $Sp(V)$.

Supposons $V \neq 0$ Shale [30] a démontré qu'il existe une et une seule représentation unitaire S de $Mp(V)$ telle que

$$(18) \qquad S(x,\phi)T(n)S(x,\phi)^{-1} = T(x(n))$$

pout tout $(x,\phi) \in Mp(V)$, et tout $n \in N$.

Nous appelerons cette représentation la représentation métaplectique. On a $S(1,-1) = -\mathrm{Id}_{\mathcal{H}}$. Lorsque $V = 0$ nous adopterons cette formule comme définition de S.

(La représentation métaplectique est aussi appelée : "représentation de Segal-Shale-Weil", "oscillator", "spinor", "harmonic").

Nous voulons donner une description concrète dans l'espace $\mathcal{H}_L$ défini dans le par. 4 . Soit donc $L \in \Lambda(V)$. Soient $x \in Sp(V)$, $\alpha \in \mathcal{H}_L$. On pose

$$(A_L(x)\alpha)(n) = \alpha(x^{-1}(n)).$$

pour tout $n \in N$. Alors A_L est un opérateur de $\mathcal{H}_L$ dans $\mathcal{H}_{xL}$, multiple d'une isométrie. On pose

$$(19) \qquad S'_L(x) = \|A_L(x)\|^{-1} F_{L,xL} A_L(x).$$

L'opérateur $S'_L(x)$ est unitaire dans $\mathcal{H}_L$, et l'on a :

$$(20) \qquad S'_L(x)T_L(n)S'_L(x)^{-1} = T_L(x(n))$$

pour tout $n \in N$.

Soit $(x,\phi) \in Mp(V)$. On pose

$$(21) \qquad S_L(x,\phi) = \phi(L)S'_L(x).$$

Alors S_L est la représentation métaplectique de $Mp(V)$ dans $\mathcal{H}_L$. (Compte-tenu de (20) et (21), il suffit de vérifier que S_L est une représentation, ce qui résulte de (14) et (16)).

I. 6. *Définition des caractères* $\rho_{\underline{l}}$:

L'espace $V_{\underline{\underline{C}}}$ est canoniquement un espace symplectique complexe. On note $\Lambda(V_{\underline{\underline{C}}})$ l'ensemble de ses sous-espaces (complexes) lagrangiens. Si $\underline{l} \in \Lambda(V_{\underline{\underline{C}}})$, on note $q_{\underline{l}}$ le nombre de valeurs propres < 0 de l'application hermitienne associée à la forme hermitienne $X \to iB(X,\bar{X})$ sur $\underline{l}$.

Si $\underline{l} \in \Lambda(V_{\underline{\underline{C}}})$, $\underline{l}$ est une sous-algèbre abélienne de $\underline{n}_{\underline{\underline{C}}}$. L'algèbre $\underline{n}_{\underline{\underline{C}}}$ opère par la représentation T dans l'espace $\mathcal{H}^\infty$ des vecteurs C^∞ de la représentation. Par restriction $\mathcal{H}^\infty$ devient un $\underline{l}$-module. On peut donc considérer les espaces vectoriels complexes $H_j(\underline{l},\mathcal{H}^\infty)$ pour $j \in \underline{\underline{N}}$. On sait que l'on a (cf [4] , [14] , ou [27]) :

$$\text{(22)} \qquad \begin{aligned} &\dim H_j(\underline{l},\mathcal{H}^\infty) = 0 \quad \text{si} \quad j \neq q_{\underline{l}} \\ &\dim H_{q_{\underline{l}}}(\underline{l},\mathcal{H}^\infty) = 1 \end{aligned}$$

Si E est un sous-espace de $V_{\underline{\underline{C}}}$, on note $Sp(V)_E$ le stabilisateur de E dans Sp(V), et $Mp(V)_E$ son image réciproque dans Mp(V).

Soit $(x,\phi) \in Mp(V)_{\underline{l}}$. Alors (x,ϕ) opère dans $\underline{l}$, et dans $\mathcal{H}^\infty$. Il opère donc dans le complexe standard $\Lambda^*\underline{l} \otimes \mathcal{H}^\infty$, et on vérifie que cette action préserve la différentielle. Donc (x,ϕ) opère dans l'espace $H_{q_{\underline{l}}}(\underline{l},\mathcal{H}^\infty)$. Comme cet espace est de dimension 1, il opère par un scalaire. Nous noterons $\rho_{\underline{l}}(x,\phi)$ ce scalaire.

I. 7. Calcul de $\rho_{\underline{l}}$ *lorsque* $\underline{l}$ *est réel :*

Soit $L \in \Lambda(V)$. Nous allons calculer $\rho_{\underline{l}}$ pour $l = L_{\underline{\underline{C}}}$.

Pour cela, on remarque que $\mathcal{H}_L^\infty$ est un espace de fonctions C^∞ sur N, et on voit immédiatement que l'application $\alpha \to \alpha(1)$ de $\mathcal{H}_L^\infty$ dans $\underline{\underline{C}}$ induit un isomorphisme de $H_0(\underline{l},\mathcal{H}^\infty)$ (qui est égal à $\mathcal{H}_L^\infty/\underline{l}\,\mathcal{H}_L^\infty$) sur $\underline{\underline{C}}$. Il résulte des formules (19) et (21) que l'on a, pour $(x, \ ) \in Mp(V)_{\underline{l}}$:

$$\text{(23)} \qquad \rho_{\underline{l}}(x,\phi) = \phi(L) \mid \det x_L \mid^{1/2}$$

I. 8. Calcul de $\rho_{\underline{l}}$ *lorsque* $\underline{l}$ *est totalement complexe :*

On suppose que l'on a $\underline{l} \cap \bar{\underline{l}} = \{0\}$. La forme hermitienne $X \to iB(X,\bar{X})$ est non dégénérée. Notons (p,q) sa signature, de sorte que $q = q_{\underline{l}}$, $p = d-q_{\underline{l}}$. L'application de restriction induit un isomorphisme de $Sp(V)_{\underline{l}}$ sur $U(p,q)$. En particulier, $Sp(V)_{\underline{l}}$ est connexe.

Nous allons démontrer l'assertion suivante : ρ_{ℓ} est l'unique caractère de $Mp(V)_{\underline{l}}$ tel que :

(24) $\rho_{\underline{l}}(1,-1) = -1$ et $\rho_{\underline{l}}(x,\phi)^2 = \det(x_{\underline{l}})$ pour tout $(x,\phi) \in Mp(V)_{\underline{l}}$.

Démonstration :

La première formule de (24) est évidente. Il en résulte que $\rho_{\underline{l}}(x,\phi)^2$ ne dépend que de x et définit un caractère de $Sp(V)_{\underline{l}} = U(p,q)$. Il existe donc $n \in \mathbb{Z}$ tel que $\rho_{\underline{l}}(x,\phi)^2 = \det(x_{\underline{l}})^n$ pour tout $(x,\phi) \in Mp(V)_{\underline{l}}$. Pour démontrer l'égalité $n = 1$, il suffit de l'établir pour $x \in U(p) \times U(q)$.

On écrit donc V comme somme directe de deux sous-espaces symplectiques : $V = V_1 + V_2$, on pose $\underline{l}_j = \underline{l} \cap V_{j\mathbb{C}}$, et on suppose que $\underline{l} = \underline{l}_1 + \underline{l}_2$, que $iB(X,\bar{X})$ est positif sur $\underline{l}_1$, et négatif sur $\underline{l}_2$.

Soient $(x_1,\phi_1) \in Mp(V_1)$, $(x_2,\phi_2) \in Mp(V_2)$. Soit (x,ϕ) l'élément de $Mp(V)$ tel que $x = x_1 + x_2$, et $\phi(L_1 + L_2) = \phi_1(L_1)\,\phi_2(L_2)$ pour tout $L_1 \in \Lambda(V_1)$, $L_2 \in \Lambda(V_2)$. L'application

(25) $(x_1,\phi_1)\ ,\ (x_2,\phi_2) \to (x,\phi)$

est un homomorphisme de $Mp(V_1) \times Mp(V_2)$ dans $Mp(V)$.

On peut considérer N_1 et N_2 (les groupes d'Heisenberg correspondant à V_1 et V_2) comme des sous-groupes de N et l'application naturelle $N_1 \times N_2 \to N$ est un homomorphisme surjectif. Il résulte du théorème de Stone-Von Neumann que, composée avec cet homomorphisme, T est isomorphe à

$T_1 \otimes T_2$, et que, composée avec (25) , S est isomorphe à $S_1 \otimes S_2$ (où T_j et S_j désignent les représentations correspondantes de N_j et $Mp(V_j)$).

C'est une partie de la démonstration de (22) que l'injection $\mathcal{H}_1^\infty \times \mathcal{H}_2^\infty \to \mathcal{H}^\infty$ induit un isomorphisme de $H_0(\underline{1}_1, \mathcal{H}_1^\infty) \times H_q(\underline{1}_2, \mathcal{H}_2^\infty)$ sur $H_q(\underline{1}, \mathcal{H}^\infty)$.

Il résulte de tout ceci qu'il suffit de démontrer (24) lorsque $V = V_1$ et lorsque $V = V_2$.

Supposons d'abord $V = V_1$. On utilise une autre réalisation de la représentation T . On note $\mathcal{H}_{\underline{1}}$ l'espace des fonctions α sur N qui sont C^∞ et qui vérifient :

$$\alpha(n \exp t E) = e^{-it} \alpha(n) \quad (n \in N,\ t \in \underline{\underline{R}})$$

$$X\alpha = 0 \qquad \text{si } X \in \underline{1}$$

$$\int_V |\alpha(\exp X)|^2 \, dX < \infty$$

On note $T_{\underline{1}}$ la représentation de N dans $\mathcal{H}_{\underline{1}}$ obtenue par translations à gauche. Elle est isomorphe à T. Notons $S_{\underline{1}}$ la représentation métaplectique dans $\mathcal{H}_{\underline{1}}$. On sait que l'on a, pour tout

$$(x,\phi) \in Mp(V)_1 \ , \ \alpha \in \mathcal{H}_{\underline{1}} \ , \ n \in N :$$

$$(26) \qquad (S_{\underline{1}}(x,\phi)\alpha)(n) = c\, \alpha(x^{-1}(n)),$$

où c vérifie $c^2 = \det(x_{\underline{1}})$. (Voir [15] par exemple). D'autre part l'application $\alpha \to \alpha(1)$ réalise un isomorphisme de $\mathcal{H}_{\underline{1}}^\infty / \underline{1}\, \mathcal{H}_{\underline{1}}^\infty$ sur $\underline{\underline{C}}$. On voit donc que l'action de (x,ϕ) dans $H_0(\underline{1}, \mathcal{H}_{\underline{1}}^\infty)$ est la multiplication par c , ce qui démontre (24) dans ce cas.

Supposons maintenant $V = V_2$. Alors on peut réaliser T (comme ci-dessus) dans l'espace $\mathcal{H}_{\bar{\underline{1}}}$ par la représentation $T_{\bar{\underline{1}}}$.

D'autre part, la dualité de Poincaré montre que $H_q(\underline{1}, \mathcal{H}^\infty)$ est isomorphe à $H^\circ(\underline{1}, \mathcal{H}^\infty) \otimes \Lambda^q\, \underline{1}$. L'espace $H^\circ(\underline{1}, \mathcal{H}_{\bar{\underline{1}}}^\infty)$ s'identifie à l'espace des fonctions

$\alpha \in \mathcal{H}^{\infty}_{\underline{l}}$ telles que $T_{\bar{\underline{l}}}(X)\alpha = 0$ pour tout $X \in \underline{l}$. Comme cet espace est de dimension 1, pour α dans cet espace, il résulte de (26) que l'on a $S_{\bar{\underline{l}}}(x,\phi)\alpha = c\alpha$, avec $c^2 = \det(x_{\bar{\underline{l}}})$. L'action de (x,ϕ) dans $H_q(\underline{\ell},\mathcal{H}^{\infty})$ est donc la multiplication par $c \det(x_{\underline{l}})$. Comme $\det(x_{\bar{\underline{l}}}) = \det(x_{\underline{l}})^{-1}$, car $\underline{l}$ est $\bar{\underline{l}}$ sont des sous-espaces lagrangiens transverses, ceci établit (24) dans ce cas . C.Q.F.D.

I. 9. Calcul de $\rho_{\underline{l}}$ *dans le cas général :*

Soit $\underline{l} \in \Lambda(V_{\underline{\underline{C}}})$. On pose $E = \underline{l} \cap V$. C'est un sous-espace isotrope de V. L'espace $E^{\perp}/E$ a une structure d'espace symplectique, et $\underline{l}/E_{\underline{\underline{C}}}$ en est un sous-espace lagrangien totalement complexe.

Soit $x \in Sp(V)_E$. Notons x^E l'élément correspondant de $Sp(E^{\perp}/E)$. Considérons $(x,\phi) \in Mp(V)_E$. Soit (x^E,ψ) un élément de $Mp(E^{\perp}/E)$ relevant x^E. Si $L \in \Lambda(V)$ est tel que $L \supset E$, le nombre $\phi(L)\psi(L/E)^{-1}$ est défini. Il ne dépend pas du choix de $L \in \Lambda(V)$ tel que $L \supset E$, comme il résulte de la formule (12) appliquée à ϕ et à ψ. On le note $\phi\,\psi^{-1}$.

Soit $(x,\phi) \in Mp(V)_{\underline{l}}$. Alors $x \in Sp(V)_E$, et on choisit un élément $(x^E,\psi) \in Mp(E^{\perp}/E)$ relevant x^E. On a $(x^E,\psi) \in Mp(E^{\perp}/E)_{\underline{l}/E_{\underline{\underline{C}}}}$. Nous allons démontrer la formule :

(27) $$\rho_{\underline{l}}(x,\phi) = \phi\psi^{-1} \,|\det x_E|^{1/2}\, \rho_{\underline{l}/E_{\underline{\underline{C}}}}(x^E,\psi).$$

Avant de faire la démonstration, remarquons que (27) ne dépend pas du choix du représentant (x^E,ψ) de x^E. Remarquons que $\rho_{\underline{l}/E_{\underline{\underline{C}}}}(x^E,\psi)$ est calculé dans le paragraphe 8. Remarquons que (23) est un cas particulier de (27) qui a été démontré à part, malgré le double emploi, à cause de sa simplicité.

Démonstration de (27) :

On choisit $L \in \Lambda(V)$ tel que L contienne E. Notons $\underline{n}_E$ l'algèbre de Lie $E^\perp/E + \underline{\underline{R}}e$. Elle s'identifie à $E^\perp + \underline{\underline{R}}e/E$. Cette algèbre de Lie opère dans $\mathcal{H}_L^\infty/E\,\mathcal{H}_L^\infty$ et on voit facilement, par restriction au sous-groupe analytique de N d'algèbre $E^\perp + \underline{\underline{R}}e$, que $\mathcal{H}_L^\infty/E\mathcal{H}_L^\infty$ est isomorphe comme $\underline{n}_E$-module à $\mathcal{H}^\infty_{L/E}$. D'autre part, soient $(x,\phi) \in Mp(V)_E$ et $(x^E,\psi) \in Mp(E^\perp/E)$ comme plus haut. Notons $\rho_E(x,\phi)$ l'action de (x,ϕ) dans $\mathcal{H}_L^\infty/E\,\mathcal{H}_L^\infty$, déduite de $S_L(x,\phi)$ par passage au quotient. Notons $S_{L/E}(x,\psi)$ l'action de (x,ψ) dans $\mathcal{H}^\infty_{L/E}$, par la représentation métaplectique. Il résulte de (21) que l'on a :

$$\rho_E(x,\phi) = \phi(L)\psi(L)^{-1}|\det x_E|^{1/2}\, S_{L/E}(x,\psi).$$

Par ailleurs, il résulte facilement des formules (22) - ou bien on le voit en les démontrant, que $H_q(\underline{l},\mathcal{H}^\infty)$ est isomorphe à $H_q(\underline{l}/E_{\underline{\underline{C}}},\mathcal{H}^\infty/E\,\mathcal{H}^\infty)$. La formule (27) s'en déduit. C.Q.F.D.

De (27) on déduit les formules suivantes :

(28) $\qquad \rho_{\underline{l}}(1,-1) = -1 \quad$ et $\quad \rho_{\underline{l}}(x,\phi)^2 = \det(x_{\underline{l}})$.

I. 10. Comparaison des $\rho_{\underline{l}}$ *:*

Soient $\underline{l}$, $\underline{l}' \in \Lambda(V_{\underline{\underline{C}}})$. Soit $(x,\phi) \in Mp(V)_{\underline{l}} \cap Mp(V)_{\underline{l}'}$. Il résulte de (28) que l'on a :

(29) $\qquad \rho_{\underline{l}}(x,\phi) = \varepsilon_{\underline{l},\underline{l}'}(x)\det(x_{\underline{l}/\underline{l}\cap\underline{l}'})\rho_{\underline{l}'}(x,\ )$

où $\varepsilon_{\underline{l},\underline{l}'}$ est un caractère du groupe $Sp(V)_{\underline{l}} \cap Sp(V)_{\underline{l}'}$ à valeurs dans $\{\pm 1\}$. Il est intéressant de calculer $\varepsilon_{\underline{l},\underline{l}'}$. Comme nous n'utiliserons pas le résultat, je le ferai avec peu de détails. On remarque tout d'abord que si V est somme directe de deux sous-espaces symplectiques V_1 et V_2 , stables par x, et tels que , posant $\underline{l}_j = \underline{l} \cap V_{j\underline{\underline{C}}}$, $\underline{l}'_j = \underline{l}' \cap V_{j\underline{\underline{C}}}$ $(j = 1,2)$, l'on ait $\underline{l} = \underline{l}_1 + \underline{l}_2$, $\underline{l}' = \underline{l}'_1 + \underline{l}'_2$, alors $\varepsilon_{\underline{l},\underline{l}'}(x) = \varepsilon_{\underline{l}_1,\underline{l}'_1}(x_1)\,\varepsilon_{\underline{l}_2,\underline{l}'_2}(x_2)$, où

x_1 et x_2 sont les restrictions de x à V_1 et V_2. D'autre part, $\varepsilon_{\underline{1},\underline{1}'}(-x) = \varepsilon_{\underline{1},\underline{1}'}(-1)\varepsilon_{\underline{1},\underline{1}'}(x)$. On est donc amené à calculer $\varepsilon_{\underline{1},\underline{1}'}(x)$ dans les cas particuliers suivants :

(i) Toutes les valeurs propres de x sont égales à -1.

(ii) x n'a pas de valeur propre < 0.

Dans le cas (ii), x est dans la composante neutre de $Sp(V)_{\underline{1}} \cap Sp(V)_{\underline{1}}$, et donc $\varepsilon_{1,1'}(x) = 1$.

Dans le cas (i), on a $\varepsilon_{\underline{1},\underline{1}'}(x) = \varepsilon_{\underline{1},\underline{1}'}(-1)$, et l'on a :

(30) $$\varepsilon_{\underline{1},\underline{1}'}(-1) = (-1)^{q_{\underline{1}}+q_{\underline{1}'}+\dim \underline{1}/\underline{1}\cap\underline{1}'}$$

<u>*Démonstration de (30)*</u> :

Soient $\underline{1}$, $\underline{1}'$, $\underline{1}'' \in \Lambda(V_{\underline{\underline{C}}})$. Soit $x \in Sp(V)$ stabilisant $\underline{1}$, $\underline{1}'$, et $\underline{1}''$. On a, d'après (29) :

(31) $$\varepsilon_{\underline{1},\underline{1}'}(x)\varepsilon_{\underline{1}',\underline{1}''}(x)\varepsilon_{\underline{1}',\underline{1}''}(x) = \det(x_{\underline{1}/\underline{1}\cap\underline{1}'})\det(x_{\underline{1}'/\underline{1}'\cap\underline{1}''})\det(x_{\underline{1}''/\underline{1}''\cap\underline{1}})$$

La formule (30) est compatible avec (39), de sorte que si on a démontré (30) pour des paires $\underline{1}$, $\underline{1}'$ et $\underline{1}$, $\underline{1}''$, on l'a démontrée pour $\underline{1}'$, $\underline{1}''$. Il suffit donc de démontrer (30) dans le cas particulier où $\underline{1}$ est réel (i.e. $\underline{1} = L_{\underline{\underline{C}}}$ avec $L \in \Lambda(V)$) et où $\underline{1}' \cap \underline{1} = \{0\}$. Par des décompositions en somme directe, on se ramène au cas où $\dim V = 2$. On doit donc examiner les cas particuliers suivants :

(i) $\underline{1}'$ est réel et $\underline{1} \cap \underline{1}' = \{0\}$.

(ii) $\underline{1}'$ est totalement complexe et $q_{\underline{1}'} = 0$.

(iii) $\underline{1}'$ est totalement complexe et $q_{\underline{1}'} = 1$.

Le premier cas résulte de (29). Le cas (iii) se ramène au cas (ii), car si $1'$ est totalement complexe, $\varepsilon_{\underline{1}',\bar{\underline{1}}'} = 1$ (en effet $Sp(V)_{\underline{1}'} = Sp(V)_{\bar{\underline{1}}'}$ est connexe, cf. paragraphe 8).

Il reste à étudier le second cas . Cela se fait en comparant les modèles $S_{\underline{1}}$ et $S_{\underline{1}'}$ de la représentation métaplectique, et nécessite un peu de calcul.

Remarque 1 :

Supposons que $x \in Sp(V)$ n'ait pas de valeur propre réelle, et que x stabilise $\underline{1}$ et $\underline{1}' \in \Lambda(V_{\underline{\underline{C}}})$. Alors

$$q_{\underline{1}} + q_{\underline{1}'} + \dim \underline{1}/\underline{1} \cap \underline{1}' = 0 \bmod 2.$$

(cela résulte de (30) et de la relation $\varepsilon_{\underline{1},\underline{1}'}(x) = \varepsilon_{\underline{1},\underline{1}'}(-x) = 1$).

Remarque 2 :

La formule (31) est compatible avec le résultat suivant, laissé en exercice au lecteur. Soit V un espace vectoriel symplectique sur un corps commutatif de caractéristique $\neq 2$. Soient L, L', L'' des sous-espaces lagrangiens de V, et soit x stabilisant L, L' et L''. Notons V_- le sous-espace dans lequel x opère avec la valeur propre généralisée -1 (i.e. $V_- = \ker(Id+x)^{\dim V}$). On pose $L_- = L \cap V_-$, etc..., et on note

$$m = \dim L_-/L_-\cap L'_- + \dim L'_-/L'_-\cap L''_- + \dim L''_-/L''_-\cap L_- .$$

On a :

$$\det x_{L/L\cap L'} \det x_{L'/L'\cap L''} \det x_{L''/L''\cap L} = (-1)^m .$$

CHAPITRE II - FORMES LINEAIRES ADMISSIBLES ET BIEN POLARISABLES

Dans tout ce chapitre, G est un groupe de Lie réel d'algèbre de Lie $\underline{g}$. Nous définissons les formes admissibles, les formes bien polarisables et terminons par quelques remarques sur la définition des représentations $T_{g,\tau}$.

II. 1. Représentation coadjointe :

Soit $g \in \underline{g}^*$. On note B_g la forme bilinéaire alternée sur $\underline{g}$ définie par la formule

(1) $\qquad B_g(X,Y) = g([X,Y]) \quad (X,Y \in \underline{g}).$

Il se trouve que le noyau de B est égal à $\underline{g}(g)$ (cf[17]) de sorte que $\underline{g}/\underline{g}(g)$ est canoniquement un espace symplectique. Le groupe $G(g)$ laisse invariant B_g.

Nous emploierons les notations suivantes. Soit V un espace vectoriel muni d'une forme bilinéaire alternée B, et soit H un groupe opérant dans V par des automorphismes linéaires conservant B. Si E est un sous-espace de V, on note $E^\perp$ le sous-espace de V orthogonal. En particulier, $V/V^\perp$ est un espace symplectique dans lequel H opère. Nous noterons H^V l'ensemble des couples $(h,m) \in H \times Mp(V/V^\perp)$, tels que h et m aient même image dans $Sp(V/V^\perp)$.

Nous dirons qu'un sous-espace L de V est lagrangien si $L = L^\perp$. Un sous-espace L de V est lagrangien si et seulement si L contient $V^\perp$, et $L/V^\perp$ est lagrangien dans $V/V^\perp$.

Compte-tenu de la description de $Mp(V/V^\perp)$ donnée en T.3, on peut décrire H^V comme l'ensemble des couples (h,ϕ), où $h \in H$, ϕ est une fonction sur l'ensemble des sous-espaces lagrangiens de V, tels que si l'on pose

$x = h_{V/V^\perp}$, et $\Phi(L/V^\perp) = \Phi(L)$ pour tout sous-espace lagrangien de V, les relations (I.11) et (I.12) soient vérifiées.

L'application $(h,\phi) \to h$ est un homomorphisme surjectif de H^V sur H dont le noyau est $\{(1,1),(1,-1)\}$. L'application $(h,\phi) \to (h_{V/V^\perp},\phi)$ est un homomorphisme dans le groupe $Mp(V/V^\perp)$.

Soit $\underline{1}$ un sous-espace lagrangien (complexe) de $V_{\underline{\underline{C}}}$. Soit $(h,\phi) \in H^V$ tel que $h\underline{1} \subset \underline{1}$. Nous poserons (cf. par. I. 6) :

(2) $$\rho_{\underline{1}}(h,\phi) = \rho_{\underline{1}/V_{\underline{\underline{C}}}^\perp}(h_{V/V^\perp},\phi) .$$

On a donc (compte-tenu de (I.28) :

(3) $$\rho_{\underline{1}}(h,\phi)^2 = \det(h_{\underline{1}/V_{\underline{\underline{C}}}^\perp}) = \det(h_{V_{\underline{\underline{C}}}/\underline{1}})^{-1}$$

(4) $$\rho_{\underline{1}}(1,-1) = -1.$$

Si X appartient à l'algèbre de Lie $\underline{h}$ de H, et si $X\underline{1} \subset \underline{1}$, on posera :

(5) $$\rho_{\underline{1}}(X) = \frac{1}{2}\,\mathrm{tr}(X_{\underline{1}/V_{\underline{\underline{C}}}^\perp})$$

Tout ceci s'applique en particulier à $G(g)$ et à $\underline{g}$ muni de la forme bilinéaire B_g. Dans toute la suite, le groupe $G(g)^{\underline{g}}$ joue un rôle fondamental. Il a été introduit dans [8].

II. 2. Formes admissibles :

Soit $g \in \underline{g}^*$. On note $X(g)$ l'ensemble des classes de représentations unitaires τ de $G(g)^{\underline{g}}$ vérifiant les propriétés suivantes :

(6) $$\tau(1,-1) = -\mathrm{Id}$$

(7) la différentielle de τ est un multiple de $ig|\underline{g}(g)$.

On dit que g est <u>admissible</u> si $X(g)$ est non vide.

Il est immédiat de voir que est admissible si et seulement s'il existe un caractère unitaire χ_g de $(G(g)_0)^{\underline{g}}$ tel que $\chi_g(1,-1) = -1$, et de différentielle $ig|\underline{g}(g)$.

S'il existe, un tel caractère est unique, car $(G(g)_0)^{\underline{g}}$ (qui est l'image réciproque de $G(g)_0$ dans $G(g)^{\underline{g}}$) a au plus deux composantes connexes. Supposons g admissible. Alors $X(g)$ est l'ensemble des classes de représentations de $G(g)^{\underline{g}}$ dont la restriction à $(G(g)_0)^{\underline{g}}$ est multiple de χ_g. Comme on a : $G(g)^{\underline{g}}/(G(g)_0)^{\underline{g}} = G(g)/G(g)_0$, on voit que $X(g)$ s'identifie aux classes de représentations projectives de $G(g)/G(g)_0$ associées au 2-cocycle défini par l'extension :

$$(G(g)_0^{\underline{g}})/\ker \chi_g \to G(g)^{\underline{g}}/\ker \chi_g \to G(g)/G(g)_0.$$

Remarque 1 :

Disons que g est <u>acceptable</u> si $(G(g)_0)^{\underline{g}}$ a deux composantes connexes, et que que g est <u>entière</u> si $ig|\underline{g}(g)$ est la différentielle d'un caractère de $G(g)_0$. On voit facilement que si g est entière, alors g est admissible si et seulement si g est acceptable.

Remarque 2 :

Supposons qu'il existe un sous-espace lagrangien $\underline{b} \subset \underline{g}_{\underline{\underline{C}}}$, stable par $\underline{g}(g)$. Rappelons la notation (5) :

(8) $$\rho_{\underline{b}}(X) = \frac{1}{2}\,\mathrm{tr}(X_{\underline{b}/\underline{g}(g)_{\underline{\underline{C}}}}) \quad (X \in \underline{g}(g)).$$

Rappelons que $\rho_{\underline{b}}$ est la différentielle d'un caractère $\rho_{\underline{b}}$ de $(G(g)_0)^{\underline{g}}$ tel que $\rho_{\underline{b}}(1,-1) = -1$.

On voit donc que g est admissible si et seulement s'il existe un caractère de $G(g)_0$ de différentielle $\rho_b + ig|\underline{g}(g)$.

Remarque 3 :

La notion d'admissibilité pour un g donné dans $\underline{g}^*$ ne dépend que du groupe G_0 .

Notations :

Si cela est nécessaire, on incorpore G dans la notation : $X(g) = X_G(g)$.
On note $X^{irr}(g)$ (resp. $X^{fac}(g)$) le sous-ensemble formé des éléments irréductibles (resp. factoriels).
Soit Γ un sous-groupe du centre de G et soit η un caractère unitaire de Γ . L'application $\gamma \to (\gamma, 1)$ identifie Γ à un sous-groupe de $G(g)^{\underline{g}}$.
On notera $X(g,\eta)$ (ou $X_{G,\Gamma}(g,\eta)$) le sous-ensemble de $X(g)$ formé des éléments dont la restriction à Γ est un multiple de η . Si $X(g,\eta)$ est non vide, on dira que g <u>est</u> η-<u>admissible</u>.

II. 3. Formes bien polarisables :

Pour un moment nous changeons les notations, et nous notons $\underline{g}$ une algèbre de Lie de dimension finie sur $\underline{\underline{C}}$. Soit $g \in \underline{g}^*$, on définit B_g comme plus haut.
Une sous-algèbre $\underline{b}$ de $\underline{g}$ est appelée une <u>polarisation</u> en g si $\underline{b}$ est un sous-espace lagrangien de $\underline{g}$.
Soit G un groupe complexe d'algèbre $\underline{g}$ et soit B le sous-groupe analytique correspondant à une polarisation $\underline{b}$ en g .
Alors Bg est un ouvert de Zariski de l'espace affine $g + \underline{b}^\perp$ (où $\underline{b}^\perp$ est l'orthogonal de $\underline{b}$ dans $\underline{g}^*$).
Les conditions suivantes sont équivalentes

(i) $Bg = g + \underline{b}^\perp$

(ii) Pour tout $\lambda \in \underline{b}^\perp$, $\underline{b}$ est une polarisation en $g + \lambda$.

(iii) Bg est fermé dans $\underline{g}^\perp$.

Lorsqu elles sont satisfaites, on dit que $\underline{b}$ vérifie la <u>condition de Pukanszky</u> (cf. [2]).

Nous dirons qu'une polarisation est <u>bonne</u> si elle est <u>résoluble</u> et si elle vérifie la condition de Pukanszky. Nous dirons que g est <u>bien polarisable</u> si g admet une bonne polarisation.

<u>Lemme 7</u> :

Supposons $\underline{g}$ semi-simple. Alors un élément $g \in \underline{g}^*$ est bien polarisable si et seulement si $\underline{g}(g)$ est une sous-algèbre de Cartan de $\underline{g}$.

<u>Démonstration</u> :

Soit r le rang de $\underline{g}$. On sait que g admet une polarisation résoluble si et seulement si $\dim \underline{g}(g) = r$, et dans ce cas, les polarisations résolubles en g sont les sous-algèbres de Borel $\underline{b}$ de $\underline{g}$ telles que $g([\underline{b},\underline{b}]) = 0$ (cf. [7] p. 60). Soit $g \in \underline{g}^*$ tel que $\dim \underline{g}(g) = r$. Soit $\underline{b}$ une polarisation résoluble en g. Comme B est un sous-groupe de Borel de G, G/B est compact, de sorte que $\underline{b}$ vérifie la condition de Pukanszky si et seulement si Gg est fermé dans $\underline{g}^*$. Identifions $\underline{g}$ et $\underline{g}^*$ par la forme de Killing. On sait que Gg est fermé si et seulement si g est semi-simple. Donc g est bien polarisable si et seulement si g est semi-simple et régulier.

C.Q.F.D.

Si $\underline{g}$ est résoluble, tout $g \in \underline{g}^*$ est bien polarisable (cf [34]). Dans l'appendice, une notion de forme linéaire "très régulière" sur une algèbre de Lie algébrique $\underline{g}$ est définie, qui généralise les formes semi-simples régulières lorsque $\underline{g}$ est semi-simple. Les formes très régulières forment un ouvert de Zariski non vide, et elles sont bien polarisables (cf [1]).

Revenons au cas où $\underline{g}$ est une algèbre de Lie réelle. Soit $g \in \underline{g}^*$. On note par la même lettre l'élement de $\underline{g}^*_{\underline{\underline{C}}}$ qui prolonge g . Une polarisation en g (resp. polarisation vérifiant la condition de Pukanszky, resp. bonne polarisation) est une sous-algèbre $\underline{b}$ de $\underline{g}_{\underline{\underline{C}}}$ qui a les mêmes propriétés pour g, considéré comme élément de $\underline{g}^*_{\underline{\underline{C}}}$.

II. 4. Remarques sur la définition des représentations $T_{g,\tau}$:

Soit G un groupe de Lie d'algèbre de Lie $\underline{g}$. Soit $g \in \underline{g}^*$ un élément admissible et bien polarisable. Soit $\tau \in X(g)$. Nous voulons associer à ces données une classe $T_{g,\tau}$ de représentations unitaires de G, vérifiant les propriétés annoncées dans l'introduction.

Je veux ici décrire une méthode naturelle pour construire $T_{g,\tau}$. Bien que ce ne soit pas celle que j'adopterai ici, pour des raisons que je développerai plus bas, elle donne une bonne idée de la paramétrisation employée.

Je vais supposer, pour simplifier, qu'il existe une bonne polarisation $\underline{b}$ en g qui soit réelle, i.e. telle que $\underline{b} = (\underline{b} \cap \underline{g})_{\underline{\underline{C}}}$. De plus je vais supposer que $\underline{b}$ est stable par G(g). On note B_0 le sous-groupe analytique de G d'algèbre $\underline{b} \cap \underline{g}$, et B le groupe $G(g)B_0$. On démontre que B est fermé dans G, et qu'il existe une représentation unitaire, uniquement déterminée, que nous noterons $\tilde{\tau}$, de B, telle que l'on ait :

(i) La restriction de $\tilde{\tau}$ à B_0 est un multiple du caractère unitaire de différentielle $ig|\underline{b}$.

(ii) Soit $x \in G(g)$. Soit $(x,\phi) \in G(g)^{\underline{g}}$ un représentant de x dans $G(g)^{\underline{g}}$. On a $\tilde{\tau}(x) = \phi\tau(x,\phi)$.

On pose alors $T_{g,\tau,\underline{b}} = \mathrm{Ind}_B^G(\tilde{\tau})$. Compte-tenu de la définition des représentations induites, rappelée dans les "principales notations", $T_{g,\tau,\underline{b}}$ est réalisée dans un espace de fonctions α sur G, à valeurs dans l'espace de τ, qui vérifient les relations suivantes.

(9) $\alpha(yx) = \tau(x,\phi)^{-1}\ \rho_{\underline{b}}(x,\phi)^{-1}\ \alpha(y)$

pour tout $y \in G$, $x \in G(g)$, $(x,\phi) \in G(g)^{\underline{g}}$.

(10) $\quad X\alpha = \langle -ig - \rho_{\underline{b}}, X\rangle \alpha$

pour tout $X \in \underline{b}$, où l'on a posé

(11) $\quad \rho_{\underline{b}}(X) = -\frac{1}{2}\,\mathrm{tr}(\mathrm{ad}\, X_{\underline{\underline{g}}_{\mathbb{C}}/\underline{b}})$ pour $X \in \underline{b}$.

Remarque 1 :

La définition (11) coïncide sur $\underline{g}(g)$ avec la définition (8), de sorte que, pour $X \in \underline{g}(g)$, les relations (9) et (10) sont compatibles.

Remarque 2 :

Rappelons que $\rho_{\underline{b}}(x,\phi)^2 = \det(\mathrm{Ad}\, x_{\underline{g}/\underline{b}_{\mathbb{C}}})^{-1}$, de sorte que la différence entre (9) et la formule qui définit les représentations induites est l'absence des valeurs absolues.

On epsère que la classe de $T_{g,\tau,\underline{b}}$ est indépendante de $\underline{b}$, et que son commutant est isomorphe à celui de τ . Lorsque ceci est vrai, il est légitime de noter $T_{g,\tau}$ cette classe. Lorsque $\underline{g}$ est nilpotente, c'est précisément la méthode inventée par Kirillov pour décrire les représentations unitaires irréductibles des groupes de Lie nilpotents connexes.

Je vais expliquer maintenant pourquoi je préjère adopter une autre méthode.

(i) Même dans le cas idéal considéré ci-dessus, où il existe une bonne polarisation réelle G(g)-invariante, on ne sait pas si $T_{g,\tau,\underline{b}}$ est indépendante de $\underline{b}$. Les résultats les meilleurs dans cette direction sont ceux d'Andler [1] qui a démontré que c'est vrai quand $\underline{g}$ est algébrique.

(ii) En général, il n'y a pas de bonne polarisation réelle. Tant qu'il y a des bonnes polarisations G(g)-invariantes, ce n'est peut-être pas fondamental, car il existe des procédés (dans le style "théorème de Borel-Weil-Bott") pour extraire des représentations de G du faisceau des fonctions sur G vérifiant (9) et (10) . Mais cela devient très vague.

(iii) En général, il n'y a pas de bonne polarisation G(g)-invariante. (Un exemple en est le produit semi-direct de SL(2,$\underline{\underline{Q}}$) avec un groupe d'Heisenberg de dimension 3. On trouvera un exemple d'un groupe algébrique complexe connexe sans polarisation réelle invariante dans [1]).

La méthode adoptée ici, esquissée dans l'introduction, et détaillée ci-dessus, est certainement moins élégante, mais elle est beaucoup plus simple, de sorte qu'on arrive à la mettre en oeuvre complètement.

CHAPITRE III - REPRESENTATIONS DES GROUPES DE LIE REDUCTIFS

II. 1. Introduction :

Dans tout ce chapitre, G est un groupe de Lie dont l'algèbre de Lie $\underline{g}$ est réductive, g un élément de $\underline{g}^*$ bien polarisable et admissible, $\tau \in X(g)$. Nous allons construire la représentation $T_{g,\tau}$. Lorsque G est connexe, et τ irréductible, nous obtenons de cette manière exactement les "représentations irréductibles tempérées de G dont le caractère infinitésimal est régulier", et notre méthode dans ce cas se réduit à dire laquelle de ces représentations (déjà construites par Harish-Chandra) nous choisissons d'appeler $T_{g,\tau}$. Dans le cas général, il faut calculer l'obstruction de Mackey à étendre une représentation de ce type de G_0 à G. Comme Kostant l'avait déjà fait dans le cas compact, on utilise une généralisation (due à D. Vogan) du théorème de Kostant-Borel-Weil-Bott.

Notations :

La classe de représentations éventuellement associée à $g \in \underline{g}^*$, admissible et bien polarisable, et $\tau \in X(g)$, sera notée $T_{g,\tau}$, ou $T^G_{g,\tau}$ s'il est utile de préciser G.

Si $G(g)$ est connexe, tous les éléments de $X(g)$ sont des multiples du caractère χ_g de $G(g)^{\underline{g}}$. Nous noterons alors T_g (ou T^G_g) la représentation T_{g,χ_g}.

Nous poserons $\underline{h} = \underline{g}(g)$. D'après le lemme II.1., c'est une sous-algèbre de Cartan de $\underline{g}$. On note $\Delta \subset \underline{h}^*_{\mathbb{C}}$ l'ensemble des racines de $\underline{g}_{\mathbb{C}}$ par rapport à $\underline{h}_{\mathbb{C}}$, $W_{\mathbb{C}}$ le groupe de Weil correspondant. Si $\alpha \in \Delta$, on note $\underline{g}^{\alpha}_{\mathbb{C}}$ le sous-espace radiciel correspondant. On identifie $\underline{h}^*$ à un sous-espace de

$\underline{g}$, en prolongeant un élément $\lambda \in \underline{h}^*$ par 0 sur tous les $\underline{g}_{\mathbb{C}}^{\alpha}$.
Si $\alpha \in \Delta$, on note $H_\alpha \in \underline{h}_{\mathbb{C}}$ la coracine correspondante (de sorte que $\alpha(H_\alpha) = 2$ et $H_\alpha \in [g_{\mathbb{C}}^{\alpha}, g_{\mathbb{C}}^{-\alpha}]$). Si $\lambda \in \underline{h}_{\mathbb{C}}^*$, on pose $\lambda_\alpha = \lambda(H_\alpha)$.

Avec ces notations, on a $g \in \underline{h}^*$, et $g_\alpha \neq 0$ pour tout $\alpha \in \Delta$. Les bonnes polarisations en g sont exactement les sous-algèbres de Borel $\underline{b}$ de $\underline{g}_{\mathbb{C}}$ contenant $\underline{h}$. Une telle sous-algèbre de Borel définit un système de racines positives $\Delta_{\underline{b}}^+$, de sorte que $\underline{b} = \underline{h}_{\mathbb{C}} + \underline{n}$, $\underline{n} = \sum_{\alpha \in \Delta_{\underline{b}}^+} \underline{g}_{\mathbb{C}}^{\alpha}$. On note $\rho_{\underline{b}}$ la demi-somme des éléments de $\Delta_{\underline{b}}^+$. On remarque que cette définition coïncide avec (II. 11).

Notons $\underline{z}$ le centre de $\underline{g}$. On écrit $\underline{h} = \underline{t} \oplus \underline{a} \oplus \underline{z}$, où $\underline{t} + \underline{a} = h \cap [\underline{g},\underline{g}]$, et $\Delta \subset i\underline{t}^* + \underline{a}^*$. Nous écrirons

(1) $$ig = \mu + i\gamma + i\lambda$$

avec $\mu \in i\underline{t}^*$, $\gamma \in \underline{a}^*$, $\lambda \in \underline{z}^*$.

Nous construirons les représentations $T_{g,\tau}$ dans l'ordre de généralité croissant suivant.

(i) G est connexe et $\underline{h} = \underline{t}$.

(ii) G est connexe et g est standard (ce qui signifie que $\mu_\alpha \neq 0$ pour tout $\alpha \in \Delta$).

(iii) g est standard.

(iv) cas général.

III. 2. Série discrète des groupes semi-simples connexes :

On suppose G connexe et $\underline{h} = \underline{t}$. Dans ce cas, notons T le sous-groupe analytique de G d'algèbre de Lie $\underline{t}$. Alors $G(g) = T$, et $\mu = ig$.
Il existe une unique involution de Cartan θ de $\underline{g}$ telle que $\theta\underline{t} = \underline{t}$.
Notons $\underline{k}$ l'ensemble des points fixes de θ , K le sous-groupe analytique correspondant. Alors $\underline{t}$ est une sous-algèbre de Cartan de $\underline{k}$. On note Δ_C le sou

le sous-ensemble de Δ formé des $\alpha \in \Delta$ tels que $\underline{\underline{g}}_{\mathbb{C}}^{\alpha} \subset \underline{k}_{\mathbb{C}}$: c'est l'ensemble des racines de $\underline{k}_{\mathbb{C}}$.

On pose $\Delta^+ = \{\alpha \in \Delta, \mu_\alpha > 0\}$ et $\Delta_C^+ = \Delta^+ \cap \Delta_C$. Bien que K ne soit pas nécessairement compact, les représentations unitaires irréductibles de K sont de dimension finie et paramétrées par leur poids dominant $\gamma \in i\underline{t}^*$ (dominant par rapport à Δ_C^+). Un élément $\gamma \in i\underline{t}^*$, dominant par rapport à Δ_C^+, est le poids dominant d'une représentation irréductible unitaire de K si et seulement si c'est la différentielle d'un caractère de T. Dans ce cas nous noterons δ_γ la représentation correspondante.

On note ρ (resp. ρ_C) la demi-somme des éléments de Δ^+ (resp. Δ_C^+). Alors l'élément $\mu + \rho - 2\rho_C$ de $i\underline{t}^*$ est dominant pour Δ_C^+ (car μ est régulier et Δ^+ dominant) et différentielle d'un caractère de T (car g est admissible).

On sait qu'il existe une et une seule représentation unitaire irréductible T de G ayant les propriétés suivantes.

(i) $\delta_{\mu+\rho-2\rho_C}$ intervient dans T.

(ii) Soit $\gamma \in it^*$, dominant pour Δ_C^+ et différentielle d'un caractère de T, vérifiant $(\gamma+2\rho_C, \gamma+2\rho_C) < (\mu+\rho, \mu+\rho)$. Alors δ_γ n'intervient pas dans T.

Nous noterons T_g cette représentation unitaire irréductible de G.

Exemple :

Si G est compact et connexe, T_g est donc la représentation irréductible de poids dominant $ig - \rho$.

L'existence d'une représentation T vérifiant (i) et (ii) est due à Harish-Chandra et à Schmid, l'unicité à Vogan (cf [37],[28],[35], ou pour plus de détails, les références données dans [9]).

Lemme 1 :

(i) Soit a un automorphisme de G. On a : ${}^{a}T_g = T_{ag}$.

(ii) Soit $g' \in \underline{g}^*$, bien polarisable, admissible, et telle que, écrivant $\underline{g}(g') = \underline{t}' + \underline{a}'$, on ait $\underline{a}' = 0$. Alors $T_g = T_{g'}$ si et seulement si $g' \in Gg$.

(iii) Les représentations unitaires irréductibles de G obtenues par ce procédé sont les représentations unitaires irréductibles de carré intégrable modulo le centre de G.

L'assertion (i) est évidente. Tout le reste est dû à Harish-Chandra (cf [37]).

III. 3. Série fondamentale pour les groupes connexes :

On suppose dans ce paragraphe que G est connexe, et que $\underline{h}$ est fondamentale, ce qui signifie que $\underline{h}$ est le centralisateur de $\underline{t}$ dans $\underline{g}$.

On note T, A, H les groupes analytiques d'algèbre de Lie $\underline{t}, \underline{a}, \underline{h}$. Alors $G(g)$ est égal à H.

On note M' le centralisateur de A dans G (nous aurons à considérer plus loin un groupe M), et $\underline{m}'$ son algèbre de Lie. On écrit $\underline{m}' = \underline{r} + \underline{s}$ où $\underline{r} = [\underline{m}', \underline{m}']$ et où $\underline{s}$ est le centre de $\underline{m}'$. On note R et S les sous-groupes analytiques correspondants de sorte que $M'_0 = RS$.

Posons $r = g|\underline{r}$, $s = g|\underline{s}$. On vérifie facilement que r est un élément bien polarisable admissible de $\underline{r}^*$, vérifiant les conditions du paragraphe III.2., de sorte que nous avons déjà défini la représentation irréductible T_r^R de R.

Comme g est admissible, il existe une (et une seule) représentation de M'_0 dont la restriction à R est T_r^R , et dont la restriction à S est le caractère de différentielle is.

Nous poserons $m' = g|\underline{m}'$. Alors m' est admissible, $M'_0(m') = M'(m') = H$.

Nous noterons $T_{m'}^{M'_0}$ la représentation de M'_0 définie ci-dessus.

D'après le lemme 1, le stabilisateur de $T_{m'}^{M'_0}$ dans M' est $M'(m')M'_0$

La représentation $\mathrm{Ind}_{M'_0}^{M'}(T_{m'}^{M'_0})$ est donc irréductible. Nous la noterons $T_{m'}^{M'}$.

On choisit un élément $X \in \underline{a}$ tel que si $\alpha \in \Delta$ vérifie $\alpha(X) = 0$, alors $\alpha(a) = 0$. On pose

$$\underline{u}' = \underline{g} \cap \sum_{\substack{\alpha(X)>0 \\ \alpha\in\Delta}} \underline{g}_{\mathbb{C}}^{\alpha}$$

et on note U' le sous-groupe analytique correspondant. Alors $M'U'$ est un sous-groupe fermé de G(c'est un sous-groupe parabolique "cuspidal", au sens de [37]).

Il est connu que la classe d'équivalence de la représentation $\mathrm{Ind}_{M'U'}^{G}(T_{m'}^{M'} \otimes \mathrm{Id}_U)$ ne dépend pas du choix de X. On la note T_g.

Lemme 2 :

(i) T_g est irréductible.

(ii) Soit a un automorphisme de G. On a ${}^{a}T_g = T_{ag}$.

(iii) Soit $g' \in \underline{g}^*$ un élément bien polarisable, admissible, tel que $\underline{g}(g)$ soit fondamental. Alors $T_g = T_{g'}$, si et seulement si $g' \in Gg$.

Tout ceci est dû à Harish-Chandra - au moins quand G est de centre fini. Le résultat (i) est particulièrement profond. Lorsque G est de centre fini, on en trouvera une démonstration dans [12]. Dans le cas général, on trouvera une démonstration dans [10], et une dans [35].

III. 4. Homologie des représentations f ndamentales :

Les hypothèses sont celles de III. 3. On fixe une polarisation $\underline{b}$ en g, totalement complexe, i.e. $\underline{b}$ est une sous-algèbre de Borel de $\underline{g}_{\mathbb{C}}$, contenant $\underline{h}$, et telle que $\underline{b} \cap \bar{\underline{b}} = \underline{h}_{\mathbb{C}}$. On note $\underline{n}$ le radical nilpotent de $\underline{b}$.

Nous noterons $\mathcal{H}$ l'espace de Hilbert dans lequel opère la représentation T_g (définie au paragraphe 3), et $\mathcal{H}^{\infty}$ le sous-espace des vecteurs C^{∞}. Alors $\mathcal{H}^{\infty}$ est un $\underline{g}_{\mathbb{C}}$-module, et, par restriction un $\underline{n}$-module.

Pour $j \in \underline{\underline{N}}$, on considère les espaces vectoriels $H_j(\underline{n}, \mathcal{H})$. Comme l'algèbre $\underline{h}$ normalise $\underline{n}$, et opère dans $\mathcal{H}^\infty$, $\underline{h}$ opère dans $H_j(\underline{n},\mathcal{H}^\infty)$.

Si $\gamma \in \underline{\underline{h}}^*_{\mathbb{C}}$, on note $H_j(\underline{n},\mathcal{H}^\infty)_\gamma$ le sous-espace propre généralisé correspondant.

On note $q_{\underline{b}}$ le nombre de valeurs propres < 0 de l'application hermitienne associée à la forme $X \to ig([X,\bar{X}])$ sur $\underline{b}$.

Lemme 3 :

Si $j \neq q_{\underline{b}}$, on a :

$$\dim H_j(\underline{n},\mathcal{H}^\infty)_{ig+\rho_{\underline{b}}} = 0 \text{ . On a } \quad \dim H_{q_{\underline{b}}}(\underline{n},\mathcal{H}^\infty)_{ig+\rho_{\underline{b}}} = 1 \text{ .}$$

Démonstration :

On choisit une involution de Cartan θ de $\underline{g}$ telle que $\theta\underline{h} = \underline{h}$. On note $\underline{k}$ l'ensemble des points fixes de θ , K le sous-groupe analytique correspondant. On note $\mathcal{H}^f$ le sous-module formé des vecteurs $\underline{k}$-finis de $\mathcal{H}^\infty$. C'est un sous-$\underline{\underline{g}}_{\mathbb{C}}$-module simple de $\mathcal{H}^\infty$. On peut donc aussi considérer les $\underline{h}$-modules $H_j(\underline{n},\mathcal{H}^f)$. Leur structure est complètement déterminée par Vogan dans [36] . En particulier, on a $\dim H_j(\underline{n},\mathcal{H}^f)_{ig+\rho_{\underline{b}}} \neq 0$ si et seulement si $j = q_{\underline{b}}$, et dans ce cas, la dimension est 1 ([36], th. 6.10). Le lemme 3 est donc corollaire du lemme 4 ci-dessous. C.Q.F.D.

Lemme 4 :

Soit T une représentation de G dans un espace de Banach $\mathcal{H}$, telle que le centre de G opère scalairement, et telle que $Z(\underline{\underline{g}}_{\mathbb{C}})$ opère scalairement dans $\mathcal{H}^\infty$. Alors l'injection $\mathcal{H}^f \to \mathcal{H}^\infty$ induit un isomorphisme de $\underline{h}$-modules $H_j(\underline{n},\mathcal{H}^f) \to H_j(\underline{n},\mathcal{H}^\infty)$.

Démonstration :

1. Rappelons l'homomorphisme d'Harish-Chandra de $Z(\underline{\underline{g}}_C)$ dans les éléments de $S(\underline{h}_{\underline{\underline{C}}})$ invariants par $W_{\underline{\underline{C}}}$. Notons le $Z \to \hat{Z}$. Il existe $\gamma \in \underline{h}^*_{\underline{\underline{C}}}$ tel que $Z(\underline{\underline{g}}_C)$ opère dans $\mathcal{H}^\infty$ par le scalaire $Z \to \hat{Z}(\gamma)$, et l'orbite $W_{\underline{\underline{C}}}\gamma$ de γ est bien déterminée (cf [7]). D'après le lemme de Casselman et Osborne [5], $H_j(\underline{n},\mathcal{H}^\infty)$ est un module $\underline{h}$-fini, et les poids de $\underline{h}$ qui interviennent sont tous de la forme $w(\gamma) + \rho_{\underline{b}}$, avec $w \in W_{\underline{\underline{C}}}$. On a un résultat analogue pour $H_j(\underline{n},\mathcal{H}^f)$, de sorte qu'il suffit de démontrer que pour tout $\eta \in \underline{t}^*_{\underline{\underline{C}}}$ l'injection $\mathcal{H}^f \to \mathcal{H}^\infty$ induit un isomorphisme $H_j(\underline{n},\mathcal{H}^f)_\eta \to H_j(\underline{n},\mathcal{H}^\infty)_\eta$.

2. On se ramène immédiatement au cas où $\underline{g}$ est semi-simple, de sorte que l'on a $\underline{t} = \underline{k} \cap \underline{h}$, et que $\underline{t}$ est une sous-algèbre de Cartan de $\underline{k}$ (parce que $\underline{h}$ est fondamentale). On pose $\underline{n}_C = \underline{k}_{\underline{\underline{C}}} \cap \underline{n}$. Il existe une suite spectrale, aboutissant à $H_*(\underline{n},\mathcal{H}^f)_\eta$, dont les flèches sont des t-morphismes, et dont le terme $E^1_{p,q}$ est $H_q(\underline{n}_C,\mathcal{H}^\infty \otimes \Lambda^p(\underline{n}/\underline{n}_C))_\eta$, (suite spectrale de Hochschild-Serre).

De même pour $H_*(\underline{n},\mathcal{H}^\infty)_\eta$. Il suffit donc de démontrer que la flèche naturelle

$$H_q(\underline{n}_C,\mathcal{H}^f \otimes \Lambda^p(\underline{n}/\underline{n}_C))_\eta \to H_q(\underline{n}_C,\mathcal{H}^\infty \otimes \Lambda^p(\underline{n}/\underline{n}_C))_\eta$$

est un isomorphisme. Comme $\Lambda^*(\underline{n}/\underline{n}_C)$ admet une filtration, par des $\underline{n}_C$-modules de dimension 1 triviaux, et compatible avec l'action de $\underline{t}$, un raisonnement standard montre qu'il suffit de prouver que pour tout $\beta \in \underline{t}^*_{\underline{\underline{C}}}$, la flèche naturelle

$$H_q(\underline{n}_C,\mathcal{H}^f)_\beta \to H_q(\underline{n}_C,\mathcal{H}^\infty)_\beta$$

est un isomorphisme.

3. Montrons que $\mathcal{H}^\infty$ est égal à l'espace des vecteurs de $\mathcal{H}$, C^∞ pour la restriction de T à K.

On choisit une base de $\underline{k}$, $X_1,\dots,X_m$, et une base de $\underline{g}$, $X_1,\dots,X_m,X_{m+1},\dots X_n$, orthonormées pour le produit scalaire $-(X,\theta X)$, où $(\ ,\ )$ est la forme de

Killing de $\underline{g}$. On pose $\Delta = \sum_1^n X_1^2$, $\Delta_C = \sum_1^m X_1^2$, $\Omega = \Delta - 2\Delta_C$.

L'opérateur Ω est dans $Z(\underline{\underline{g}}_C)$ et donc opère scalairement dans $\mathcal{H}^\infty$. Notons D et D_C les extensions self-adjointes de Δ et Δ_C opérant dans $\mathcal{H}^\infty$ (on sait que sur $\mathcal{H}^\infty$, Δ et Δ_C sont essentiellement self-adjoints). Alors pour tout $k \in \underline{\underline{N}}$, D^k et D_c^k ont même domaine de définition. Comme l'espace des vecteurs C^∞ pour G (resp. pour K) est l'intersection des domaines de définition des D^k (resp. D_c^k), notre assertion est démontrée.

Le lemme 4 est donc conséquence du lemme 5 ci-dessous. C.Q.F.D.

Lemme 5 :

Soit T une représentation de K dans un espace de Banach $\mathcal{H}$. Soient $\mathcal{H}^\infty$ l'espace des vecteurs C^∞ de $\mathcal{H}$, $\mathcal{H}^f$ le sous-espace de $\mathcal{H}^\infty$ formé des vecteurs $\underline{k}$-finis. Soit $\eta \in \underline{\underline{t}}_C$. On suppose qu'il existe un sous-groupe discret Z du centre de K tel que K/Z soit compact, et la restriction de T à Z soit scalaire.

Alors la flèche naturelle $H_j(\underline{n}_c,\mathcal{H}^f)_\eta \to H_j(\underline{n}_c,\mathcal{H}^\infty)_\eta$

est un isomorphisme.

Démonstration :

L'espace $\mathcal{H}^f$ est dense dans $\mathcal{H}^\infty$. Soit δ une représentation irréductible de K intervenant dans $\mathcal{H}^f$, et soit $\mathcal{H}_\delta$ le sous-espace de $\mathcal{H}$ dans lequel K opère de manière isomorphe à un multiple de δ . Alors $\mathcal{H}_\delta$ est fermé dans $\mathcal{H}$, contenu dans $\mathcal{H}^\infty$, et admet un supplémentaire fermé invariant. De plus, le nombre de δ pour lesquels on a : $H_j(\underline{n}_c,\mathcal{H}_\delta)_r \neq 0$ est fini d'après le théorème de Kostant [19] .

Il suffit donc de prouver l'assertion suivante. On suppose que pour toute représentation irréductible S de K de dimension finie, on a

$H_j(\underline{n}_c,\mathcal{H}_\delta)_\eta = 0$. Alors $H_j(\underline{n}_c,\mathcal{H}^\infty)_r = 0$.

Comme K/Z est compact, $\underline{t}$ opère de manière semi-simple dans le complexe $\Lambda^*\underline{n}_c \otimes \mathcal{H}^\infty$. On considère un élément $\omega \in (\Lambda^j \underline{n}_c \otimes \mathcal{H}^\infty)_\eta$ tel que $d\omega = 0$. Notons, pour tout δ, ω_δ la composante de ω dans $\Lambda^j\underline{n}_c \otimes \mathcal{H}_\delta$. On a $d\omega = 0$. On peut choisir $v_\delta \in (\Lambda^{j+1} \underline{n}_c \otimes \mathcal{H}_\delta)_\eta$ tel que $dv_\delta = \omega_\delta$. Le problème est de choisir les v_δ de telle sorte qu'il existe $v \in \Lambda^{j+1}n_c \otimes \mathcal{H}^\infty$ dont les composantes soient les v_δ.

Il existe un opérateur d^* de degré $+1$ dans le complexe $\Lambda^* \underline{n}_c \otimes \mathcal{H}^\infty$, commutant à l'action du commutant de $\underline{k}$ dans $\mathcal{H}^\infty$, commutant à l'action de $\underline{h}$, et ayant les propriétés suivantes. Posons $\square = dd^* + d^*d$. Dans $(\Lambda^j \underline{n}_c \otimes \mathcal{H}_\delta)_\eta$ l'opérateur $\square$ est inversible si et seulement si $H_j(\underline{n}_c, \mathcal{H}_\delta)_\eta = 0$. On peut donc choisir $v_\delta = \square^{-1} d\omega_\delta^*$. L'opérateur $\square$ est calculé et défini dans [19]. On y trouve en particulier que dans $(\Lambda^* \underline{n}_c \otimes \mathcal{H}_\delta)_\eta$, $\square$ est un scalaire. Notons le $c_{\delta,\eta}$. Choisissons un produit scalaire défini négatif, invariant sur $\underline{k}$, et notons encore δ le poids dominant de δ. On peut choisir d^* de telle sorte que :

$$2c_{\delta,\eta} = (\delta+\rho_c, \delta+\rho_c) - (\eta+\rho_c, \eta+\rho_c).$$

Soit $\| \ \|$ une norme définissant la topologie de $\Lambda^*\underline{n}_c \otimes \mathcal{H}$. L'opérateur de Casimir de $\underline{k}$ opère dans δ par le scalaire $(\delta+\rho_c, \delta+\rho_c)$, de sorte qu'un élément u de $\Lambda^*n \otimes \mathcal{H}^\infty$, dont les composantes sont notées u_δ, vérifie :

$$\Sigma(\delta+\rho_c, \delta+\rho_c)^N \| u_\delta \| < \infty$$

pour tout $N \in \underline{\underline{N}}$. Appliquant ceci à $u = d^*\omega$, on en conclut que l'on a : $\Sigma \| v_\delta \| < \infty$. Posons $v = \Sigma v_\delta$. On a $\Sigma(\delta+\rho_c, \delta+\rho_c)^N \| v_\delta \| < \infty$ pour tout $N \in \underline{\underline{N}}$. Donc $v \in \Lambda^*\underline{n}_c \otimes \mathcal{H}^\infty$. C.Q.F.D.

Remarque :

Lorsque G est compact, le lemme 3 est un résultat de Kostant [19], qui entraîne le théorème de Borel-Weil-Bott. Lorsque $\underline{h} = \underline{t}$, c'est un résultat de Schmid, qui entraîne la "conjecture de Langlands" sur les réalisations des séries discrètes [29].

III. 5. Formes linéaires standard :

Nous ne supposons plus que G est connexe. Dans ce paragraphe, nous supposons g standard, i.e. nous supposons que, avec les notations (1), on a : $\mu_\alpha \neq 0$ pour tout $\alpha \in \Delta$.

Ceci implique que $\underline{h}$ est fondamental. On a défini (paragraphe III. 3) la représentation $T_g^{G_0}$ de G_0, et il résulte du lemme 2 que son stabilisateur dans G est égal à $G(g)G_0$.

On considère l'algèbre $\underline{n} = \sum_{\substack{\alpha\in\Delta \\ \mu_\alpha>0}} \underline{\underline{g}}_{\mathbb{C}}^{\alpha}$. On pose $\underline{b} = \underline{h}_{\mathbb{C}} + \underline{n}$. C'est une polarisation totalement complexe, et elle est stable par $G(g)$.

Notons $\mathcal{H}$ l'espace de $T_g^{G_0}$, et $\mathcal{H}^\infty$ le sous-espace des vecteurs C^∞. Supposons que S soit une représentation unitaire de $G(g)^{\underline{g}}$ dans $\mathcal{H}$ telle que, pour tout $(x,\phi) \in G(g)^{\underline{g}}$, et tout $y \in G_0$, l'on ait :

$$(2) \qquad S(x,\phi)T_g^{G_0}(y)S(x,\phi)^{-1} = T_g^{G_0}(xyx^{-1}) .$$

Alors S laisse stable $\mathcal{H}^\infty$, et, avec les notations du paragraphe 4, on en déduit que $G(g)^{\underline{g}}$ opère dans l'espace (de dimension 1) $H_{q_{\underline{b}}}(\underline{n},\mathcal{H}^\infty)_{ig+\rho_{\underline{b}}}$.

Lemme 6 :

Il existe une, et une seule, représentation unitaire de $G(g)^{\underline{g}}$ dans $\mathcal{H}$, vérifiant (2), telle que l'action de $G(g)^{\underline{g}}$ dans $H_{q_{\underline{b}}}(\underline{n},\mathcal{H}^\infty)_{ig+\rho_{\underline{b}}}$ soit égale au caractère $\rho_{\underline{b}}$.

(Rappelons que le caractère de $\rho_{\underline{b}}$ de $G(g)^{\underline{g}}$ est défini pour tout sous-espace $G(g)$-invariant lagrangien de $\underline{\underline{g}}_{\mathbb{C}}$, cf. chapitre II. formule (2)).

Nous noterons S_g la représentation de $G(g)^{\underline{g}}$ définie dans le lemme 6.

Démonstration du lemme 6 :

Soit S' une représentation projective de $G(g)$ dans $\mathcal{H}$, par des opérateurs unitaires, et vérifiant (2). Une telle représentation existe. Soit $c(x)$ le scalaire qui s'en déduit représentant l'action de $x \in G(g)$ dans $H_{q_{\underline{b}}}(\underline{n},\mathcal{H}^\infty)_{ig+\rho_{\underline{b}}}$. Alors on définit $S_g(x,\phi)$ par la formule

$$S_g(x,\phi) = c(x)^{-1}\ \rho_{\underline{b}}(x,\phi)S'(x).$$

Comme c'est l'unique possibilité, on voit que l'on a démontré l'existence et l'unicité de la représentation S_g.

Supposons que x soit dans $G_0(g)$. Alors $S_g(x,\phi)$ doit être proportionnel à $T_g^{G_0}(x)$. Comme on a $G_0(g) = H$, l'action de $T_g^{G_0}(x)$ dans $H_{q_{\underline{b}}}(\underline{n},\mathcal{H}^\infty)_{ig+\rho_{\underline{b}}}$ est par définition la multiplication par le caractère de différentielle $ig + \rho_{\underline{b}}$. On a donc :

$$(3) \qquad S_g(x,\phi) = \chi_g(x,\phi)^{-1}\ T_g^{G_0}(x)$$

pour tout $(x,\phi) \in G(g)_0^{\underline{g}}$.

En particulier, $S_g(x,\phi)$ est unitaire pour $(x,\phi) \in G(g)_0^{\underline{g}}$.

Notons $Z_{\underline{g}}$ le centralisateur de $\underline{g}$ dans G. Alors le groupe $Z_{\underline{g}}G(g)_0$ est d'indice fini dans $G(g)$. Si $x \in Z_{\underline{g}}$, on a $S_g(x,\phi) = \rho_{\underline{b}}(x,\phi) = \phi$ $(=\pm 1)$. Le groupe $\|S_g(G(g)^{\underline{g}})\|$ est un sous-groupe fini de $]0,\infty[$, et donc trivial. Donc S_g est unitaire. C.Q.F.D.

Soit $\tau \in X(g)$. Soient $x \in G(g)$, $y \in G_0$, $(x,\phi) \in G(g)^{\underline{g}}$ représentant x. On pose

$$(4) \qquad (\tau \otimes S_g T_g^{G_0})(xy) = \tau(x,\phi) \otimes S_g(x,\phi)T_g^{G_0}(y).$$

C'est un opérateur unitaire dans l'espace produit tensoriel de l'espace de τ et de celui de $T_g^{G_0}$. On vérifie (grâce à (2) et (3)) qu'il ne dépend que de xy et que $\tau \otimes S_g T_g^{G_0}$ est une représentation de $G(g)G_0$.

Sa restriction à G_0 est un multiple de $T_g^{G_0}$, et son commutant est isomorphe à celui de τ.

On pose

(5) $$T_{g,\tau} = \mathrm{Ind}_{G(g)G_0}^{G}(\tau \otimes S_g\, T_g^{G_0})$$

Lemme 7 :

(i) Le commutant de τ est isomorphe à celui de $T_{g,\tau}$.

(ii) Soit a un automorphisme de G. Alors ${}^aT_{g,\tau} = T_{ag,{}^a\tau}$.

(iii) Soit $g' \in \underline{g}^*$ un élément admissible, bien polarisable, et standard. Si $g' \notin Gg$, les représentations $T_{g,\tau}$ et $T_{g',\tau'}$ sont disjointes pour tout $\tau' \in X(g')$.

(iv) Soient $\tau, \tau' \in X(g)$. Les espaces d'entrelacement entre τ et τ', et $T_{g,\tau}$ et $T_{g,\tau'}$ sont isomorphes.

Démonstration :

Cela résulte du lemme 2 et de la théorie de Mackey.

Remarque 1 :

Supposons G compact. Il est facile de voir que les représentations $T_{g,\tau}$ avec $\tau \in X^{irr}(g)$ forment l'ensemble des représentations unitaires irréductibles de G. Il est facile de voir que cette description du dual unitaire de G est équivalente à celle de Kostant [19].

Enfin, il est facile aussi de voir que $T_{g,\tau}$ est isomorphe à la représentation obtenue par translations à gauche dans l'espace des fonctions α, C^∞ sur G, à valeurs dans l'espace de τ, qui vérifient :

(6) $$\alpha(yx) = \tau(x,\phi)^{-1}\, \rho_{\underline{b}}^{-}(x,\phi)^{-1}\, \alpha(y)$$

pour $y \in G$, $x \in G(g)$, $(x,\phi) \in G(g)^{\underline{g}}$.

(7) $X\alpha = 0$ pour $X \in \bar{\underline{n}}$, où $\underline{b}$ est comme au début du paragraphe.

Nous ne nous servirons pas de ce résultat qui est à comparer aux formules (9) et (10), chapitre II.

Remarque 2 :

Il arrive que l'on ait $G(g) = Z_{\underline{g}}\, G(g)_0$ (nous en verrons un exemple important au paragraphe 7). Dans ce cas la construction se simplifie énormément. On a :

(8) $$T_{g,\tau} = \mathrm{Ind}_{Z_{\underline{g}}G_0}^{G} (\tau \otimes T_g^{G_0})$$

où l'on a noté $\tau \otimes T_{\underline{g}}^{G_0}$ la représentation de $Z_{\underline{g}}G_0$ définie par la formule :

(9) $$(\tau \otimes T_g^{G_0})(xy) = \tau(x,1)T_g^{G_0}(y)$$

pour $x \in Z_g$, $y \in G_0$.

III. 6. Le cas général :

Nous ne faisons plus d'hypothèses restrictives, ni sur G, ni sur g - sauf celles expliquées en III. 1. Rappelons que $\nu \in \underline{a}^*$ est défini en (1). On note M le centralisateur de ν dans G, $\underline{m}$ son algèbre de Lie, $m = g|\underline{m}$. On a $M(m) = G(g)$, et on remarque que m est standard.

On pose $\underline{u} = (\sum_{\substack{\alpha\in\Delta \\ \nu_\alpha>0}} \underline{g}_{\mathbb{C}}^{\alpha}) \cap \underline{\underline{g}}$, et on note U le sous-groupe analytique de G correspondant. Alors MU est un sous-groupe fermé de G.

Bien que l'on ait $M(m) = G(g)$, $M(m)^{\underline{m}}$ et $G(g)^{\underline{g}}$ ne sont pas en général égaux. Il y a cependant une bijection naturelle entre $X(g)$ et $X(m)$. En effet, soit $x \in G(g)$, et soient $(x,\phi) \in G(g)^{\underline{g}}$, $(x,\psi) \in M(m)^{\underline{m}}$ des représentants de x. Soit L un sous-espace lagrangien de $\underline{m}$. Alors $L + \underline{u}$ est un sous-espace lagrangien de $\underline{g}$. Il résulte de la formule (12) du chapitre I que le nombre $\phi(L+\underline{u})\psi(L)^{-1}$ ne dépend pas du choix de L. On le note $\phi\psi^{-1}$.

Soit $\tau \in X(g)$. On définit un élément $\sigma \in X(m)$ en posant

(10) $\psi^{-1}\,\tau(x,\phi) = \sigma(x,\psi)$

pour tout $(x,\psi) \in M(m)^{\underline{m}}$. Cela ne dépend pas du choix de $(x,\phi) \in G(g)^{\underline{g}}$ représentant x.

La représentation $T^{M}_{m,\sigma}$ a été définie au paragraphe 5. On pose :

(11) $T_{g,\tau} = \mathrm{Ind}^{G}_{MU}(T^{M}_{m,\sigma} \otimes \mathrm{Id}_U)$.

Lemme 8 :

Les représentations $T_{g,\tau}$ de G ont les propriétés (i), (ii), (iii), (iv), (v) de l'introduction.

Le lemme 8 sera démontré dans les paragraphes suivants.

III. 7. Démonstration du lemme 8 lorsque G *est connexe* :

Dans ce paragraphe, nous supposons que G est connexe. Les notations sont celles du paragraphe 6.

Rappelons la méthode d'Harish-Chandra pour construire des représentations de G. On note M' le centralisateur de $\underline{a}$ dans G, $\underline{m}'$ son algèbre de Lie, On pose $m' = g|\underline{m}'$. Soit H le centralisateur de $\underline{h}$ dans G. Comme G est connexe, on a $H = G(g) = M'(m')$. De manière analogue au paragraphe 3, on choisit une sous-algèbre $\underline{u}'$ de $\underline{g}$, on note U' le sous-groupe analytique correspondant, de sorte que $M'U'$ est un sous-groupe parabolique "cuspidal" de G.

De manière analogue à (10), on définit un élément σ' de $X(m')$ en posant

(12) $\psi'^{-1}\tau(x,\phi) = \sigma'(x,\psi')$

pour tout $(x,\psi') \in M'(m')^{\underline{m}'}$. Cela ne dépend pas du choix du représentant (x,ϕ) de x dans $G(g)^{\underline{g}}$. On définit la représentation $T^{M'}_{m',\sigma'}$ de M' comme au paragraphe 5 (la remarque 2 de ce paragraphe s'applique).

On pose $T'_{g,\tau} = \mathrm{Ind}_{M'U'}^{G}(T^{\mu'}_{m',c'} \otimes \mathrm{Id}_{U'})$.

Lemme :

La représentation $T'_{g,\tau}$ ne dépend pas du choix de $\underline{u}'$. Les propriétés (i), (ii), (iii), (iv), de l'introduction sont vérifiées pour les représentations $T'_{g,\tau}$.

Démonstration :

Tout ceci est essentiellement dû à Harish-Chandra. Donnons quelques détails. Tout d'abord, $X(g)$ est de type I, et la construction de $T'_{g,\tau}$ commute aux sommes boréliennes de représentations. Il suffit donc de considérer le cas où $\tau \in X^{irr}(g)$, ce que nous faisons ci-dessous.

Dans ce cas $T^{M'}_{m',\sigma'}$ est une représentation irréductible de M', de carré intégrable module le centre.

Montrons que $T^{M'}_{m',\sigma'}$ ne dépend pas du choix de $\underline{u}'$. Cela revient à démontrer que τ' ne dépend pas du choix de $\underline{u}'$, et donc que si $x \in G(g)$, $(x,\phi) \in G(g)^{\underline{g}}$, $(x,\psi') \in M'(m')^{\underline{m}'}$, et si L est un sous-espace lagrangien de $\underline{m}'$, le nombre $\phi(L+u')\psi'(L)^{-1}$ ne dépend pas du choix de $\underline{u}'$. Il faut donc voir que $\phi(L+\underline{u}')$ ne dépend pas du choix de $\underline{u}'$. Cela résulte de la formule (12) du chapitre I, et de ce que $\underline{u}'$ est invariant par x.

La première assertion du lemme est un résultat d'Harish-Chandra (cf [37]).

La propriété (i) signifie que $T'_{g,\tau}$ est irréductible. Lorsque $[G,G]$ est de centre fini, cela résulte de la théorie d'Harish-Chandra [12]. Le point clé est que γ_α est non nul pour toute racine α réelle, i.e. pour tout $\alpha \in \Delta$ tel que $\alpha|\underline{t} = 0$ (ceci vient de ce que g_α est non nul pour tout $\alpha \in \Delta$). Cependant, le résultat n'est explicité que lorsque $\underline{h}$ est fondamental. On trouvera le fait que $T'_{g,\tau}$ est irréductible, complètement explicité, et sans l'hypothèse de centre fini, avec une démonstration différente, dans [32].

$$T_m^{M_0} = \mathrm{Ind}_{M_0'U''}^{M_0}(T_{m'}^{M_0'} \otimes \mathrm{Id}_{U''}) .$$

Mais ceci est vrai par définition même de $T_m^{M_0}$ (paragraphe 3).

C.Q.F.D.

Remarque :

Lorsque G est connexe, et $\underline{h}$ fondamental, nous avions défini $T_{\underline{g}}$ deux fois (paragraphe 3 et paragraphe 5). Le lemme 9 montre que les deux définitions coïncident.

III. 3. Fin de la démonstration du lemme 8 :

On emploie les notations du paragraphe 7. Comme les constructions faites commutent aux sommes boréliennes, il suffit de démontrer le lemme 8 quand τ est factorielle. Nous supposons ci-dessous que τ est factorielle.

Comme $Z_{\underline{g}}G_0(g)$ est d'indice fini dans $G(g)$, il existe une sous-représentation irréductible dans la restriction de τ à $G_0(g)^{\underline{g}}$. On en choisit une, et on l'appelle τ_1. On note $G(g)^{\underline{g}}_{\tau_1}$ le stabilisateur de τ_1 dans $G(g)^{\underline{g}}$, et $G(g)_{\tau_1}$ son image dans $G(g)$. On choisit une représentation projective de $G(g)^{\underline{g}}_{\tau_1}$ qui prolonge τ_1, que nous noterons encore τ_1, et qui vérifie les relations :

(13) $\tau_1(\hat{x}\hat{y}) = \tau_1(\hat{x})\tau_1(\hat{y})$ et $\tau_1(\hat{y}\hat{x}) = \tau_1(\hat{y})\tau_1(\hat{x})$ pour tout $\hat{x} \in G(g)^{\underline{g}}_{\tau_1}$, $\hat{y} \in G_0(g)^{\underline{g}}$.

Soient $x, x' \in G(g)_{\tau_1}$ et soient $\hat{x}, \hat{x}'$ des représentants dans $G(g)^{\underline{g}}_{\tau_1}$. On pose

(14) $\tau_1(\hat{x}\,\hat{x}') = c(x,x')\tau_1(\hat{x})\tau_1(\hat{x}')$.

On a ainsi défini un 2-cocycle sur $G(g)_{\tau_1}/G_0(g)$.

Il existe une unique représentation projective de $G(g)_{\tau_1}/G_0(g)$, que nous noterons τ_2, telle que l'on ait $\mathrm{Ind}_{G(g)^{\underline{g}}_{\tau_1}}^{G(g)^{\underline{g}}} (\tau_2 \otimes \tau_1) = \tau$. Elle vérifie

$$(15) \qquad \tau_2(xx') = c(x,x')^{-1}\, \tau_2(x)\tau_2(x')$$

pour $x, x' \in G(g)_{\tau_1}$.

On considère la représentation $T^{G_0}_{g,\tau_1}$ de G_0 . Comme le lemme 8 est valable pour G_0, d'après le paragraphe 7, cette représentation est irréductible, et son stabilisateur dans G est le groupe $G(g)_{\tau_1} G_0$.

Nous allons construire une représentation projective de $G(g)_{\tau_1} G_0$, que nous noterons T_1, et vérifiant les propriétés suivantes

$$(15) \qquad T_1(y) = T^{G_0}_{g,\tau_1}(y) \quad \text{si} \quad y \in G_0 .$$

$$(16) \qquad T_1(xy) = T_1(x)T_1(y) \quad \text{et}$$

$$T_1(yx) = T_1(y)T_1(x) \quad \text{si} \quad x \in G(g)_{\tau_1} ,\ y \in G_0 .$$

$$(17) \qquad T_1(xx') = c(x,x')T_1(x)T_1(x') \quad \text{pour} \quad x, x' \in G(g)_{\tau_1} .$$

De plus nous montrerons que l'on a

$$(18) \qquad T^G_{g,\tau} = \mathrm{Ind}^G_{G(g)_{\tau_1} G_0}(\tau_2 \otimes T_1) .$$

Le lemme 8 résulte immédiatement du cas particulier des groupes connexes, de la formule (18), et de la théorie du Mackey.

Construction de T_1 :

La construction de T_1 est tout à fait similaire à celle de $T^G_{g,\tau}$. On introduit le groupe $M_1 = M \cap G_0$. On note σ_1 la représentation de $M_1(m)^{\underline{m}}$ définie de manière analogue à σ (formule 10). Le stabilisateur $M(m)^{\underline{m}}_{\sigma_1}$ de σ_1 dans $M(m)^{\underline{m}}$ est l'image réciproque de $G(g)_{\sigma_1}$ dans $M(m)$. On utilise encore (10) pour définir une représentation projective, notée encore σ_1, de

$M(m)\frac{m}{\sigma_1}$, et vérifiant les propriétés analogues à (13) et (14), avec le même cocycle c.

Nous définissons une représentation projective, que nous noterons $\sigma_1 \otimes S_g T_g^0$, du groupe $M(m)_{\sigma_1} M_0$, en posant

$$(19) \qquad (\sigma_1 \otimes S_m T_m^{M_0})(xy) = \sigma_1(\hat{x}) \otimes S_m(\hat{x}) T_m^{M_0}(y)$$

pour $x \in M(m)_{\sigma_1}$, $\hat{x} \in M(m)\frac{m}{\sigma_1}$ représentant x, $y \in M_0$.

Nous définissons une représentation projective, que nous noterons R, dans l'espace de $T_{m,\sigma_1}^{M_1}$, du groupe $M(m)_{\sigma_1} M_1$. Si $y \in M_1$, on pose $R(y) = T_{m,\sigma_1}^{M_1}(y)$. Si $x \in M(m)_{\sigma_1}$, on définit $R(x)$ de la manière suivante. Rappelons que T_{m,σ_1} est induite à partir de la représentation $\sigma_1 \otimes S_m T_m^0$ du groupe $M_1(m)M_0$. Un élément de l'espace de $T_{m,\sigma_1}^{M_1}$ est donc une fonction α sur M_1 , à valeurs dans l'espace de $\sigma_1 \otimes S_m T_m^{M_0}$, vérifiant certaines relations. Pour une telle fonction, on pose

$$(R(x)\alpha)(y) = (\sigma_1 \otimes S_m T_m^{M_0})(x)\alpha(x^{-1}yx)$$

pour tout $y \in M_1(m)M_0$. On vérifie que lorsque $x \in M_1(m)$, les deux définitions de $R(x)$ coïncident. Enfin, si $x \in M(m)_{\sigma_1}$ et si $y \in M_1$, on pose $R(xy) = R(x)R(y)$. On vérifie sans difficulté que l'on a $R(yx) = R(y)R(x)$, pour $x \in M(m)_{\sigma_1}$, $y \in M_1$, et $R(xx') = c(x,x')R(x)R(x')$ pour $x,x' \in M(m)_{\sigma_1}$. Rappelons que la représentation $T_{g,\tau_1}^{G_0}$ est induite par la représentation $T_{m,\sigma_1}^{M_1} \otimes Id_U$ de M_1U . Elle est réalisée dans un espace de fonctions sur G_0 à valeurs dans l'espace de $T_{m,\sigma_1}^{M_1}$. Soit α une telle fonction. On pose, si $x \in G(g)_{\sigma_1} = M(m)_{\sigma_1}$, $(T_1(x)\alpha)(y) = R(x)\alpha(x^{-1}yx)$ pour tout $y \in G_0$, et l'on pose $T_1(xy) = T_1(x)T_{g,\sigma_1}^{G_0}(y)$ pour $x \in G(g)_{\sigma_1}$, $y \in G_0$.

Les formules (15), (16), (17) sont faciles à vérifier. Il reste à démontrer la formule (18).

Pour cela, nous remarquons que, d'après le théorème d'induction par étages, $T^G_{g,\tau}$ est induite à partir de la représentation $(\sigma\otimes S_m T_m^{M_0}) \otimes Id_U$ du groupe $M(m)M_0U$. D'autre part, par construction de T_1, il est facile de voir que la représentation $Ind^G_{G(g)_{\sigma_1} G_0}(\tau_2 \otimes T_1)$ est induite à partir de la représentation $\tau_2 \otimes (\sigma_1 \otimes S_m T_m^{M_0}) \otimes Id_U$ du groupe $M(m)_{\sigma_1}M_0U$. Il nous suffit donc de démontrer que l'on a :

(20) $$Ind^{M(m)M_0}_{M(m)_{\sigma_1}M_0}(\tau_2 \otimes (\sigma_1 \otimes S_m T_m^{M_0})) = \sigma \otimes S_m T_m^{M_0} .$$

Pour cela nous décrivons un opérateur d'entrelacement entre ces deux espaces. Il est facile de voir que σ est isomorphe à la représentation induite par la représentation $\tau_2 \otimes \sigma_1$ de $M(m)^{\underline{m}}_{\sigma_1}$. Un élément β de l'espace de la représentation $\sigma \otimes S_m T_m^{M_0}$ est donc une fonction sur $M(m)^{\underline{m}}$, à valeurs dans $\mathcal{K} \otimes \mathcal{L} \otimes \mathcal{H}$, où l'on a noté $\mathcal{K}$ l'espace de τ_2, $\mathcal{L}$ l'espace de σ_1, $\mathcal{H}$ l'espace de $T_m^{M_0}$. La fonction β vérifie

$$\beta(\hat{x}\hat{y}) = (\tau_2(y)^{-1} \otimes \sigma_1(\hat{y})^{-1} \otimes Id)\beta(\hat{x})$$

pour $\hat{x} \in M(m)^{\underline{m}}$, $\hat{y} \in M(m)^{\underline{m}}_{\sigma_1}$. A β on associe une fonction α sur $M(m)M_0$, à valeurs dans $\mathcal{K} \otimes \mathcal{L} \otimes \mathcal{H}$, en posant, pour $x \in M(m)$, $y \in M_0$, $\hat{x} \in M(m)^{\underline{m}}$ représentant x :

$$\alpha(xy) = (1 \otimes 1 \otimes T_m^{M_0}(y^{-1})S_m(\hat{x})^{-1})\beta(\hat{x}).$$

On vérifie que l'application $\beta \to \alpha$ est un opérateur d'entrelacement entre les deux représentations figurant dans (20). C.Q.F.D.

Remarque :

Les formules (15), (16), (17) calculent l'obstruction de Mackey à étendre la représentation $T^{G_0}_{g,\tau_1}$ à son stabilisateur dans G. Elle est isomorphe à l' l'obstruction qu'il y a à étendre τ_1 à son stabilisateur dans $G(g)^{\underline{g}}$.
Peut-on décrire de manière aussi simple l'obstruction de Mackey pour les au-

tres représentations irréductibles de G^0, plus particulièrement les représentations irréductibles tempérées dont le caractère infinitésimal n'est pas régulier ?

CHAPITRE IV - TECHNIQUES DE RECURRENCE

Dans ce chapitre, nous étudions ce que devient une forme linéaire admissible, ou bien polarisable, quand on la restreint à un idéal, ou à l'algèbre de Lie du "petit groupe" de la théorie de Mackey. Nous n'utiliserons les résultats ci-dessous que dans le cas d'un idéal nilpotent. Cependant, en vue d'applications ultérieures, j'ai traité une situation plus générale.

4. 1. Formes bien polarisables :

Dans ce paragraphe, $\underline{g}$ est une algèbre de Lie complexe de dimension finie, et g est un élément de $\underline{g}^*$.

Lemme 1 :

Soit $\underline{q}$ un idéal de $\underline{g}$ contenu dans $\ker g$. Soit g' l'élément de $\underline{g}/\underline{q}$ obtenu par passage au quotient. On suppose $\underline{q}$ résoluble. Alors g est bien polarisable si et seulement s'il en est de même de g'.

Démonstration : c'est évident.

Ci-dessous, on considère un idéal $\underline{l}$ de $\underline{g}$. On pose $l = g|\underline{l}$, $\underline{h} = \underline{g}(l)$, $h = g|\underline{h}$. Nous notons L et H les groupes analytiques correspondants.

Lemme 2 :

La forme g est bien polarisable si et seulement s'il en est de même de l et de h. Dans ce cas il existe une bonne polarisation $\underline{b}$ en g telle que l'on l'on ait : $\underline{b} = \underline{b} \cap \underline{l} + \underline{b} \cap \underline{h}$.

Dans la démonstration, nous aurons à utiliser le résultat suivant.

Lemme 3 :

On a $L(1)_0 g = g + (\underline{h} + \underline{l})^{\perp}$ ou $(\underline{h} + \underline{l})^{\perp}$ est l'orthogonal de $\underline{h} + \underline{l}$ dans $\underline{g}^*$.

Démonstration :

Voir [24] p. 500.

Démonstration du lemme 2 :

1. Soit $\underline{t}$ une sous-algèbre de $\underline{g}$. Posons $t = g|\underline{t}$. On suppose qu'il existe une bonne polarisation en g contenue dans $\underline{t}$. Soit $\underline{b}$ une sous-algèbre de $\underline{t}$. C'est une bonne polarisation en t si et seulement si elle a la même propriété pour g.

2. Soit $\underline{b}$ un sous-espace lagrangien de $\underline{g}$. Les conditions suivantes sont équivalentes :

(i) $\underline{b} \subset \underline{k}$.

(ii) $\underline{b} \cap \underline{l}$ est lagrangien dans $\underline{l}$.

(iii) $\underline{b} \cap \underline{h}$ est lagrangien dans $\underline{h}$.

Quand l'une de ces conditons est vérifiée, on a $\underline{b} = \underline{b} \cap \underline{l} + \underline{b} \cap \underline{h}$. (cf [7] p. 57).

Supposons que $\underline{b}$ soit une bonne polarisation en g contenue dans $\underline{k}$. Alors $\underline{b} \cap \underline{l}$ est une bonne polarisation en l, et $\underline{b} \cap \underline{h}$ une bonne polarisation en h.

Réciproquement, soient $\underline{a}$ une bonne polarisation en l, $\underline{c}$ une bonne polarisation en h telles que $\underline{c}$ normalise $\underline{a}$. Alors $\underline{a} + \underline{c}$ est une bonne polarisation en g. (La condition de Pukanszky vient du lemme 3).

3. Soit $\underline{r}$ une algèbre de Lie résoluble, soit $\underline{s}$ un idéal de $\underline{r}$, soit $r \in \underline{r}^*$, et posons $s = r|\underline{s}$. Alors il existe une bonne polarisation $\underline{b}$ en r telle que $\underline{b} \cap \underline{s}$ soit une bonne polarisation en s(cf [33]).

4. On suppose g bien polarisable. On note $\underline{r}$ le plus grand idéal résoluble de $\underline{l}$, et on pose $r = g|\underline{r}$. Soit $\underline{b}$ une bonne polarisation en g. En appliquant 1 et 3 à l'algèbre $\underline{b} + \underline{r}$, on voit que g a une bonne polarisation $\underline{b}$ telle que $\underline{b} \cap \underline{r}$ soit une bonne polarisation en r.
On considère l'algèbre $\underline{g}(r)$. On pose $\underline{q} = \underline{r}(r) \cap \mathrm{Ker}\, r$. C'est un idéal de $\underline{g}(r)$. On pose $\underline{g}_1 = \underline{g}(r)/\underline{q}$, $\underline{l}_1 = \underline{l}(r)/\underline{q}$, et on note g_1, l_1 les formes linéaires obtenues par passage au quotient sur $\underline{g}_1^*$ et $\underline{l}_1^*$.
Il résulte du lemme 1 et de 2 que g_1 est bien polarisable.
Nous allons démontrer qu'il existe une bonne polarisation $\underline{b}$ en g telle que $\underline{b} \cap \underline{l}$ soit une bonne polarisation en l. On le démontre par récurrence sur la dimension de $\underline{g}$. C'est clair si $\dim \underline{g} = 0$. On suppose le résultat établi pour les algèbres de Lie de dimension strictement inférieure à $\dim \underline{g}$. On considère deux cas :.

i) $\dim \underline{g}_1 < \dim \underline{g}$. Alors il existe une bonne polarisation $\underline{b}_1$ en g_1 telle que $\underline{b}_1 \cap \underline{l}_1$ soit une bonne polarisation en l_1. Soit $\underline{b}'$ l'image réciproque de $\underline{b}_1$ dans $\underline{g}(r)$. D'après 3, comme $\underline{b}'$ est résoluble il existe une bonne polarisation $\underline{a}$ en r normalisée par $\underline{b}'$. D'après 2 (appliqué à $\underline{r}$), $\underline{b}' + \underline{a}$ est une bonne polarisation en g et $(\underline{b}' \cap \underline{l}(r)) + \underline{a}$ une bonne polarisation en l . Notre assertion est démontrée dans ce cas.

ii) $\dim \underline{g}_1 = \dim \underline{g}$. Cela implique que $\underline{r}$ est de dimension au plus un, contenu dans le centre de $\underline{g}$. D'autre part, posons $\underline{s} = [\underline{l},\underline{l}]$. Alors $\underline{s}$ est un idéal semi-simple de $\underline{g}$. Soit $\underline{g}'$ le centralisateur de $\underline{s}$ dans $\underline{g}$, soient $s = g|\underline{s}$, $g' = g|\underline{g}'$. Alors $\underline{g}$ est somme directe de $\underline{s}$ et $\underline{g}'$. Soit $\underline{b}$ une bonne polarisation en g. Soit $\underline{a}$ la projection de $\underline{b}$ sur $\underline{s}$.
C'est une sous-algèbre résoluble de $\underline{s}$, et l'on a $\underline{a}^\perp \subset \underline{a}$, où $\underline{a}^\perp$ désigne l'orthogonal par rapport à B_s . Choisissons une polarisation $\underline{a}' \subset \underline{a}$ pour $s|\underline{a}$. Alors $\underline{a}'$ est aussi une polarisation pour s. Comme $\underline{a}'$ est résoluble, il en résulte que $\underline{a}'$ est une sous-algèbre de Borel de $\underline{s}$ ([7] p. 60). On a

donc $\underline{a} = \underline{a}'$, ce qui entraîne que l'on a : $\underline{a} = \underline{b} \cap \underline{s}$. D'après 2, $\underline{a}$ est une bonne polarisation en s, et notre assertion est démontrée dans ce cas. Comme il existe une bonne polarisation $\underline{b}$ en g telle que $\underline{b} \cap \underline{l}$ soit une bonne polarisation en l, il résulte de 2 que $\underline{b} \cap \underline{h}$ est une bonne polarisation en h et que $\underline{b} = \underline{b} \cap \underline{l} + \underline{b} \cap \underline{h}$.

5. On suppose g bien polarisable. Soit $\underline{d}$ une algèbre de Lie de dérivations de $\underline{s}$, résoluble, et stabilisant g. Nous allons montrer que g a une bonne polarisation stable par $\underline{d}$.
On note $\underline{r}$ le plus grand idéal résoluble de $\underline{g}$, et on définit $\underline{g}_1$, g_1 comme en 4. Notons que g_1 est bien polarisable et que $\underline{d}$ opère comme algèbre résoluble de dérivations de $\underline{g}_1$, stabilisant g_1 .
On raisonne par récurrence sur la dimension de $\underline{g}$. Comme en 4, on considère deux cas.

i) dans $\underline{g}_1 <$ dim $\underline{g}$. La démonstration est analogue a celle faite en 4.

ii) dim $\underline{g}_1$ = dim $\underline{g}$. Alors $\underline{g}$ est réductive , et $\underline{g}(g)$ est une sous-algèbre de Cartan de $\underline{g}$. Le centre de $\underline{g}$ est de dimension au plus 1 et $\underline{d}$ opère trivialement dans ce centre. On peut donc identifier $\underline{d}$ à une sous-algèbre de $\underline{s}(g)$, et n'importe quelle sous-algèbre de Borel de $\underline{g}$ contenant $\underline{g}(g)$ convient.

6. On suppose que l et h sont bien polarisables. Soit $\underline{c}$ une bonne polarisation en h. D'après 5, appliqué à $\underline{c}$ et $\underline{l}$, il existe une bonne polarisation $\underline{a}$ en l , stable par $\underline{c}$. D'après 2, $\underline{c} + \underline{a}$ est une bonne polarisation en g. C.Q.F.D.

Corollaire du lemme 2 :

i) On suppose $\underline{l}$ résoluble. Alors g est bien polarisable si et seulement si h est bien polarisable.

ii) On suppose $\underline{g}/\underline{l}$ résoluble. Alors g est bien polarisable si et seulement si l est bien polarisable.

Les cas les plus importants sont les suivants : $l = [\underline{g},\underline{g}]$, et $\underline{l} = \underline{u}$ (le plus grand idéal nilpotent).

IV. 2. Formes admissibles :

Dans ce paragraphe, G est un groupe de Lie d'algèbre de Lie $\underline{g}$, $g \in \underline{g}^*$, Γ un sous-groupe du centre de G, η un caractère unitaire de Γ.

Lemme 4 :

Soit $\underline{q}$ un idéal G-invariant de $\underline{g}$ contenu dans $\ker g$, tel que le sous-groupe analytique Q correspondant soit fermé. On pose $G' = G/Q$, $\underline{g}' = \underline{g}/\underline{q}$, $\Gamma' = \Gamma/\Gamma \cap Q$, et on note g' l'élément de $\underline{g}'^*$ obtenu par passage au quotient. On suppose que g est η-admissible. Alors η est trivial sur $\Gamma \cap Q$. On note η' le caractère de Γ obtenu par passage au quotient. Alors g' est η'-admissible.

Plus précisément, l'application $x \to (x,1)$ identifie Q à un sous-groupe de $G(g)^{\underline{g}}$, et $G'(g')^{\underline{g}'}$ est isomorphe à $G(g)^{\underline{g}}/Q$. La composition avec la projection $G(g)^{\underline{g}} \to G'(g')^{\underline{g}'}$ donne un isomorphisme de $X(g',\eta')$ sur $X(g,\eta)$.

Démonstration :

C'est évident.

Dans la suite, $\underline{l}$ est un idéal G-invariant de $\underline{g}$. On pose $l = g|\underline{l}$, $h = \underline{g}(l)$, $h = g|\underline{h}$. Le groupe $G(l)$ opère dans $\underline{l}$ en conservant B_l. On note H son revêtement à deux feuillets $H = G(l)^{\underline{l}}$. On note Γ' l'image réciproque de Γ dans H, et η' le caractère de Γ' défini par la formule

$$\eta'(\gamma, \pm 1) = \pm\eta(\gamma)$$

pour $\gamma \in \Gamma$. On note L le sous-groupe analytique d'algèbre de Lie $\underline{l}$.

De manière analogue au lemme 3, on a :

Lemme 5 :

On a : $L(1)_0 g = g + (\underline{h}+\underline{1})^{\perp}$, et $G(1)(h) = G(g)L(1)_0$.

Démonstration :

Cf [24] p. 500.

Rappelons que $G(g)^{\underline{g}}$ est décrit comme un ensemble de couples (x,ϕ) avec $x \in G(g)$. De même H est décrit comme un ensemble de couples (x,ψ) avec $x \in G(1)$, et $H(h)^{\underline{h}}$ comme un ensemble de triplets (x,ψ,θ), avec $x \in G(1)(h)$, et $(x,\psi) \in H(h)$.

Soient L' un sous-espace lagrangien de $\underline{1}$ et L'' un sous-espace lagrangien de $\underline{h}$. Alors $L' + L''$ est un sous-espace lagrangien de $\underline{g}$. Soit $x \in G(g)$, et soient (x,ϕ) et (x,ψ,θ) des représentants dans $G(g)^{\underline{g}}$ et $H(h)^{\underline{h}}$ respectivement. On peut considérer le nombre

$$\phi(L'+L'')\psi(L')^{-1}\theta(L'')^{-1}$$

Il résulte des formules (12) qu'il ne dépend pas du choix de L' et L''. On le note $\phi\psi^{-1}\theta^{-1}$.

Lemme 6 :

La forme g est η-admissible si et seulement si h est η'-admissible. Dans ce cas, étant donné $\tau \in X_{G,\Gamma}(g,\eta)$, il existe un unique élément $\sigma \in X_{H,\Gamma'}(h,\eta')$ tel que l'on ait :

(1) $$\sigma(x,\psi,\theta) = \phi\psi^{-1}\theta^{-1}\tau(x,\phi)$$

pour tout $(x,\phi) \in G(g)^{\underline{g}}$ et tout représentant (x,ψ,θ) de x dans $H(h)^{\underline{h}}$.

L'application $\tau \to \sigma$ est une bijection de $X_{G,\Gamma}(g,\eta)$ sur $X_{H,\Gamma'}(h,\eta')$ préservant le type.

De plus, 1 est admissible.

Démonstration :

1. Supposons g η-admissible, et soit $\tau \in X_{G,\Gamma}(g,\eta)$. Montrons que h est admissible. Posons $A = (L(1)^{\underline{1}})_0$. C'est un sous-groupe de $H(h)$, et l'application $a \to (a,1)$ identifie A à un sous-groupe de $H(h)^{\underline{h}}$. Notons B l'image réciproque de $G(g)$ dans $H(h)^{\underline{h}}$. D'après le lemme 5, on a $H(h)^{\underline{h}} = BA$, et $A/A \cap B$ est homéomorphe à $(\underline{h}+\underline{l})$. Donc $A \cap B$ est connexe, et $A/A \cap B$ simplement connexe. Comme g est admissible la restriction de τ à $A \cap B$ fournit un caractère de différentielle $ig|\underline{l}(g)$. Ce caractère se prolonge uniquement en un caractère χ de A de différentielle $ig|\underline{l}(1)$.

Définissant σ sur A par la formule (1), on pose $\sigma(ab) = \sigma(a)\chi(b)$ pour $a \in A$, $b \in B$. Ceci ne dépend pas des choix faits et fournit la représentation cherchée dans $X_{H,\Gamma'}(h,\eta')$.

Il existe un unique prolongement de χ à $(L(1)_0)^{\underline{1}}$ tel que $\chi(1,-1) = -1$, et donc 1 est admissible.

2. Supposons h η'-admissible. Soit $\sigma \in X_{H,\Gamma'}(h,\eta')$. La formule (1) définit un élément de $X_{G,\Gamma}(g,\eta)$. Donc g est η-admissible.

C.Q.F.D.

CHAPITRE V - EXTENSIONS DES REPRESENTATIONS DES GROUPES DE LIE NILPOTENTS

Dans ce chapitre, U est un groupe de Lie nilpotent connexe d'algèbre de Lie $\underline{u}$, et u un élément admissible de $\underline{u}^*$. Nous rappelons la construction de Kirillov de la représentation unitaire irréductible associée à u, et le calcul de l'obstruction de Mackey à étendre une telle représentation. Ce chapitre est essentiellement destiné à fixer les notations.

V. 1. - La théorie de Kirillov :

On sait que $U(u)$ est connexe. D'autre part, l'application $x \to (x,1)$ est un isomorphisme de $U(u)$ sur son image dans $U(u)^{\underline{u}}$, de sorte que dire que u est est admissible, c'est dire que $iu|\underline{u}(u)$ est différentielle d'un caractère de $U(u)$.

(Remarquons que le centre Z de U est connexe, et qu'une forme linéaire u' sur $\underline{u}$ est admissible si et seulement si, notant $\underline{z}$ l'algèbre de Lie de Z, $iu'|\underline{z}$ est différentielle d'un caractère de Z).

Il existe des polarisations réelles $\underline{b}$ en u, i.e. des polarisations telles que $\underline{b} = (\underline{b} \cap \underline{u})_{\underline{\underline{C}}}$. Soit $\underline{b}$ une telle polarisation, soit B le sous-groupe analytique d'algèbre $\underline{b} \cap \underline{u}$. Il est fermé, et il existe un caractère unitaire $\chi_{\underline{b}}$ de différentielle $iu|\underline{b} \cap \underline{u}$. On pose $T_{u,\underline{b}} = \mathrm{Ind}_B^U(\chi_{\underline{b}})$.

Lemme 1 : (Kirillov [17]).

La représentation $T_{u,\underline{b}}$ est irréductible, et sa classe ne dépend pas de $\underline{b}$. On la note T_u.

Si u et u' sont deux éléments admissibles de $\underline{u}^*$, on a $T_u = T_{u'}$ si et seulement si $u' \in Uu$.

V. 2. Opérateurs d'entrelacement :

Notons $\mathcal{H}_{\underline{b}}$ l'espace de la représentation induite $T_{u,\underline{b}}$. Soit $\underline{b}'$ une autre polarisation réelle. Comme les représentations $T_{u,\underline{b}}$ et $T_{u,\underline{b}'}$ sont équivalentes, il existe un opérateur d'entrelacement $\mathcal{H}_{\underline{b}} \to \mathcal{H}_{\underline{b}'}$. Il existe un choix canonique d'un tel opérateur d'entrelacement. Nous le noterons $F_{\underline{b}',\underline{b}}$. Il est caractérisé par la propriété suivante : il existe une mesure de positive invariante sur $B'/B \cap B'$ telle que, pour tout élément $y \in U$, et tout vecteur C^{∞}, α, de $\mathcal{H}_{\underline{b}}$, on ait

(1) $$(F_{\underline{b}',\underline{b}}\alpha)(y) = \int_{B'/B\cap B'} \alpha(yx)\chi_{\underline{b}}(x)dy$$

(voir Lion [21]).

V. 3. La représentation S_u :

Soit A un groupe d'automorphismes de U stabilisant u. On peut donc définir le revêtement $A^{\underline{u}}$ de A.

On choisit une polarisation réelle $\underline{b}$ en u. Soit $x \in A$. On définit un opérateur $A_{\underline{b}}(x)$ de $\mathcal{H}_{\underline{b}}$ dans $\mathcal{H}_{x\underline{b}}$ en posant, pour $\alpha \in \mathcal{H}_{\underline{b}}$ et $y \in U$:

(2) $$(A_{\underline{b}}(x)\alpha)(y) = \alpha(x^{-1}(y)).$$

On pose

(3) $$S'_{\underline{b}}(x) = \|A_{\underline{b}}(x)\|^{-1} F_{\underline{b},x\underline{b}} A_{\underline{b}}(x).$$

C'est un opérateur unitaire dans $\mathcal{H}_{\underline{b}}$, et l'on a

(4) $$S'_{\underline{b}}(x)T_{u,\underline{b}}(y)S'_{\underline{b}}(x)^{-1} = T_{u,\underline{b}}(x(y))$$

pour tout $y \in U$.

Soit $(x,\phi) \in A^{\underline{u}}$. On pose

(R) $$S_{\underline{b}}(x,\phi) = \phi(\underline{b})S'_{\underline{b}}(x)$$

Lemme 2 :

(i) $S_{\underline{b}}$ est une représentation unitaire de $A^{\underline{u}}$ dans $\mathcal{K}_{\underline{b}}$.

(ii) $S_{\underline{b}}$ ne dépend pas de $\underline{b}$, dans le sens suivant : si F entrelace $T_{u,\underline{b}}$ et $T_{u,\underline{b}'}$ alors F entrelace $S_{\underline{b}}$ et $S_{\underline{b}'}$.

Nous noterons S_u la représentation de $A^{\underline{u}}$ dans l'espace de T_u ainsi définie .

$$\text{(iii)}\quad S_u(x,\phi)T_u(y)S_u(x,\phi)^{-1} = T_u(x(y))$$

pour tout $y \in U$, $(x,\phi) \in A^{\underline{h}}$.

(La représentation S_u a été décrite de manière différente dans [9] . On trouvera dans [13] une bonne explication à l'existence d'une représentation de $A^{\underline{u}}$ vérifiant (iii). La présentation adoptée ici est due à Lion [21]).

V. 4. *Extension des représentations des groupes de Lie nilpotents* :

Dans ce paragraphe, G est un groupe de Lie d'algèbre de Lie $\underline{g}$, $\underline{u}$ est un idéal nilpotent G-invariant de $\underline{g}$, U le sous-groupe analytique de G d'algèbre de Lie $\underline{u}$, u un élément admissible de $\underline{u}^*$. On suppose U fermé. Nous noterons $\underline{h} = \underline{g}(u)$, $H = G(u)^{\underline{u}}$, $\underline{q} = \underline{u}(u) \cap \ker u$, Q le sous-groupe analytique d'algèbre de Lie $\underline{q}$. L'application $x \to (x,1)$ est un isomorphisme du groupe (connexe) $U(u)$ sur son image dans $G(u)^{\underline{u}}$, et Q est la composante neutre du noyau du caractère χ_u de $U(u)$ de différentielle $iu|\underline{u}(u)$.
Le groupe Q est donc fermé et invariant dans H .
On pose $\underline{g}_1 = \underline{h}/\underline{q}$, $G_1 = H/Q$.
Soit Γ un sous-groupe du centre de G. Soit η un caractère unitaire de Γ qui a même restriction à $\Gamma \cap U(u)$ que χ_u. On note Γ' l'image réciproque de Γ dans H, et on pose $\Gamma_1 = \Gamma' U(u)/Q$. C'est un sous-groupe de centre de G_1. On note η_1 le caractère de Γ_1 qui provient par passage au quotient de caractère η' de $\Gamma'U(u)/Q$ qui prolonge χ_u , et tel que

$\eta'(\gamma,\pm 1) = \pm\eta(\gamma)$ pour $\gamma \in \Gamma$.

Soit T_1 une représentation de G_1 dont la restriction à Γ_1 est un multiple de η_1. On définit une représentation, notée $T_1 \otimes S_u\, T_u$ du groupe $G(u)U$ de la manière suivante. Soient $x \in G(u)$ et $y \in U$.

Soit $(x,\psi) \in H$ un représentant de x. On note encore (x,ψ) son image dans G_1. On pose :

(6) $$(T_1 \otimes S_u\, T_u)(xy) = T_1(x,\psi) \otimes S_u(x,\psi)T_u(y).$$

On vérifie que cette définition ne dépend pas du choix de $x,y,(x,\psi)$, et que cela donne une représentation.

Lemme 3 :

i) Le stabilisateur de T_u ~~dans G~~ est le groupe $G(u)U$. Il est fermé.

ii) Soit T_1 une représentation de G_1 dont la restriction à Γ_1 est un multiple du caractère η_1. On pose

(7) $$T = \mathrm{Ind}_{G(u)U}^{G}(T_1 \otimes S_u\, T_u)$$

La restriction de T à Γ est un multiple de η. La restriction de T à U est portée par l'orbite (sous G) de T_u dans le dual unitaire de U. L'application $T_1 \to T$ est une bijection des ensembles de représentations de G_1 et G décrits ci-dessus. Cette bijection induit un isomorphisme des espaces d'opérateurs d'entrelacement et des commutants.

Démonstration :

Tout cela résulte des lemmes 1 et 2 et de la théorie de Mackey (cf [8]).

C.Q.F.D.

CHAPITRE 6 - CONSTRUCTION DES REPRESENTATIONS $T_{g,\tau}$

Dans ce chapitre, G est un groupe de Lie d'algèbre de Lie $\underline{g}$, Γ un sous-groupe du centre de G, η un caractère unitaire de Γ , $g \in \underline{g}^*$ un forme linéaire τ-admissible, $\tau \in X(g,\eta)$.

Nous allons construire les représentations $T_{g,\tau}$ par récurrence sur dim $\underline{g}$. Supposons d'abord dim $\underline{g} = 0$. Alors $g = 0$, $G(g) = G$, et l'on pose $T_{g,\tau}(x) = \tau(x,1)$ pour tout $x \in G$. Les propriétés (i) à (iiiii) de l'introduction sont vérifiées.

On suppose la construction faite pour tous les groupes de Lie de dimension strictement inférieure, de telle sorte que les propriétés (i) à (iiiii) soient vérifiées.

On note $\underline{u}$ le plus grand idéal nilpotent de $\underline{g}$. On pose $u = g|\underline{u}^*$. Le sous-groupe analytique U de G est fermé et invariant ; de plus u est admissible (ch. IV lemme 6). On emploie les notations $\underline{h}$, $\underline{g}_1$, G_1, Γ_1, η_1 , etc... du paragraphe V. 4. De plus, on pose $h = g|\underline{h}$, et on note g_1^* l'élément de $\underline{g}_1$ déduit de h par passage au quotient. Il résulte du chapitre IV, lemmes 1 et 2, que g_1 est bien polarisable. Il résulte du chapitre IV, lemmes 4 et 6 , que g_1 est η_1-admissible, et qu'il y a une bijection canonique $\tau \to \tau_1$ entre $X_{G,\Gamma}(g,\eta)$ et $X_{G_1,\Gamma_1}(g_1,\eta_1)$.

Nous allons considérer deux cas.

i) On suppose que l'on a : dim $\underline{g}$ = dim $\underline{g}_1$. Alors $\underline{u}$ est de dimension au plus 1, u est injectif, $\underline{g}$ est réductive de centre $\underline{u}$. On définit $T_{g,\tau}$ comme au chapitre III.

ii) On suppose que l'on a : dim $\underline{g}_1$ < dim $\underline{g}$. Alors l'hypothèse de récurrence nous permet de construire la représentation $T^{G_1}_{g_1,\tau_1}$ de G_1. D'après la propriété (iiiii), sa restriction à Γ_1 est le caractère . η_1. Le lemme 3, chapitre V, nous permet de poser :

(1) $$T^G_{g,\tau} = \mathrm{Ind}^G_{G(u)U}(T^{G_1}_{g_1,\tau_1} \otimes S_u\, T_u).$$

Remarque :

Si $\underline{g}$ est réductive, mais si u n'est pas injective, nous avons défini deux fois $T^G_{g,\tau}$: une fois dans le cas ii) ci-dessus, et une fois au chapitre III. On vérifie facilement que les deux définitions coïncident.

Théorème 1 :

Les représentations $T_{g,\tau}$ vérifient les propriétés (i) à (iiiii) de l'introduction.

Démonstration :

Les propriétés (i), (iiii) et (v) résultent immédiatement du lemme 8, chapitre III, dans le premier cas, de l'hypothèse de récurrence et du lemme 3 chapitre V dans le second cas.

Démontrons (iii). La restriction de $T_{g,\tau}$ à U est portée par l'orbite de T dans $\hat{U}$. Cela résulte de la construction de $T_{g,\tau}$ au chapitre III lorsque $\underline{g}$ est réductif (1er cas), et de la formule (1) sinon. Soient $g' \in \underline{g}^*$, $\tau' \in X(g')$, $u' = g'|\underline{u}$. Supposons $T_{g,\tau}$ et $T_{g',\tau'}$ non disjointes. Alors les restrictions à U sont non disjointes, et il en résulte que l'on a $u' \in Gu$. Quitte à remplacer g' par un conjugué, on peut supposer que l'on a : $u = u'$.

La définition de $\underline{g}_1$, G_1, etc... ne dépend que de u. Si $\underline{g} = \underline{g}_1$, (iii) résulte du lemme 8, chapitre III.

Si $\underline{g} \neq \underline{g}_1$, il résulte du lemme 3, chapitre V, que T_{g_1,τ_1} et $T_{g'_1,\tau'_1}$ sont non disjointes. L'hypothèse de récurrence montre que g_1 et g'_1 sont conjugués par G_1, et donc par $G(u)$.

Quitte à remplacer g' par un conjugué, on peut supposer que l'on a : $u = u'$, $g_1 = g_1'$. Il en résulte que g et g' ont même restriction à $\underline{g}(u) + \underline{u}$. Il résulte du lemme 5, chapitre 4, que g et g' sont dans la même G-orbite. C.Q.F.D.

CHAPITRE VII - APPLICATIONS

Comme je l'ai dit dans l'introduction, je ne donne ici que des applications simples de la construction de représentations $T_{g,\tau}$. Elles sont de deux ordres : une classification de l'ensemble des représentations unitaires irréductibles d'un groupe de Lie moyennable de type I, et une application à la représentation régulière des groupes de Lie localement algébriques connexes, avec en particulier la classification des représentations irréductibles de carré intégrable.

VII. 1. Groupes de Lie moyennables :

Dans ce paragraphe, on suppose que G est moyennable. Ceci est équivalent aux deux conditions suivantes : G/G_0 est moyennable, et le radical résoluble de G est cocompact.

Les résultats de ce paragraphe sont essentiellement contenus dans l'article de Pukanszky [26] .

Théorème 1 :

On suppose G moyennable. Soit I un idéal primitif de la C^*-algèbre de G. Il existe une forme linéaire admissible et bien polarisable $g \in \underline{g}^*$, et $\tau \in X^{irr}(g)$ tels que I soit le noyau de $T_{g,\tau}$ dans $C^*(G)$.

Démonstration :

On raisonne par récurrence sur la dimension de G. Lorsque la dimension de G est nulle, le résultat est évident. On suppose ci-dessous que la dimension de $\underline{g}$ est strictement positive et que le résultat est établi pour tous les groupes de Lie moyennables de dimension strictement inférieure. Soit $\underline{u}$ le plus grand idéal nilpotent de $\underline{g}$, et soit U le sous-groupe analytique correspondant de G. D'après le théorème 4.3. de Gootman et Rosenberg [11] , il existe une représentation unitaire irréductible T de G, de noyau I, dont la restriction à U est portée par une quasi-orbite transitive de G dans le dual unitaire de U, car U est de type I et car G/U est moyennable. Soit $u \in \underline{u}^*$ un élément admissible tel que l'orbite de la représentation T_u de U porte la restriction de T à U . D'après le lemme 3, par. V.4., avec les notations de ce lemme, et posant $\Gamma = \{1\}$, il existe une représentation unitaire irréductible T_1 de G_1 , dont la restriction à Γ_1 est η_1 , et telle que l'on ait :

$$T = \mathrm{Ind}_{G(u)U}^{G} (T_1 \otimes S_u T_u) .$$

Comme dans le chapitre VI, on considère deux cas.

1er cas :

$\dim \underline{g}_1 = \dim \underline{g}$. Alors $\underline{g}$ est réductive de centre $\underline{u}$, et $\underline{g}/\underline{u}$ est semi-simple compacte. Comme le groupe des automorphismes de $\underline{g}/\underline{u}$ est fini, modulo les automorphismes intérieurs, la restriction de T à G_0 est portée par l'orbite sous G d'une représentation unitaire irréductible de G_0 . Cette représentation de G_0 est de la forme $T_g^{G_0}$ (où g est une forme linéaire admissible bien polarisable sur $\underline{g}$, cf. par. III.1., exemple). Il résulte du par. III. 5. qu'il existe $\tau \in X^{irr}(g)$ tel que l'on ait $T = T_{g,\tau}$, et donc I est le noyau de $T_{g,\tau}$.

2ème cas :

$\dim \underline{g}_1 < \dim \underline{g}$. Comme G_1 est moyennable (comme quotient d'un sous-groupe fermé de G) il existe un élément admissible bien polarisable $g_1 \in \underline{g}_1^*$ et $\tau_1 \in X^{irr}(g_1)$ tels que T_1 et $T_{g_1,\tau_1}^{G_1}$ aient même noyau dans la C^*-algèbre de G_1. Comme la restriction de T_1 à Γ_1 est le caractère η_1, il en est de même de T_{g_1,τ_1}, ce qui implique que l'on a $\tau_1 \in X^{irr}(g_1,\eta_1)$. Il existe donc un élément admissible et bien polarisable $g \in \underline{g}^*$, et $\tau \in X^{irr}(g)$ tels que les éléments G_1, g_1, τ_1 définis ci-dessus soient les mêmes que ceux qui sont définis au chapitre VI à partir de g et de τ. Par définition, on a (cf. chapitre VI) :

$$T_{g,\tau} = \mathrm{Ind}_{G(u)U}^{G} (T_{g_1,\tau_1}^{G_1} \otimes S_u T_u) .$$

Comme les représentations T_1 et T_{g_1,τ_1} ont même noyau, il en est de même de T et $T_{g,\tau}$, d'après les théorèmes de continuité de Fell. C.Q.F.D.

Corollaire :

On suppose G moyennable. Soit T une représentation unitaire irréductible normale de G. Il existe une forme linéaire admissible bien polarisable $g \in \underline{g}^*$ et $\tau \in X^{irr}(g)$ tels que $T = T_{g,\tau}$.

Démonstration :

En effet, une représentation unitaire irréductible de G ayant même noyau que T dans la C^*-algèbre de G est équivalente à G. C.Q.F.D.

Remarque 1 :

Ce corollaire, joint au théorème 1 du chapitre VI, donne une paramétrisation du dual unitaire des groupes de Lie moyennables de type I.

Remarque 2 :

Pukanszky [26] a obtenu ces résultats dans le cas des groupes connexes. La démonstration donnée ici n'est pas fondamentalement différente, mais l'emploi du théorème "marteau pilon" de [11] permet un exposé plus simple et un énoncé plus général. Notons que la paramétrisation du dual unitaire et la définition de l'admissibilité d'une forme linéaire, sont différentes dans [26].

Remarque 3 :

Lorsque G est connexe et moyennable - ou plus généralement lorsque G vérifie une certaine hypothèse (H) définie par Charbonnel et Khalgui [6] , on peut réaliser toutes les représentations $T_{g,\tau}$ comme "représentation holomorphes induites". C'est-à-dire qu'on peut trouver une polarisation $\underline{b}$ en g telle que $T_{g,\tau}$ soit réalisée, par translations à gauche, dans un sous-espace approprié de l'espace des fonctions vérifiant les relations (9) et (10) du paragraphe II.4. (Ceci est implicite dans [16] et dans [23]). On voit donc que toutes les représentations unitaires irréductibles normales d'un groupe moyennable connexe (ou plus généralement de type (H)) sont obtenues pas induction holomorphe (ceci est bien connu pour les groupes résolubles connexes - cf [2] et [24], et est démontré pour les groupes moyennables connexes simplement connexes dans [23]).

Remarque 4 :

Supposons G moyennable et connexe. Soit $g \in \underline{g}^*$ un élément admissible et bien polarisable, et soit $\tau \in X^{irr}(g)$. Pukanszky a démontré [26] que la représentation irréductible $T_{g,\tau}$ est normale si et seulement si G_g est un sous-ensemble localement fermé de $\underline{g}^*$ et si τ est dimension finie. Notons aussi que, dans le cas général, une application immédiate des résultats de Thoma [33] donne des conditions nécessaires pour que G soit de type I en fonction de la structure de G(g) , pour chaque $g \in \underline{g}^*$ admissible et bien polarisable.

VII. 2. Groupes de Lie algébriques :

Nous nous intéressons dans ce paragraphe aux groupes de Lie connexes localement algébriques, (c'est-à-dire localement isomorphes au groupe des points réels d'un groupe algébrique affine défini sur $\mathbb{R}$).
Pour les nécessités de la récurrence, nous considérerons une classe plus vaste de groupes.

Définition :

Un groupe de Lie G est dit presque algébrique s'il vérifie les conditions suivantes

i. Il est localement algébrique.

ii. Il existe un sous-groupe G' de G d'indice fini tel que, notant $Z_{G'}$ le centre de G', on ait $Z_{G'} G_0 = G'$.

Nous aurons besoin des lemmes suivants. On note $\underline{u}$ le plus grand idéal nilpotent de $\underline{g}$ et U le sous-groupe analytique correspondant.

Lemme 1 : Soit G un groupe de Lie presque algébrique. Alors G est de type I, et G opère régulièrement dans U (i.e. les orbites de G dans le dual unitaire de U sont localement fermées).

Démonstration :

Les deux résultats se déduisent immédiatement du cas où G est connexe, où ils sont dus à Dixmier et Pukanszky (cf. e.g. [26] p. 85-86). C.Q.F.D.

Lemme 2 : Soit G un groupe de Lie presque algébrique. Soit $u \in \underline{u}^*$ un élement admissible. Le groupe G_1 défini au paragraphe V. 4. est presque algébrique.

Démonstration :

Il suffit visiblement de le démontrer quand G est connexe, et pour cela, de le démontrer quand G est connexe et simplement connexe. Soit Γ un sous-groupe discret du centre de G tel que G/Γ soit égal à la composante neutre d'un groupe algébrique. Alors $G(u)/\Gamma$ a un nombre fini de composantes connexes. Donc $\Gamma G(u)_0$ est d'indice fini dans $G(u)$. D'autre part $G(u)$ est localement algébrique. On voit donc que $G(u)$ et $G(u)^{\underline{u}}$ sont presque algébriques. Rappelons que l'on a $G_1 = G(u)^{\underline{u}}/Q$, où Q est le sous-groupe analytique d'algèbre $\underline{q} = \underline{u}(u) \cap \ker u$. La condition ii/ est vérifiée pour G_1 . Pour montrer que G_1 est localement algébrique, il suffit de montrer que G est localement isomorphe à un groupe algébrique G tel que le sous-groupe analytique Q correspondant soit algébrique. Pour cela il suffit que le sous-groupe U correspondant à $\underline{u}$ soit unipotent, car tout sous-groupe de Lie connexe d'un groupe unipotent est algébrique. Soit donc G un groupe algébrique localement isomorphe à G. Soit Z le centre de G, et soit T le maximal de Z. Alors G est localement isomorphe à $G/T \times T$. Remplaçant T par un groupe unipotent abélien de même dimension, on voit que G est localement isomorphe à un groupe algébrique G tel que la composante neutre Z_0 de Z soit unipotente. Alors Z_0 est contenu dans U, et U/Z_0 est unipotent, donc U est unipotent. C.Q.F.D.

Dans la suite, G est un groupe de Lie presque algébrique, Γ est un sous groupe fermé du centre de G, et η un caractère unitaire de Γ . On note $\hat{G}_\eta$ l'ensemble des classes de représentations unitaires irréductibles de G dont la restriction à Γ est le caractère η . Comme G est de type I, il existe une unique classe μ_η de mesures boréliennes sur l'espace borélien standard $\hat{G}_\eta$ tel que la représentation $\operatorname{Ind}_\Gamma^G(\eta)$ soit quasi-équivalente à la représentation $\int_{\hat{G}_\eta} T \, d\mu_\eta(T)$. On dira que μ_η est la classe de la mesure de Plancherel de $\hat{G}_\eta$.

Théorème 2 :

Le complémentaire dans $\hat{G}_\eta$ de l'ensemble des représentations de la forme $T_{g,\tau}$, où g parcourt l'ensemble des formes η-admissibles bien polarisables, et τ l'ensemble $X^{irr}(g,\eta)$, est de mesure nulle pour μ_η.

Démonstration :

On raisonne par récurrence sur la dimension de G. Le résultat est évident pour les groupes discrets. On suppose la dimension de G strictement positive, et le résultat établi pour les groupes presque algébriques de dimension inférieure. Notons $\hat{U}_\eta$ l'ensemble des classes de représentations unitaires irréductibles de U dont la restriction à $\Gamma \cap U$ est η. Notons ν une mesure de volume fini dans la classe de Plancherel de $\hat{U}_\eta$, et notons $\bar{\nu}$ l'image de ν sur $\hat{U}_\eta/G$. On choisit une section borélienne s de $\hat{U}_\eta/G$ dans $\underline{u}^*$ (de sorte que pour toute orbite ω de G dans $\hat{U}_\eta$, on a : $T_{s(\omega)} \in \omega$). Pour chaque $u \in \underline{u}^*$, admissible pour la restriction de η à $\Gamma \cap U$, on définit G_1, Γ_1, η_1 comme au paragraphe V. 4. Le groupe Γ_1 est fermé. En effet, il suffit de voir que $\Gamma U(u)$ est fermé dans $G(u)$. Comme $\Gamma \cap U = \Gamma \cap U(u)$, il suffit de voir que ΓU est fermé dans G. Il suffit de le faire quand G est connexe et simplement connexe. Comme U contient la composante neutre du centre Z de G il suffit de prouver que ZU est fermé. Divisant par un sous-groupe de Z, il suffit de le montrer quand G est un ouvert d'un groupe algébrique. Alors ZU/U est fini et le résultat est clair.

Notons μ_{η_1} la mesure de Plancherel correspondante sur $\hat{G}_{1,\eta_1}$. On peut former la représentation R^u de G définie par la formule :

$$R^u = \int \mathrm{Ind}_{G(u)U}^{G} (\pi \otimes S_u T_u) d\mu_{\eta_1}(\pi).$$

Pour tout $\omega \in \hat{U}_\eta/G$, notant $u = s(\omega)$, nous poserons $R^\omega = R^u$.

Compte-tenu du lemme 1, une légère généralisation de Kleppner et Lipsman [18] (qui considèrent le cas $\Gamma = \{1\}$) montre que la représentation $\mathrm{Ind}_{\Gamma}^{G}(\eta)$ est quasi-équivalente à la représentation

$$\int_{\hat{U}_\eta/G} R^{\omega}\, d\bar{\nu}(\omega).$$

Pour démontrer le théorème, il suffit donc de démontrer que pour tout $u \in \underline{u}^*$ comme ci-dessus, l'ensemble des représentations de la forme T_{g_1,τ_1}, avec $\tau_1 \in X^{irr}(g_1,\eta_1)$ a un complémentaire dans $\hat{G}_{\eta_1}$, qui est de mesure nulle pour μ_{η_1}. En effet, si l'on choisit $g \in \underline{g}^*$ et $\tau \in X^{irr}(g, \ )$ tels que les éléments g_1 et τ_1 correspondant définis au chapitre VII soient égaux aux éléments g_1 et τ_1 ci-dessus, on a, d'après le chapitre VI :

$$\mathrm{Ind}_{G(u)U}^{G}(T_{g_1,\tau_1} \otimes S_u\, T_u) = T_{g,\tau}$$

La représentation $\mathrm{Ind}_{\Gamma}^{G}(\eta)$ est donc désintégrée en représentations $T_{g,\tau}$, ce qui prouve notre assertion.

Nous considérons deux cas.

1er cas : $\dim \underline{g}_1 = \dim \underline{g}$. Alors $\underline{g}$ est réductive de centre $\underline{u}$.

Pour le groupe G_0 le théorème 2 est vrai : il est dû à Harish-Chandra [12] lorsque $[G_0,G_0]$ est de centre fini et a été étendu au cas général par Wolf [38].

Appliquant comme ci-dessus le résultat de Kleppner et Lipoman [18], on est ramené à l'assertion suivante : soit $g \in \underline{g}^*$ un élément η-admissible, et soit $\sigma \in X^{irr}_{G_0,\Gamma\cap G}(g,\eta')$ (où η' est la restriction de η à $\Gamma \cap G_0$).

Toutes les représentations unitaires irréductibles de G dont la restriction à G_0 est portée par l'orbite sous G de la représentation $T^{G_0}_{g,\sigma}$ sont de la forme $T^{G}_{g,\tau}$. Ceci résulte du paragraphe III. 8.

2ème cas : dim $\underline{g}_1$ < dim $\underline{g}$. On applique l'hypothèse de récurrence (grâce au lemme 2). C.Q.F.D.

On trouvera dans l'appendice des idées qui devraient être utiles pour décrire précisément la classe de la mesure de Plancherel. J'espère revenir sur cette question. Je donne ci-dessous un résultat partiel, généralisant des résultats bien connus de Harish-Chandra (pour les groupes semi-simples), Moore et Wolf (pour les groupes nilpotents) et Charbonnel (pour les groupes résolubles).

Théorème 3 : (Les notations sont celles du théorème 2).

Une représentation unitaire irréductible T de G intervient discrètement dans $\mathrm{Ind}_\Gamma^G(\eta)$ si et seulement s'il existe un élément η-admissible bien polarisable $g \in \underline{g}^*$ tel que $G(g)/\Gamma$ soit compact, et $\tau \in X^{irr}(g,\eta)$, tels que T soit isomorphe à $T_{g,\tau}$.

Démonstration : D'après le théorème 2, il suffit de démontrer l'assertion suivante. Soit $g \in \underline{g}^*$ un élément η-admissible bien polarisable. Soit $\tau \in X^{irr}(g,\eta)$. Alors $T_{g,\tau}$ intervient discrètement dans $\mathrm{Ind}_\Gamma^G(\eta)$ si et seulement si $G(g)/\Gamma$ est compact.

Nous démontrons cette assertion par récurrence sur la dimension de G. Elle est claire pour les groupes de dimension 0 (rappelons que, par hypothèse, ceux-ci ont un sous-groupe abélien d'indice fini). On suppose donc la dimension de G strictement positive, et le résultat établi pour tous les groupes presque algébriques de dimension inférieure. On pose $u = g|\underline{u}$, et on emploie les notations du chapitre VI. On considère deux cas.

1er cas : dim $\underline{g}_1$ = dim $\underline{g}$. Appliquant la méthode de Kleppner et Lipsman comme dans la démonstration du théorème 2, on voit que $T_{g,\tau}$ intervient discrètement dans $\mathrm{Ind}_\Gamma^G(\eta)$ si et seulement si les deux conditions suivantes sont réunies.

i. Soit $T_{g,\sigma}^{G_0}$ une représentation irréductible de G_0 dont le G-orbite porte la restriction de $T_{g,\tau}$ à G_0 (voir paragraphe III.8). Alors $T_{g,\sigma}^{G_0}$ intervient discrètement dans $\mathrm{Ind}_{\Gamma\cap G_0}^{G_0}(\eta')$, où η' est la restriction de η à $\Gamma \cap G_0$.

Cela implique, d'après Harish-Chandra (cf [37]), que $G_0(g)/\Gamma\cap G_0(g)$ est compact.

On suppose donc que $G_0(g)/\Gamma\cap G_0(g)$ est compact. Cela entraîne que $G_0(g)$ est connexe (et donc σ est l'unique élément de $X_G^{irr}(g)$).

ii. La représentation τ intervient discrètement dans la représentation de $G(g)^{\underline{g}}$ induite par la représentation de $\Gamma(G(g)_0)^{\underline{g}}$ dont la restriction à Γ est le caractère η, et la restriction à $(G(g)_0)^{\underline{g}}$ le caractère χ_g.

Comme $G(g)^{\underline{g}}$ contient un sous-groupe abélien d'indice fini, on voit que cela implique que le groupe $G(g)/\Gamma G(g)_0$ est compact.

Les conditions i. et ii. ensemble sont équivalentes à la compacité de $G(g)/\Gamma$.

2ème cas : $\dim \underline{g}_1 < \dim \underline{g}$.

Il résulte de la démonstration du théorème 2 que $T_{g,\tau}$ intervient discrètement dans $\mathrm{Ind}_\Gamma^G(\eta)$ si et seulement si les deux conditions suivantes sont réalisées.

i. Soit ω l'orbite de T_u dans $\hat{U}_\eta$. Alors on a : $\bar{\nu}(\{\omega\}) > 0$.

ii. T_{g_1,τ_1} intervient discrètement dans $\mathrm{Ind}_{\Gamma_1}^{G_1}(\eta_1)$.

Etudions la condition i. Soit D le sous-groupe de U, connexe, algébrique, engendré par $\Gamma \cap U$, et soit $\underline{d}$ son algèbre de Lie. Alors $D/\Gamma \cap U$ est compact, et l'ensemble des formes linéaires sur $\underline{u}$ qui sont admissibles pour la restriction de η à $\Gamma \cap U$ est une réunion localement finie de sous-espaces affines $\underline{u}_j^*$ (indexée par les caractères de $D/\Gamma\cap U$) parallèles à l'orthogonal $\underline{d}^\perp$ de $\underline{d}$ dans $\underline{u}^*$. On en déduit une partition de $\hat{U}_\eta$ en sous-ensembles isomorphes (pour la structure borélienne) à $\underline{u}_j^*/U$, et la mesure de Plancherel sur $\underline{u}_j^*/U$ est équivalente à l'image d'une mesure finie sur $\underline{u}_j^*$ équivalente à la mesure de Lebesgue.

On voit donc que la condition i. est équivalente à la condition i'.

i'. Le sous-ensemble $G_0 u$ est ouvert dans $u + \underline{d}^\perp$.

Cette condition est encore équivalente à l'égalité $\underline{g}u = \underline{d}^\perp$, et, passant à l'orthogonal, à la condition $(\underline{g}u)^\perp = \underline{d}$. Mais on a $(\underline{g}u)^\perp = \underline{u}(g)$. La condition i. est donc équivalente à la condition: i". Le groupe $U(g)/\Gamma\cap U(g)$ est compact.

Etudions la condition ii. Par l'hypothèse de récurrence, elle est équivalente à la compacité de $G(g_1)/\Gamma_1$. D'après le lemme 5, paragraphe IV. 2., ceci est équivalent, à la compacité du groupe $G(g)U(u)/\Gamma U(u) = G(g)/\Gamma U(g)$.

Donc i. et ii. ensemble sont équivalentes à la compacité de $G(g)/\Gamma$. C.Q.F.D.

Remarque :

Les groupes presque algébriques unimodulaires pour lesquels il existe des représentations unitaires irréductibles intervenant discrètement dans $\mathrm{Ind}_\Gamma^G(\eta)$ ont une structure très particulière (voir l'article d'Anh cité dans l'appendice).

BIBLIOGRAPHIE

[1] M. ANDLER. Sur des représentations construites par la méthode des orbites. C. R. Acad. Sc. Paris 290 (1980) 873-875.

[2] L. AUSLANDER et B. KOSTANT. Polarization and unitary representations of solvable groups. Invent. Math. 14 (1971) 255-354.

[3] P. BERNAT et al. Représentations des groupes de Lie résolubles. Dunod, Paris 1972.

[4] J. CARMONA. Représentations du groupe de Heisenberg dans les espaces de (0,q)-formes. Math. Ann. 205 (1973) 89-112.

[5] W. CASSELMAN-M. S. OSBORNE. The n-cohomology of representations with an infinitesimal character. Compositio Math. 31 (1975) 219-227.

[6] J. Y. CHARBONNEL et M. S. KHALGUI. Polarisations pour un certain type de groupes de Lie. C. R. Acad. Sc. Paris 287 (1978) 915-917.

[7] J. DIXMIER. Algèbres enveloppantes. Gauthier-Villars, Paris 1974.

[8] M. DUFLO. Sur les extensions des représentations irréductibles des groupes de Lie nilpotents. Ann. Sc. Ecole Norm. Sup. 5 (1972) 71-120.

[9] M. DUFLO. Représentations de carré intégrable des groupes semi-simples réels. Sem. Bourbaki exp. 508, 1977-1978.

[10] T. J. ENRIGHT. On the fundamental series of a real semi-simple Lie algebra. Their irreducibility, resolutions and multiplicity formulae. Annals of Math. 110 (1979) 1-82.

[11] E. C. GOOTMAN et J. ROSENBERG. The structure of Crossed product C^*-algebras : a proof of the generalized Effros-Hahn conjecture. Invent. Math. 52 (1979) 283-298.

[12] HARISH-CHANDRA. Harmonic analysis on real reductive groups. Annals of Math. 104 (1976) 117-201.

[13] R. HOWE. On the character of Weil's representation. Trans. Amer. Math. Soc. 177 (1975) 287-298.

[14] N. E. HURT. Proof of an analogue of a conjecture of Langlands for the Heisenberg-Weyl group. Bull. London Math. Soc. 4 (1972) 127-129.

[15] M. KASHIWARA et M. VERGNE. On the Segal-Shale-Weil Representations and Harmonic Polynomials. Invent. Math. 44 (1978) 1-47.

[16] M. S. KHALGUI. Sur les caractères des groupes de Lie à radical cocompact. Preprint 1980.

[17] A. A. KIRILLOV. Représentations unitaires des groupes de Lie nilpotents. Uspekhi Mat. Nauk . 17 (1962) 57-110.

[18] A. KLEPPNER et R. L. LIPSMAN. The Plancherel formula for group extensionsII. Ann. Sci. Ec. Norm. Sup. 6 (1973) 103-132.

[19] B. KOSTANT. Lie algebra cohomology and the generalized Borel-Weil theorem. Ann. of Math. 74 (1961) 329-387.

[20] G. LION. Indices de Maslov et représentation de Weil. Publ. Université Paris 7, N° 2, 1978.

[21] G. LION. Extension de représentations de groupes de Lie nilpotents. C. R. Acad. Sc. Paris 288 (1979) 615-618.

[22] G. LION et M. VERGNE. The Weil representation, Maslov index and theta series. Birkhäuser , Boston 1980.

[23] R. L. LIPSMAN. Orbit theory and representations of Lie groups with co-compact radical. Preprint, Maryland 1980.

[24] L. PUKANSZKY. Unitary representations of solvable Lie groups. Ann. Sc. E.N.S. 4 (1971) 435-491.

[25] L. PUKANSZKY. Characters of connected Lie groups. Acta Mathematica 133 (1974) 81-137.

[26] L. PUKANSZKY. Unitary representations of Lie groups with co-compact radical and applications. Trans. Amer. Math. Soc. 236 (1978) 1-50.

[27] I. SATAKE. Unitary representations of a semi-direct product of Lie groups on $\bar{\partial}$-cohomology spaces. Math. Ann. 190 (1971) 177-202.

[28] W. SCHMID. Some properties of square integrable representations of semisimple Lie groups. Ann. of Math. 102 (1975) 535-564.

[29] W. SCHMID. L^2-cohomology and the discrete series. Annals of Math. 103 (1976) 375-394.

[30] D. SHALE. Linear symmetries of free boson fields. Amer. Math. Soc. 103 (1962) 149-167.

[31] J.M. SOURIAU. Construction explicite de l'indice de Maslov et applications. Fourth international colloquium on group theoritical methods in physics, University of Nijmegen, 1975.

[32] B. SPEH et D. VOGAN. Reducibility of generalized principal series representations , Preprint 1978.

[33] E. THOMA. Uber unitare Darstellungen abzahlarer diskreter Gruppen. Math. Ann. 153 (1964) 111-138.

[34] M. VERGNE. Construction de sous-algèbres subordonnées à un élément du dual d'une algèbre de Lie résoluble. C. R. Acad. Sc. Paris 270 (1970) 173-175 et 704-707.

[35] D. VOGAN. The algebraic structure of representations of semi-simple Lie groups. I Ann. of Math. 109 (1979) 1-60. II. Preprint 1977.

[36] D. VOGAN. Irreducible characters of semi-simple Lie groups II. Duke Math. Journal 46 (1979) 805-859.

[37] G. WARNER. Harmonic analysis on semi-simple Lie groups . Springer-Verlag, New-York 1972.

[38] J. WOLF. The action of a real semi-simple Lie group on a complex manifold II. Unitary representations on partially holomorphic cohomology spaces. Mem. Amer. Math. Soc. 138 (1974).

APPENDICE : PARAMETRIZATION OF THE SET OF REGULAR ORBITS OF THE COADJOINT REPRESENTATION OF A LIE GROUP

(Un texte préparé pour une conférence à l'Université de Maryland, décembre 1978).

We consider only algebraic groups and algebras. It is likely that all the results can be formulated for non algebraic groups as well, but this introduces various kinds of complications which I prefer to avoid in these lectures. I shall discuss theses complications in the last paragraph.

I. *Lie algebras over an algebraicaly closed field k of characteristic 0 :*

Let k be as above, G an affine algebraic group defined over k, $\underline{g}$ its Lie algebra, $\underline{g}^*$ the dual space of $\underline{g}$. If $g \in \underline{g}^*$ we denote by $G(g)$ the stabilizer of g and by $\underline{g}(g)$ its Lie algebra. If H is an algebraic group we denote by H_0 the connected component of 1 in H.

Suppose $g \in \underline{g}^*$ is regular (i.e. the orbit G is of maximum dimension). Then $G(g)_0$ is commutative [6]. We denote by $S(g)$ its maximal torus. We say that g is strongly regular if g is regular and $S(g)$ of maximum dimension. We denote by $\underline{g}^*_s$ the set of strongly regular elements. It follows from [11] (or it can be proved in an elementary way) that $\underline{g}^*_s$ is an open set, and that if g and g' are strongly regular, $S(g)$ and $S(g')$ are conjugate (even by an element of (G_0, G_0)).

Let us fix a torus S with Lie algebra $\underline{s}$ such that $S(g)$ is conjugate to S for all $g \in \underline{g}^*_s$. We use the following notations : H = centralizer of S in G, H' = normalizer of S in G, $\underline{h}$ = Lie algebra of both H and H', $W = H'/H$, $\underline{p} = [\underline{s},\underline{g}]$. It is clear that $\underline{g} = \underline{h} \oplus \underline{p}$. We identify $\underline{g}^*$ and $\underline{h}^* \oplus \underline{p}^*$.

Lemma 1 :

(i) Every G-orbit in $\underline{g}_s^*$ intersects $\underline{h}$.

(ii) In $\underline{h}^*$, every regular element is strongly regular. The set $\underline{h}^* \cap \underline{g}_s^*$ is equal to the set of regular elements $h \in \underline{h}^*$ such that $\pi_\omega(h) \neq 0$, where π_ω is an homogeneous polynomial on $\underline{h}^*$, defined below.

(iii) There is a natural bijection of $\underline{g}_s^*/G$ onto $(\underline{h}^* \cap \underline{g}_s^*)/H'$.

Corollary :

$k(\underline{g}^*)^G$ is isomorphic to $k(\underline{h}^*)^{H'}$.

Definition of π_ω *:*

The dimension of $\underline{p}$ is even, say 2d. We choose a non zero exterior 2d-form ω on $\underline{p}$. On $\underline{p}$, we consider for every $h \in \underline{h}^*$ the 2-forms B_h such that $B_h(X,Y) = h([X,Y])$ $(X,Y \in \underline{p})$. We define $\pi_\omega(h)$ by the formula

$$(d!)^{-1} B_h \wedge \ldots \wedge B_h = \pi_\omega(h)\omega$$ (there are d factors B_h, so that $\pi_\omega(h)$ is the Pfaffian).

Example 1 :

We suppose G semi-simple and connected. Then $g \in \underline{g}^*$ is strongly regular if and only if it is regular, and semi-simple (when identified to an element of $\underline{g}$ using the Killing form). This is the case if and only if G(g) is a Cartan subgroup of G. Here W is the Weyl group, and the corollary read : $k(\underline{g}^*)^G$ $k(\underline{h}^*)^W$, which is due to Chevalley. Note that the most interesting part in Chevalley's theorem is the isomorphism $k[\underline{g}^*]^G$ $k[\underline{h}^*]^W$, but this does not generalize in the situation of the corollary.

Remark 1 :

Let us say that $g \in \underline{g}^*$ is <u>very regular</u> if there exist $\phi_1,\ldots,\phi_p$, elements of $k(\underline{g}^*)^G$, defined at g , whose differentials at g are linearly independant, with $p = \dim G(g)$. The set of very regular elements is open, and non empty by a result of Rosenlicht. It is contained in the set of regular elements. Moreover, if g is very regular, $G(g)$ centralizes $G(g)_0$ [6] . Thus, for the very regular elements in $\underline{h}^* \cap \underline{g}_s^*$, we have $G(g) \subset H$. This proves that $k(\underline{h}^*)^H$ is an extension of $k(\underline{h}^*)^{H'}$ with Galois group W.

Remark 2 :

If g is very regular, the set $\Sigma(g)$ of semi-simple elements of $G(g)$ is a subgroup. Consider the set of $g \in \underline{g}^*$ which are very regular, strongly regular, and such that $G(g)$ has the maximum number of connected components (among very regular and strongly regular elements). Then this set is open, and for g in this set the $\Sigma(g)$ are mutually conjugate in G. This follows from [11]or it can be proved in an elementary way.

Remark 3 :

If G is connected, so is H . When G is connected, $k(\underline{g}^*)^G$ is called by Dixmier the heart of $\underline{g}$. Thus the heart of $\underline{h}$ is an extension of the heart of $\underline{g}$ with Galois group W. If moreover G is solvable, W = 1.

Remark 4 :

Corresponding to the restriction mapping from $k[\underline{g}^*]^G$ into $k[\underline{h}^*]^{H'}$, one can define a "Harish-Chandra mapping" from $U(\underline{g})^G$ into $U(\underline{h})^{H'}$. This involves the choice of a "system of positive roots" of $\underline{g}$ in $\underline{s}^*$, but the result is independant of this choice.

Example 2 :

G is the semi-direct product of $k^* \times k^*$ by a unipotent group whose Lie algebra has a basis x, y, z, w with brackets $[x,y] = z$. The action of $(t,u) \in k^* \times k^*$ is given by : $(t,u)x = tux$, $(t,u)y = t^{-1}uy$, $(t,u)z = u^2z$, $(t,u)w = u^{-4}w$. Let a,b be the canonical basis of the Lie algebra of $k^* \times k^*$. We can choose $\underline{s} = k\,a$. Then $\underline{h} = k\,a + kb + kz + kw$. Note that $\underline{g}$ is unimodular, but that $\underline{h}$ is not. For generic g , $\Sigma(g)$ is conjugate to the subgroup $k^* \times \{\pm 1\}$ of $k^* \times k^*$.

2. *Real Lie algebras :*

a. *Strongly regular orbits :*

We consider a Lie group G with Lie algebra $\underline{g}$. We denote by $\underline{g}_{\underline{\mathbb{C}}}$ the complexified Lie algebra. We assume that G has a finite number of connected components, and that G_0 ; the connected component of 1, is the analytic subgroup with Lie algebra $\underline{g}$ of an algebraic group $G_{\underline{\mathbb{C}}}$ with Lie algebra $\underline{g}_{\underline{\mathbb{C}}}$.

Lemma 2 :

Let g and g' be in the same connected component (for the Hausdorff topology) of $\underline{g}^*_{\underline{\mathbb{C}},s} \cap \underline{g}^*$. Then S(g) and S(g') are conjugate by an element of (G_0, G_0).

We denote by T(g) the maximal compact subgroup of $G(g)_0$, when g is regular in $\underline{g}^*$ (recall that $G(g)_0$ is commutative). From lemma 2, we get :

Theorem 1 :

There exist a finite number $T_1, \ldots, T_q$ of connected compact abelian subgroups of G such that if $g \in \underline{g}^*_{\underline{\mathbb{C}},s} \cap \underline{g}^*$,T(g) is conjugate in G to one and only one T_j .

Consider one T_j . We introduce some notations : $\underline{t}_j$ is the Lie algebra of T_j , $H_j = Z_G(T_j)$, $H_j' = N_G(T_j)$, $\underline{p}_j = [\underline{t}_j, \underline{g}]$, $\underline{g}_j^*$ is the set of $g \in \underline{g}^*_{\underline{\mathbb{C}},s} \cap \underline{g}^*$ such that T(g) is conjugate to T_j, $\underline{h}_j$ the Lie algebra of

H_j .

We choose Haar measures on G and on H_j . There correspond Haar measures on $\underline{g}$ and on $\underline{h}_j$. On $\underline{p}_j$ we choose the quotient measure, which is defined by a differential form ω_j . As above we define an homogeneous polonomial $\pi_j = \pi_{\omega_j}$ on $\underline{h}_j^*$, and we identify $\underline{g}^*$ and $\underline{h}_j^* \oplus \underline{p}_j^*$.

Lemma 3 :

(i) The intersection of a G-orbit in $\underline{g}_j^*$ with $\underline{h}_j^*$ is a non-empty $h_j^!$-orbit. Thus $\underline{g}_j^*/\Gamma$ is isomorphic to $(\underline{h}_j^* \cap \underline{g}_j^*)/H_j^!$.

(ii) $\underline{g}_j^* \cap \underline{h}_j^*$ is the set of $h \in \underline{h}_j^*$ which are strongly regular, such that $\pi_j(h) \neq 0$, and $T(h) = T_j$.

Remark 1 :

$\underline{g}_j^* \cap \underline{h}_j^*$ is the union of a certain number of connected components (for the Hausdorff topology) of a Zariski open subset of $\underline{h}_j^*$. If T_j is "fundamental" (i.e. the dimension of T_j is maximal), then it is Zariski open.

Remark 2 :

The case $q = 1$ is specially interseting. This happens for instance if $\underline{g}$ is solvable (cf. [3]), or more generally if the semi-simple part of $\underline{g}$ is compact. This happens also if $\underline{g}$ has a complex structure, or if $\underline{g}$ is reductive with one conjugacy class of Cartan subalgebra.

b. *Admissible orbits* :

Let $g \in \underline{g}^*$. Then $G(g)$ has a canonical two-fold covering group (*) $G(g)^{\underline{g}}$. We denote by $(G(g)^{\underline{g}})_0$ the inverse image in $G(g)^{\underline{g}}$ of $G(g)_0$, and

(*) défini au chapitre II.

by $(1, -1)$ the non trivial element of the kernel of $G(g)^{\underline{g}} \to G(g)$. We say that g is <u>admissible</u> if there is a character χ_g of $(G(g)_0)^{\underline{g}}$ with differential ig and such that $\chi_g(1, -1) = -1$. If it exists, such a character is unique and unitary. If g is admissible, we denote by $X^{irr}(g)$ the set of irreductible classes of unitary representations of $G(g)^{\underline{g}}$ whose restriction to $G(g)_0)^{\underline{g}}$ is a multiple of χ_g.

We denote by $\underline{g}^*_{aj}$ the set of admissible elements in $\underline{g}^*_j$. The purpose of these lectures is in fact the description of $\underline{g}^*_{aj}/G$. Let X be an element of $i\underline{t}_j$ such that $\alpha(X)$ is non zero for every non zero root α of $\underline{t}_{\underline{\underline{C}}}$ in $\underline{g}_{\underline{\underline{C}}}$ and let P be the set of non zero roots which are positive on X. We denote by ρ the element $\frac{1}{2} \sum_{\alpha \in P} (\dim \underline{g}^{\alpha}_{\underline{\underline{C}}})\alpha$, with evident notations. We denote by R_j the set of $h \in \underline{h}^*_j$ such that $\rho + ih|\underline{t}_j$ is the differential of a character of T_j. The set R_j does not depend on the choice of P, and is the union of affine spaces, translated from the orthogonal $\underline{t}^{\perp}_j$ of $\underline{t}_j$ in $\underline{h}^*_j$. The connected components of R_j are indexed by elements l of a lattice L_j. We denote by L'_j the set of $l \in L_j$ such that π_j is not zero on R^l_j. Let $l \in L'_j$. Then $R^l_j \cap \underline{g}_j$ is non empty, and an union of connected components (for the Hausdorff topology) of a Zariski open set of R^l_j.

Theorem 2 :

The set $\underline{g}^*_{aj}/G$ is isomorphic to $\{\bigcup_{l \in L'_j} (R^l_j \cap \underline{g}^*_j)/H_j\}$ $/W_j$ (where $W_j = H'_j/H_j$).

We use theorem 2 to describe a measure of $\underline{g}^*_{aj}/G$. To simplify, we suppose that $\underline{g}$ is unimodular (if not we have to use measures with values in a suitable line bundle, cf. [3]). Recall that we have chosen measures on G and on H_j. We put on T_j the normalized Haar measure, on $\underline{t}^{\perp}_j$ the corresponding Haar measure (for the duality given by $\exp(i < g, X >)$). On each R^l_j $(l \in L_j)$ we put the translated Haar measure. This gives a measure on R_j, and thus on the open subset $R_j \cap \underline{g}_j^*$. We multiply this measure by $(2\pi)^{-d}|\pi_j|$ (where $2d = \dim \underline{p}_j$ and $\pi = 3,14\ldots$). The result is a measure on $R_j \cap \underline{g}_j^*$ which is $H_j^!$-invariant, and depends only on the Haar measure on G. Each $H_j^!$-orbit in $\underline{h}_j^*$ has a canonical invariant measure (we choose the normalization described in [2] p. 20). On $\underline{g}_{aj}^* = (R_j \cap \underline{g}_j^*)/H_j^!$, we put the quotient measure, and we denote it by μ_j.

c. Plancherel formula :

The group G is as in a. Moreover we suppose it is unimodular, and provided with a Haar measure. Let $\underline{g}_a^*$ be the set of strongly regular admissible elements of $\underline{g}^*$. For each $\Omega \in \underline{g}_a^*/G$, we choose an element $g_\Omega \in \Omega$. Let $\hat{G}$ be the unitary dual of G.

Conjecture :

There exist an injective mapping $(g_\Omega, \tau) \to T_{g_\Omega,\tau}$ from $\bigcup_{\Omega \in \underline{g}_a^*/G} X^{irr}(g_\Omega)$ into $\hat{G}$, with the following properties.

(i) (Infinitesimal character). If $u \in U(\underline{g})$, and if $\hat{u}$ is the corresponding element of $k[\underline{g}^*]$ (under the isomorphism defined in [5]), then $T_{g_\Omega,\tau}(u)$ is a multiple of $\hat{u}(ig)$.

(ii) (Kirillov's character formula). Let $\Omega \in \underline{g}_a^*/G$ and $\tau \in X^{irr}(g\ )$ Then Ω is tempered, $T_{g_\Omega,\tau}$ is of trace class, and for $\phi \in C_c^\infty(\underline{g})$ with support in a sufficiently small neigborhood of 0, we have

$$\mathrm{tr}[T_{g_\Omega,\tau}(\phi)] = \dim \tau \int_\Omega (j\phi)^\wedge(g)\,d\beta_\Omega(g)$$

where $^\wedge$ is the Fourier transform, β_Ω the canonical measure on Ω, $T_{g_\Omega,\tau}(\phi) = \int \phi(X)\ \ T_{g_\Omega,\tau}(\exp X)dX$, and $j(X) = \left|\frac{d(\exp X)}{dX}\right|^{1/2}$.

(iii) (Plancherel formula). For each $j = 1,\ldots,p$ there exists a function ρ_j on $\bigcup_{\Omega \in g^*_{aj}/G} X^{irr}(g_\Omega)$ with values in $]0, \infty[$ such that

$$(1) \qquad \phi(1) = \sum_{j=1}^{p} \int_{g^*_{aj}/G} \sum_{\tau \in X^{irr}(g_\Omega)} [\mathrm{tr}\, T_{g_\Omega,\tau}(\phi)]\, \rho_j(g_\Omega,\tau)\, d\mu_j(\Omega)$$

for $\phi \in C^\infty_c(G)$.

(iv) (Plancherel formula, continued). Suppose T_j is fundamental, then

$$\rho_j(g_\Omega,\tau) = (\, (G(g_\Omega)/G(g_\Omega)_0))^{-1} \dim\ \tau \ .$$

Examples :

1. Assume $\underline{g}$ is nilpotent. Then G_0 is the direct product of a compact connected abelian group by a simply connected nilpotent group. If G is connected the conjecture reduces to Kirillov's thesis [8] . If G is not connected then the conjecture is still valid, as it is seen using [4] and [5].

2. Assume that G is connected and solvable. Then, except perhaps for the fact that all relevant orbits are tempered, the conjecture is true (cf. [2] ch. 9 for (ii) and [3] for (iv)).

3. Assume that G is compact. If G is connected (ii) is proved in [9] , and (iv) can be obtained in the same way. Then the case of disconnected G can be reduced to the connected case.

Let us explain the construction of $T_{g,\tau}$. Here, since g is regular , $G'(g)_0$ is a Cartan subgroup of G_0. Let $\underline{t}$ be its Lie algebra. Let $\underline{b}$ be the Borel subalgebra of $\underline{g}_{\mathbb{C}}$ which contains $\underline{t}$ and the root spaces corresponding to roots $\alpha \in i\underline{t}^*$ such that $(ig, \alpha) < 0$. There is a well defined character ρ of $G(g)^{\underline{g}}$ such that $\rho(x)^2 = (\det \mathrm{Ad}_{\underline{g}/\underline{b}}(x))^{-1}$ (note that $\underline{b}$ is $G(g)$-stable, and that this is essentially the definition of $G(g)^{\underline{g}}$). Then $T_{g,\tau}$ is the representation by left translations in the space of functions ϕ on with values in the space of τ which satisfy $\phi(xy) = \rho(y)^{-1}\tau(y)^{-1}\phi(x)$

for $x \in G$, $y \in G(g)$ (note that $\rho(y)^{-1}\ \tau(y)^{-1}$ is well defined on $G(g)$) , and $\rho * X = 0$ for X in the nilradical of $\underline{b}$. (By the Borel-Weil theorem, if G is connected, T_g is the dual of irreducible representation with lowest weight $ig + \rho$).

4. Assume that $\underline{g}$ is reductive, and G in the Harish- Chandra class (which means here that Ad x is for any $x \in G$ an inner automorphism of $\underline{g}_{\mathbb{C}}$). Then, the conjecture is true : (iii), (iv) and the irreducibility of $T_{g,\tau}$ are due to Harish-Chandra [7] , and (ii) to [12] . Note that one has to be careful when comparing our parametrization of the relevant part of $\hat{G}$ with Harish-Chandra's. Ours seems more natural (although more complicated) because it does not involve a choice of positive roots as for instance in [7] , I par. 27. It is interesting to remark that the validity of (iv) contains for instance [7] III cor. p. 164 , including the exact value of the constant c_G . It is reasonable to guess that the validity of Langland's conjecture (cf. [13]) will allow us to extend these results to the case where G is reductive, but not necessarily in the Harish-Chandra class (as we have done for compact groups in example 3).

3. *Non algebraic groups* :

The results of paragraphs 1 and 2 b are easy to generalize to non algebraic algebras. It is known how to generalize those of paragraphe 3 c to connected solvable groups (cf. [3]).

Consider first the case of a Lie group G with a finite number of connected components, and locally isomorphic to an algebraic group. Then one has to make two modifications in the formula (1). First, it may happen that the compact part of the connected center of G_0 is not a direct factor. In this case, one decomposes the regular representation of G using the Fourier transform on this subgroup, and for each character of this subgroup one writes a "projective" Plancherel formula analoguous to (1). Secondly, it may happen that $X^{irr}(g)$ is not finite. Never the less it is provided with a canonical Plancherel formula (because $G(g)/G(g)_0$ is discrete) which replaces the sum in formula (1).

Remark that (iii) and (iv), modified as above, imply for locally algebraic groups with a finite number of connected components precise conjectures for the existence, parametrization, and the formal degree of square integrable representations. They are coherent with the results of N. H. Anh [1] . When G is not locally algebraic with a finite number of connected components, there are serious problems due in particular to the fact that some groups are not necessarily of type I, and orbits not locally closed. See [3] to understand what happens in a concrete example (connected solvable groups).

COMMENTAIRES

1. Le lemme 1 (i) et son corollaire sont un cas particulier d'un résultat de V. Kac ("Infinite root systems, representations of graphes and invariant theory", Invent. Math. 56 (1980) 57-92), valable pour toutes les représentations linéaires des groupes algébriques.

2. Examinons ce qui a été fait sur la "conjecture" du paragraphe 2. c.

(0) Comme toutes les formes linéaires fortement régulières sont bien polarisables, ces notes fournissent les représentations $T_{g,\tau}$.

(i) On peut démontrer que ces représentations vérifient (i). C'est même vrai en supposant seulement g bien polarisable.

(ii) Il n'est pas vrai que toutes les orbites $\Omega \in \underline{g}_a^*/G$ soient tempérées. Il faut se restreindre à un ouvert de Zariski de $\underline{g}^*$ que rencontre tous les R_j . Avec cette restriction elle est vraie dans le cas résoluble (comme on le déduit de résultats de Pukanszky et de Charbonnel-Dixmier). Lorsque $\Omega \in \underline{g}_a^*/G$ est tempérée, la formule pour la trace de $T_{g,\tau}$ est vérifiée par M.S. Khalgui dans le cas des groupes connexes à radical co-compact, et est bien probablement vraie en général compte-tenu des résultats de W. Rossmann [12] .

(iii) La démonstration du théorème 2 chapitre VII peut être modifiée pour montrer que les représentations $T_{g,\tau}$ avec $g \in \underline{g}_a^*$ et $\tau \in X^{irr}(g)$ suffisent à décomposer $L^2(G)$, ce qui est un premier pas vers la formule (1).

(iv) Cette partie de la conjecture peut être généralisée de la manière suivante aux T_j non fondamentaux : la formule

$$\rho_j(g_\Omega,\tau) = (\ (G(g_\Omega)/G(g_\Omega)_0))^{-1} \dim \tau$$

est asymptotiquement vraie quand g_Ω tend vers ∞ . Sous cette forme, elle donne une interprétation intéressante des constantes apparaissant dans les cas connus de la formule de Plancherel - en particulier pour les groupes de Lie semi-simples et la formule d'Harish-Chandra. Dans le cas des séries principales sphériques des groupes de Lie semi-simples réels, R. Mneimnei a vérifié que cette interprétation est correcte.

REFERENCES

1. N. H. Anh. Lie groups with square integrable representations. Annals of Math. 104 (1976) 431-458

2. P. Bernat and al. . Représentations des groupes de Lie résolubles. Dunod, Paris 1972.

3. J. Y. Charbonnel. La formule de Plancherel pour un groupe résoluble connexe II. Math. Annalen. 250 (1980) 1-34.

4. M. Duflo. Sur les extensions des représentations irréductibles des groupes de Lie nilpotents. Ann. Sci. Ec. Norm. Sup. 5 (1972) 71-120.

5. M. Duflo. Opérateurs différentiels bi-invariants sur un groupe de Lie. Ann. Sci. Ec. Norm. Sup. 10 (1977) 265-288.

6. M. Duflo and M. Vergne. Une propriété de la représentation co-adjointe d'une algèbre de Lie. C.R. Acad. Sci. Paris. 268 (1969) 583-585.

7. Harish-Chandra. Harmonic analysis on real reductive groups
   I. J. of Functional Analysis 19 (1975) 104-204.
   III. Annals of Math. 104 (1976) 117-201.

8. A. A. Kirillov. Représentations unitaires des groupes de Lie nilpotents. Uspekhi Mat. Nauk 17 (1962) 57-110.

9. A. A. Kirillov. The characters of unitary representations of Lie groups. Functional Analysis and its applications 2 (1968) 40-55.

10. A. Kleppner and R. L. Lipsman. The Plancherel formula for group extensions II. Ann. Sci. Ec. Norm. Sup. 6 (1973) 103-132.

11. R. W. Richardson. Deformations of Lie subgroups and the variation of the isotropy subgroups. Acta Math. 129 (1972) 35-73.

12. W. Rossmann. Kirillov's character formula for reductive Lie groups. Inv. Math. 48 (1978) 207-220.

13. W. Schmid. L^2-cohomology and the discrete series. Ann. of Math. 103 (1976) 375-394.

CENTRO INTERNAZIONALE MATEMATICO ESTIVO

(C.I.M.E.)

ON A NOTION OF RANK FOR UNITARY REPRESENTATIONS OF THE CLASSICAL GROUPS

ROGER HOWE
Yale University

Research supported partially by NSF grant MCS79-05018

1. Review of the Heisenberg group and the oscillator representation.

The goal of these lectures is to introduce some general concepts concerning unitary representations of locally compact groups, and to apply these concepts to the study of representations of semisimple Lie groups, especially the symplectic group. Historically, what we may call "general representation theory", on one hand, and the representation theory of semisimple Lie groups on the other, have, with some notable exceptions, tended to go their separate ways; general concepts have not proven very powerful in semisimple harmonic analysis, which has needed its own special methods. The key phenomenon permitting at least a partial merger here is the oscillator representation of the symplectic group. Although this representation has received increasing attention in recent years, it may be still unfamiliar to some. Because of that, and because it will play such a pervasive role in the present study, I will begin by reviewing the basic definitions and salient properties of the oscillator representation. General references for the following material are [Cr], [H1], [Wi1].

Let F be a local field - $\mathbb{R}$, $\mathbb{C}$, or non-Archimedean. We will assume F is not of characteristic 2. We let W be a vector space of dimension $2m$ over F, on which is defined a symplectic form $\langle\ ,\ \rangle$. We may choose a basis $\{e_i, f_i\}_{i=1}^m$ for W, such that

$$\langle e_i, e_j\rangle = 0 = \langle f_i, f_j\rangle \qquad \langle e_i, f_j\rangle = \delta_{ij} \tag{1.1}$$

where δ_{ij} is Kronecker's δ. We will suppose such a basis chosen once and for all. We will refer to it as our <u>standard symplectic basis</u>. Taking coordinates with respect to the standard basis identifies W with F^{2m}, and we will make this identification when convenient.

The group of linear isometries of the form $<\ ,\ >$ is the symplectic group of W and $<\ ,\ >$. We will denote it variously as $Sp(W, <\ ,\ >)$, or $Sp(W)$, or Sp_{2m}, or $Sp_{2m}(F)$, or simply Sp, generally keeping the designation as short as is consistent with precision of reference.

Define a two-step nilpotent group $H(W)$, the Heisenberg group attached to W by the recipe:

$$H(W) = W \oplus F \tag{1.2}$$

as set, and has group law

$$(w,t)(w',t') = (w+w',\ t+t'+(\tfrac{1}{2})<w,w'>)$$

with $w, w' \in W$ and $t, t' \in F$.

The group $H(W) = H$ has a natural locally compact topology, and we may consider its unitary representation theory. It is very well understood. Observe that the center $Z(H)$ of H is also the commutator subgroup of H. Therefore all representations of H trivial on $Z(H)$ factor to the abelian group

$$H/Z(H) \simeq W$$

and are thus identified to one-dimensional characters of W.

Let ρ then be an irreducible unitary representation of H which is non-trivial on $Z(H)$. Then the restricted representation $\rho|Z(H)$ must be a multiple of a single unitary character $\chi(\rho)$, the central character of ρ. The basic result about ρ is the Stone-von Neumann Theorem, which says that ρ is determined up to unitary equivalence by $\chi(\rho)$. In other words, given a non-trivial character χ of $Z(H)$, there is a unique irreducible ρ , up to unitary equivalence, such that $\chi = \chi(\rho)$.

For a general locally compact group, let $\hat{G}$ denote the unitary dual of G, the collection of equivalence classes of irreducible representations of G. Then according to our remarks just above,

$$\hat{H} \simeq \hat{W} \cup (Z(H)^{\wedge} - \{1\})$$

where 1 here is the trivial character of $Z(H)$.

It is convenient to introduce coordinates on $Z(H)^{\wedge}$. In the description (1.2) of H, the center $Z(H)$ is identified to F, so that $Z(H)^{\wedge}$ is identifiable with $\hat{F}$. We choose once and for all a non-trivial character χ_1 of F, called a <u>basic character</u> of F. Then an arbitrary character of F has the form

$$\chi_t(s) = \chi_1(ts) \qquad t, s \in F \tag{1.3}$$

Write $F - \{0\} = F^x$ for the multiplicative group of non-zero elements of F. Then with the identifications just made we can write

$$\hat{H} \simeq \hat{W} \cup F^x$$

We can denote the element of $\hat{H}$ with central character χ_t by ρ_t.

Here is a realization of the representation ρ_t. Let Y be the subspace of W spanned by the f_i's of the standard basis (1.1), and let X be the span of the e_i's. We can realize ρ_t on the space $L^2(Y)$. The formulas for elements of H acting on $f \in L^2(Y)$ are

(1.4) $$\rho_t(y',0)f(y) = f(y+y') \qquad y,y' \in Y$$

$$\rho_t(x,0)f(y) = \chi_t(\langle y,x \rangle)f(y) \qquad x \in X$$

$$\rho_t(0,s)f(y) = \chi_t(s)f(y) \qquad s \in F$$

The symplectic group $Sp(W)$ acts on $H(W)$ by automorphisms in the obvious way:

$$g(w,t) = (g(w),t) \qquad g \in Sp, \ (w,t) \in H(W)$$

Whenever one has a group G acting by automorphisms on a locally compact group H, one has also an action of G on $\hat{H}$, denoted Ad^*, by the recipe

$$Ad^*g(\rho)(h) = \rho(g^{-1}(h)) \qquad \rho \in \hat{H}, \ g \in G, \ h \in H$$

If $\rho \in \hat{H}$ is fixed by G, it means $Ad^*g(\rho)$ is unitarily equivalent to ρ for each $g \in G$. Hence there is a unitary operator U_g, uniquely defined up to a scalar multiple, such that

$$U_g \, \rho(h) U_g^{-1} = \rho(g(h)) \qquad g \in G, \ h \in H.$$

Since U_g is well-defined up to scalars, the transformation it induces on the projective space associated to the space of ρ is well-defined period. Also, it is easy to see that the map $g \to U_g$ is multiplicative up to scalar multiples. Hence the map $g \to U_g$ is called a projective representation. One can then show (see e.g. [My]; the general proof is difficult but under certain simplifying assumptions which apply here, it is relatively easy) that there is a central extension $\tilde{G}$ of G, and a bonafide unitary representation $\tilde{U}$ of $\tilde{G}$ on the space of ρ, such that

$$\tilde{U}(g) \ \rho(h) \ \tilde{U}(g)^{-1} = \rho(g(h)) \qquad g \in G, \ h \in H$$

where we have used g to denote both an element of $\tilde{G}$ and its image in G. We will generally abuse notation in this way. If G is perfect (equal to its own commutator subgroup) then $\tilde{G}$ may also be taken to be perfect, and it is in this case uniquely defined, and so is $\tilde{U}$.

We may specialize these remarks to the action of $Sp(W)$ on $H(W)$. Since Sp acts trivially on $Z(H)$ we see the representations $\rho_t \in \hat{H}$ are fixed by $Ad^* Sp$. Since Sp is perfect, there is a unique perfect central extension $\tilde{Sp}$, and a unitary representation, which we will denote ω_t, of $\tilde{Sp}$, such that

$$(1.5) \qquad \omega_t(g)\ \rho_t(h)\ \omega_t(g)^{-1} = \rho_t(g(h)) \qquad g \in \tilde{Sp}(W),\ h \in H(W)$$

We will call ω_t (or the projective representation of Sp that gives rise to it) an <u>oscillator representation</u> of $\tilde{Sp}$. Nominally ω_t depends on t, but in fact this dependence is rather weak, and there are only a finite number of mutually non-equivalent ω_t's. We will denote a typical one of them simply by ω.

Of course, the group $\tilde{Sp}$ could conceivably depend on t. However, it is a theorem of Shale [Sh] for $F = \mathbb{R}$ and Weil [Wil] otherwise that it does not, and furthermore, except for the case $F = \mathbb{C}$, when $\tilde{Sp} = Sp$, the group $\tilde{Sp}$ is a two-fold cover of Sp, so that we have a diagram

$$1 \to \mathbb{Z}_2 \to \tilde{Sp} \to Sp \to 1$$

Under ω, the group $\mathbb{Z}_2$ is represented by ± 1. In any irreducible representation of $\tilde{Sp}$, the group $\mathbb{Z}_2$ will either act by ± 1, or just by 1. In the latter case, the representation factors to define a representation of Sp.

It is not easy to give formulas for $\omega_t(g)$ for all $g \in \tilde{S}p$. However, on a certain subgroup of $\tilde{S}p$, formulas can be given consistent with the formulas (1.4). Let $P_m(W) = P_m$ be the subgroup of $Sp(W)$ that preserves X, the span of the e_i's. We have

$$(1.6) \qquad P_m \simeq GL(Y) \cdot N_m(W)$$

where $N_m(W) = N_m$ is the subgroup of P_m that leaves X pointwise fixed. Furthermore, we can identify N_m with the space of symmetric bilinear forms on Y,

$$N_m(W) \xrightarrow{\beta} S^{2*}(Y)$$

$$n \to \beta_n$$

by the rule

$$(1.7) \qquad \beta_n(y_1,y_2) = \langle y_1,ny_2\rangle \qquad n \in N_m,\ y_i \in Y.$$

We let $\tilde{P}_m$ denote the inverse image in $\tilde{S}p$ of P_m, and similarly for $\tilde{G}L(Y)$. However, N_m may be lifted in a unique manner to a subgroup of $\tilde{S}p$, and we will continue to denote this subgroup of $\tilde{S}p$ also by N_m. We can then write, for $f \in L^2(Y)$

$$(1.8) \qquad \omega_t(g)f(y) = \gamma(g)\,|\det_Y g|^{-1/2}\, f(g^{-1}(y)) \qquad g \in \tilde{G}L,\ y \in Y$$

$$\omega_t(n)f(y) = \chi_t(-\tfrac{1}{2}\ \beta_n(y,y))f(y) \qquad n \in N_m$$

where γ is a factor of absolute value 1, and $\det_Y g$ denotes the determinant of g acting on Y, and $|\ |$ indicates the standard absolute value on F ([Wi2], Chap. 1, §3).

The multiplicative group F^x also acts by automorphisms on H by the rule

(1.9) $$s(w,t) = (sw, s^2t)$$

For this action we see that

(1.10) $$Ad^*s(\rho_t) = \rho_{s^{-2}t}$$

Since this action of F^x on H obviously commutes with the action of Sp, we conclude, by the naturality of the extension process, that

(1.11) $$\omega_t \simeq \omega_{s^2t} \qquad s,t \in F^x$$

Hence the set of oscillator representations ω is parametrized by the finite set F^x/F^{x2}.

Furthermore, the set of ω_t is closed under taking contragredients. Given a representation ρ of a group G on a Hilbert space $\mathcal{H}$, let σ^* denote the contragredient representation of G on the dual space $\mathcal{H}^*$. Since $\chi_t^* = \chi_{-t}$, it is clear that $\rho_t^* = \rho_{-t}$ for $\rho_t \in \hat{H}$, and hence

(1.12) $$\omega_t^* = \omega_{-t}$$

The representations ω_t have some important hereditary properties which we will now detail. Let $W = W^1 \oplus W^2$ be a decomposition of W into two orthogonal subspaces. Then it is obvious that

(1.13) $$H(W) \simeq (H(W^1) \times H(W^2))/\Delta^-$$

where Δ^- denotes the antidiagonal of $Z(H^1) \times Z(H^2) \simeq F \times F$. Here and below, if x denotes some object attached to W in the discussion above, then x^1 and x^2 denote the similar objects attached to W^1 and W^2.

From (1.13) it is clear that

(1.14)
$$\rho_t \simeq \rho_t^1 \check{\otimes} \rho_t^2 \qquad \text{(outer tensor product)}$$

Further, we have embeddings

$$i_j : Sp(W^j) \to Sp(W)$$

which may be lifted to maps of $\tilde{S}p$. Then

(1.15)
$$\omega_t \circ (i_1 \times i_2) \simeq \omega_t^1 \check{\otimes} \omega_t^2 \qquad \text{(outer tensor product)}$$
$$\omega_t \circ i_1 \simeq \infty \, \omega_t^1$$

Finally we come to one of the most remarkable properties of ω. Let W be a symplectic vector space as usual, and let V be a vector space on which an inner product (a symmetric, non-degenerate, bilinear form) is defined. Put

(1.16)
$$W^V = W \otimes V$$

Both W and V may be identified with their dual spaces by means of the bilinear forms defined on them. This leads to isomorphisms

(1.17)
$$W \otimes V \simeq \text{Hom}(W,V) \simeq \text{Hom}(V,W)$$

We may use one of these alternative forms for W^V when convenient.

The space W^V is naturally a symplectic vector space, with symplectic form given by the tensor product of the forms on W and V. Let $O(V) = O$ denote the isometry group of the given inner product on V. There are obvious embeddings of $Sp(W)$ and $O(V)$ into $Sp(W^V)$, and the images of the two groups clearly commute. In fact, it is not hard to see that $O(V)$ is the centralizer of $Sp(W)$ in $Sp(W^V)$, and vice versa. Thus the pair

$(Sp(W), O(V))$ form what I have called a dual pair in $Sp(W^V)$.

Consider an oscillator representation ω_t^V of $\tilde{Sp}(W^V)$. Let $\tilde{Sp}(W)$ and $\tilde{O}(V)$ denote the inverse images in $\tilde{Sp}(W^V)$ of $Sp(W)$ and $O(V)$. Note that $\tilde{Sp}(W)$ as here defined may not be the same as what was defined earlier. When it is necessary to distinguish them, we will denote the present one by $\tilde{Sp}(W)_V$. The difference between the two is fairly mild. According to the Shale-Weil description of $\tilde{Sp}(W)$, we have

$$\tilde{Sp}(W)_V \simeq \tilde{Sp}(W) \quad \text{if } \dim V \text{ is odd;}$$

$$\tilde{Sp}(W)_V \simeq Sp(W) \times \mathbb{Z}_2 \quad \text{if } \dim V \text{ is even.}$$

Using formulas (1.11) and (1.15) it is not difficult to see that the restriction $\omega_t^V|\tilde{Sp}(W)$ is a tensor product of $\dim V$ oscillator representations. Explicitly, if $\{v_i\}_{i=1}^{\ell}$ is an orthogonal basis for V, and the inner product of v_i with itself is t_i, then

$$(1.18) \qquad \omega_t^V|\tilde{Sp}(V) \simeq \omega_{tt_1} \otimes \omega_{tt_2} \otimes \cdots \otimes \omega_{tt_\ell}$$

We will call $\omega_t^V|\tilde{Sp}(V)$ the <u>Weil representation</u> of $\tilde{Sp}(W)$ <u>associated to V</u>.

The groups $\tilde{Sp}(W)$ and $\tilde{O}(V)$ commute with one another and so form a dual pair in $\tilde{Sp}(W^V)$. The reduction of ω_t^V on either group would be provided by the following

<u>Conjecture</u>: The groups $\omega_t^V(\tilde{Sp}(W))$ and $\omega_t^V(\tilde{O}(V))$ generate each other's commutants, in the sense of von Neumann algebras.

Although proving this conjecture is at present probably more a matter of hard work than insight, a proof has not been written down. However, for certain V, relatively direct proofs are available [H1], [H2].

Theorem 1.1: The above conjecture is true if $2 \dim V \leq \dim W$, or if $O(V)$ is compact (i.e., anisotropic). Furthermore, in the case that

$2 \dim V \leq \dim W$, the image of the parabolic subgroup $\tilde{P}_m(W)$ of $\widetilde{Sp}(W)$ already generates the commutant of $\omega_t^V(\tilde{O}(V))$.

2. The N_m-rank of representations of Sp.

In this section we introduce a simple-minded general method for studying representations, and we apply it in a particular way to the study of representations of $\widetilde{Sp}(W)$ for W a symplectic vector space. The rigidity of the representation theory of the Heisenberg group makes the application fruitful. In the next section we will prove some auxiliary results which show that the considerations of this section are less <u>ad hoc</u> than they perhaps seem at first. An analysis similar to that given here applies to other classical groups. The exceptional groups are significantly different.

Let G be a separable locally compact group, and let $H \subseteq G$ be a closed subgroup. We will assume G and H are type I. If the representation theory of H is quite well understood while that of G is relatively mysterious, we might attempt to study representations of G by restricting them to H. In this connection, we make three definitions. Take a representation $\rho \in \hat{G}$ and consider $\rho|H$, its restriction to H. Since H is type I, we know by the direct integral theory [Dx], [Nk], that $\rho|H$ is defined up to unitary equivalence by a projection valued measure on $\hat{H}$. We will call this projection-valued measure, and the unitary equivalence class it defines, the <u>H-spectrum</u> of ρ. The H-spectrum is obviously a unitary invariant of ρ and therefore provides a potential means of classifying $\rho \in \hat{G}$. We note that in fact we may define the H-spectrum for any representation of G, not only for irreducible ones.

Useful information about ρ might follow from knowledge about ρ considerably cruder than its exact H-spectrum. For example, the dual space $\hat{H}$ has the structure of a T_0 topological space defined by the Fell topology [F1]. The (closed) support of the H-spectrum of ρ will be called the <u>geometric</u> <u>H-spectrum</u>. A still cruder piece of information about ρ is the following. Given a representation σ of H, we let $n\sigma$, for n a positive integer or ∞, denote the direct sum of n copies of σ. Recall that two representations σ and σ' are called <u>quasi-equivalent</u> if $\infty\sigma$ and $\infty\sigma'$ are equivalent. We will say that the representation ρ of G is <u>H-regular</u> if $\rho|H$ is quasi-equivalent to some subrepresentation of the regular representation of G. Otherwise we will say $\rho|H$ is <u>H-singular</u>.

An obvious class of candidates for mysterious groups G are the semisimple groups. A class of candidates for subgroups whose representation theory is well-understood are the unipotent radicals of parabolic subgroups. We will consider the case $G = Sp(W)$ and $H = N_m$ where N_m is as defined in (1.6), the unipotent radical of the stabilizer of the span of the e_i's, with the e_i's as in formula (1.1). As noted in (1.7), we have $N_m \simeq S^{2*}(Y)$, where Y is the span of the f_i's. Since X and Y are in duality via the form $<\ ,\ >$, we may also write

$$N_m \simeq S^2(X),$$

with $S^2(X)$ denoting the second symmetric power of X. The dual space of $S^2(X)$ is $S^{2*}(X)$, the space of symmetric bilinear forms on X. We may identify $S^{2*}(X)$ with $\hat{N}_m$, by associating to $\beta \in S^{2*}(X)$ the character χ_β defined by

$$(2.1) \qquad \chi_\beta(n) = \chi_1(\beta(n)) \qquad n \in N_m \simeq S^2(X) .$$

Here χ_1 is the basic character of F chosen in §1.

We have $N_m \subseteq P_m \subseteq Sp$. Consider the N_m-spectrum of a representation σ of P_m. This is a projection-valued measure on $\hat{N}_m$. For a Borel set

$U \subseteq \hat{N}_m$, let $\pi_\sigma(U)$ denote the associated projection. Since π_σ comes from a representation of P_m, it allows $Ad^* P_m$ as a group of automorphisms. Specifically, we have the formula

$$\pi_\sigma(Ad^* p(U)) = \sigma(p)\, \pi_\sigma(U)\, \sigma(p)^{-1} \qquad p \in P_m \tag{2.2}$$

The action of P_m on $\hat{N}_m$ via Ad^* is identified via (2.1) to the natural action of $GL(X)$ on $S^{2*}(X)$. Thus the $Ad^* P_m$ orbits in $\hat{N}_m$ are naturally parametrized by the isomorphism classes of symmetric bilinear forms over F of rank less than or equal to $\dim X$. In particular, since F is not of characteristic 2, the number of orbits is finite. If β is a symmetric bilinear form on X, let $\mathcal{O}_\beta$ denote the $Ad^* P_m$-orbit through β. We define the rank of $\mathcal{O}_\beta$ to be the rank of β.

Each $\mathcal{O}_\beta$ is an analytic variety over F, and a such it carries a well-defined measure class, which is locally represented by Haar measure in any local coordinate system. Although the $\mathcal{O}_\beta$ do not generally carry $Ad^* P_m$-invariant measures, the canonical measure class just described for each orbit is invariant under $Ad^* P_m$.

It follows from the transformation law (2.2) that the restriction of the spectral measure π_σ to an orbit $\mathcal{O}_\beta \subseteq \hat{N}_m$ must be absolutely continuous with respect to the canonical measure class on $\mathcal{O}_\beta$. Also, the restriction of π_σ to $\mathcal{O}_\beta$ must be of uniform multiplicity. Summarizing this discussion yields

Proposition 2.1: Given a representation σ of P_m, the N_m-spectrum of σ is determined by the multiplicities $n(\sigma,\beta)$ of π_σ restricted to $\mathcal{O}_\beta$ for each $Ad^* P_m$-orbit $\mathcal{O}_\beta \subseteq \hat{N}_m$.

The multiplicities $n(\sigma,\beta)$ are non-negative integers or $+\infty$. We will say the orbit $\mathcal{O}_\beta$ occurs in σ if $n(\sigma,\beta) > 0$. We will say σ is supported on the $\mathcal{O}_\beta$ which occur in σ.

Given a representation σ of Sp, we may restrict it to P_m, and all the above notions may then be applied. Further, they apply just as well to $\tilde{S}p$ as to Sp.

Definition 2.2: Given a representation σ of $\tilde{S}p(W)$, the $\underline{N_m\text{-rank}}$ of σ is the maximum of the ranks of the $\mathrm{Ad}^* P_m$ orbits in $\hat{N}_m$ occurring in σ. We will say that σ is of $\underline{\text{pure } N_m\text{-rank}}$ ℓ if all orbits occurring in σ have rank ℓ.

Example 2.3: On F, define the rank one symmetric bilinear form β_s by

$$\beta_s(r_1, r_2) = sr_1r_2 \qquad s, r_1, r_2 \in F.$$

The formula (1.8) shows that the oscillator representation ω_t of $\tilde{S}p$ is supported on the orbit $\mathcal{O}_{\beta_s}$ with $s = -(\frac{1}{2})t$. In particular, ω_t is of pure N_m-rank 1.

This example shows the concept of rank is non-trivial. We will establish some basic properties of the N_m-spectrum, and particularly of N_m-rank.

Given sets $U_1, U_2 \subseteq \hat{N}_m$, we use the conventional notation

$$U_1 + U_2 = \{u_1 + u_2 \colon u_i \in U_i\}.$$

Lemma 2.4: Let σ_1 and σ_2 be two representations of $\tilde{P}_m$. An orbit $\mathcal{O}_\beta \subseteq \hat{N}_m$ occurs in $\sigma_1 \otimes \sigma_2$ if and only if there are orbits $\mathcal{O}_{\beta_i}$ occurring respectively in σ_i such that $\mathcal{O}_\beta$ is open in $\mathcal{O}_{\beta_1} + \mathcal{O}_{\beta_2}$.

Proof: First consider the outer tensor product $\sigma_1 \check{\otimes} \sigma_2$ as a representation of $\tilde{P}_m \times \tilde{P}_m$. Then the $N_m \times N_m$-spectrum of $\rho_1 \otimes \rho_2$ is the direct product, in the obvious sense, of the N_m-spectra of σ_1

and σ_2. Explicitly, for sets $U_1, U_2 \subseteq \hat{N}_m$, we have

$$\pi_{\sigma_1 \check{\otimes} \sigma_2}(U_1 \times U_2) = \pi_{\sigma_1}(U_1) \otimes \pi_{\sigma_2}(U_2)$$

Taking the inner tensor product of σ_1 and σ_2 amounts to restricting the outer tensor product to the diagonal. In particular, the N_m-spectrum of the inner tensor product $\sigma_1 \otimes \sigma_2$ will be given by

$$\pi_{\sigma_1 \otimes \sigma_2}(U) = \pi_{\sigma_1} \times \pi_{\sigma_2}(\delta^{*-1}(U))$$

for $U \subseteq \hat{N}_o$, where

$$\delta^*: \hat{N}_m \times \hat{N}_m \to \hat{N}_m$$

is the dual of the diagonal map

$$\delta: N_m \to N_m \times N_m$$

$$n \to (n,n)$$

Since the spectral measure of σ_1 or σ_2 is the sum of its restrictions to the various orbits, we may as well assume for purposes of this lemma that σ_i is supported on a single orbit $\mathcal{O}_{\beta_i}$. It is clear that

$$\delta^*(\mathcal{O}_{\beta_1} \times \mathcal{O}_{\beta_2}) = \mathcal{O}_{\beta_1} + \mathcal{O}_{\beta_2}$$

If $\mathcal{O}_\beta$ is an orbit disjoint from $\mathcal{O}_{\beta_1} + \mathcal{O}_{\beta_2}$, then $\delta^{*-1}(\mathcal{O}_\beta)$ will be disjoint from $\mathcal{O}_{\beta_1} \times \mathcal{O}_{\beta_2}$, so clearly

$$\pi_{\sigma_1 \otimes \sigma_2}(\mathcal{O}_\beta) = 0$$

If $\mathcal{O}_\beta$ is contained but not open in $\mathcal{O}_{\beta_1} + \mathcal{O}_{\beta_2}$, then $\delta^{*-1}(\mathcal{O}_\beta)$

will be a subvariety of positive codimension in $\mathcal{O}_{\beta_1} \times \mathcal{O}_{\beta_2}$, and will have zero measure with respect to the canonical measure class on $\mathcal{O}_{\beta_1} \times \mathcal{O}_{\beta_2}$. Hence again $\pi_{\sigma_1 \otimes \sigma_2}(\mathcal{O}_\beta) = 0$.

Lemma 2.5: Let β_1 and β_2 be two symmetric bilinear forms on X. Recall $m = \dim X$.

a) All orbits which are open in $\mathcal{O}_{\beta_1} + \mathcal{O}_{\beta_2}$ have rank equal to $\min(m, \operatorname{rank} \beta_1 + \operatorname{rank} \beta_2)$

b) Suppose $\operatorname{rank}(\beta_1 + \beta_2) = \operatorname{rank} \beta_1 + \operatorname{rank} \beta_2$. Let R_i be the radical of β_i. Then the radical of $\beta_1 + \beta_2$ is $R_1 \cap R_2$, and $\beta_1 + \beta_2$ factored to $X/(R_1 \cap R_2)$ is isomorphic to the direct sum of the β_i factored to X/R_i. Hence $\mathcal{O}_{\beta_1} + \mathcal{O}_{\beta_2}$ contains only a single orbit of rank equal to $\operatorname{rank} \beta_1 + \operatorname{rank} \beta_2$.

Proof: Part a) is obvious since the condition that $\beta_1' + \beta_2'$ have rank less than the maximum possible, as specified in a), is a non-trivial polynomial condition on $(\beta_1', \beta_2') \in \mathcal{O}_{\beta_1} \times \mathcal{O}_{\beta_2}$.

Suppose $\beta_1 + \beta_2$ are as in b). Clearly $R_1 \cap R_2$ is contained in the radical of $\beta_1 + \beta_2$. On the other hand we have the standard formula

$$\operatorname{rank} \beta_1 + \dim R_1 = m = \operatorname{rank} \beta_2 + \dim R_2$$

Hence

$$\dim(R_1 \cap R_2) \geq \dim R_1 + \dim R_2 - m = m - (\operatorname{rank} \beta_1 + \operatorname{rank} \beta_2)$$
$$= m - \operatorname{rank}(\beta_1 + \beta_2)$$

By dimension counting, then, we see $R_1 \cap R_2$ is indeed the radical of $\beta_1 + \beta_2$. We may divide by $R_1 \cap R_2$ and reduce to the case when $R_1 \cap R_2 = \{0\}$, so that

$$\operatorname{rank}(\beta_1 + \beta_2) = \operatorname{rank} \beta_1 + \operatorname{rank} \beta_2 = \dim X$$

In this case we see that $X = R_1 \oplus R_2$ is a direct sum decomposition of X. Since R_1 is complementary to R_2, the restricted form $\beta_2|R_1$ is non-degenerate. Similarly $\beta_1|R_2$ is non-degenerate. Thus we see that $R_1 \oplus R_2$ is an orthogonal direct sum decomposition for the form $\beta_1 + \beta_2$, and exhibits as isomorphic the direct sum of the β_i factored to X/R_i. This proves the lemma.

Lemma 2.6: Let σ_1 and σ_2 be two representations of $\tilde{P}_o$. Then

a) N_m-rank $(\sigma_1 \otimes \sigma_2) = \min(m, N_m\text{-rank } \sigma_1 + N_m\text{-rank } \sigma_2)$;

b) If σ_1 and σ_2 are of pure rank, then so is $\sigma_1 \otimes \sigma_2$;

c) If each of σ_1 and σ_2 is supported on a single $\mathrm{Ad}^* P_m$ orbit in $\hat{N}_m$, and if the sum of the N_m-ranks of the σ_i is at most m, then $\sigma_1 \otimes \sigma_2$ is also supported on a single $\mathrm{Ad}^* P_m$-orbit.

Proof: These statements are immediate from the two preceding lemmas.

We need also to understand how N_m-rank behaves under restriction to smaller symplectic groups. Let $\{e_i, f_i\}_{i=1}^m$ be the standard symplectic basis of formula (1.1). Let W_k denote the span of $\{e_i, f_i\}$ for $i \leq k$. Thus $\dim W_k = 2k$ and $W = W_m$. Set $X_k = X \cap W_k$ and $Y_k = Y \cap W_k$. Let $P_k(W_k)$ be the stabilizer of X_k in $Sp(W_k)$. Then $P_k(W_k) = P_m(W) \cap Sp(W_k)$. Here we are considering $Sp(W_k)$ to be the subgroup of $Sp(W)$ leaving $W_k^{\perp}$, the orthogonal subspace to W_k, spanned by e_i and f_i for $i > k$, pointwise fixed. Let $N_k(W_k)$ denote the unipotent radical of $P_k(W_k)$. Then $N_k(W_k) = N_m(W) \cap Sp(W_k)$. Also, as in formula (1.7), we have $N_k(W_k) \simeq S^2(X_k)$. Thus we have a diagram

(2.3)
$$\begin{array}{ccc} N_k(W_k) & \xrightarrow{i} & N_m(W) \\ \wr\| & & \wr\| \\ S^2(X_k) & \hookrightarrow & S^2(X) \end{array}$$

The vertical isomorphisms are given by dualizing formula (1.7). The top horizontal map is just inclusion, and the bottom horizontal map is the symmetric square of the inclusion $X_k \subseteq X$. It is clear from formula (1.7) that diagram (2.3) commutes. It follows that the dual diagram

(2.4)
$$\begin{array}{ccc} \hat{N}_m(W) & \xrightarrow{i^*} & \hat{N}_k(W_k) \\ \wr\| & & \wr\| \\ S^{2*}(X) & \longrightarrow & S^{2*}(X_k) \end{array}$$

also commutes. In diagram (2.4) the top map is just restriction of a character from $N_m(W)$ to $N_k(W_k)$, and the bottom map is given by restriction of bilinear forms from X to X_k.

Lemma 2.7: Let σ be a representation of $\tilde{P}_m(W)$, and consider the restriction $\sigma|\tilde{P}_k(W_k)$. Let $\beta' \in S^{2*}(X_k)$. In order that the $Ad^*P_k(W_k)$-orbit $\mathcal{O}_{\beta'} \subseteq \hat{N}_k(W_k)$ occur in $\sigma|\tilde{P}_k(W_k)$, it is necessary and sufficient that there is an $Ad^*P_k(W)$-orbit $\mathcal{O}_\beta$ in $\hat{N}_m(W)$ which occurs in σ and such that $\mathcal{O}_{\beta'}$ is open in $i^*(\mathcal{O}_\beta)$, with i^* as in diagram (2.4).

Proof: This is analogous to lemma 2.4. Let π_σ be the $N_m(X)$-spectrum of σ. Let $(\pi_\sigma)_k$ denote the $N_k(W_k)$-spectrum of $\sigma|\tilde{P}_k(W_k)$. Then for a set $U' \subseteq \hat{N}_k(W_k)$ we have

$$(\pi_\sigma)_k(U') = \pi_\sigma(i^{*-1}(U'))$$

As in lemma 2.4, we may assume for purposes of this proof that σ is supported on a single $Ad^*P_m(W)$-orbit $\mathcal{O}_\beta$. Then we see if $\mathcal{O}_{\beta'} \subseteq \hat{N}_k(W_k)$ is an $Ad^*P_k(W_k)$-orbit, and $\mathcal{O}_{\beta'}$ is not contained in $i^*(\mathcal{O}_\beta)$, obviously $i^{*-1}(\mathcal{O}_{\beta'})$ is disjoint from $\mathcal{O}_\beta$, so $(\pi_\sigma)_k(\mathcal{O}_{\beta'}) = 0$. If $\mathcal{O}_{\beta'} \subseteq i^*(\mathcal{O}_\beta)$, but is not open, then $i^{*-1}(\mathcal{O}_{\beta'}) \cap \mathcal{O}_\beta$

is a subvariety of positive codimension in $\mathcal{O}_\beta$, and therefore has measure zero for the canonical measure class. Hence in this case too we have $(\pi_\sigma)_k(\mathcal{O}_{\beta'}) = 0$.

Lemma 2.8: Consider $\beta \in S^{2*}(X)$, and $\beta' \in S^{2*}(X_k)$

a) If $\mathcal{O}_{\beta'}$ is open in $i^*(\mathcal{O}_\beta)$, then

$$\text{rank } \beta' = \min(k, \text{rank } \beta)$$

b) If $\text{rank}(\beta|X_k) = \text{rank } \beta$, then the radical of $\beta|X_k$ is $R \cap X_k$, where R is the radical of β, and β factored to X/R is naturally isomorphic to $\beta|X_k$ factored to $X_k/(R \cap X_k)$.

Proof: This is analogous to lemma 2.5. Part a) holds because the condition that $\beta_1|X_k$ have rank less than that specified is a non-trivial polynomial condition on $\beta_1 \in \mathcal{O}_\beta$. For part b) observe that the radical of $\beta|X_k$ has dimension

$$k - \text{rank}(\beta|X_k) = k - (m - \dim R) = k + \dim R - m$$
$$\leq \dim (R \cap X_k)$$

But clearly $R \cap X_k$ is contained in the radical of $\beta|X_k$, so they must be equal. Dividing out by $R \cap X_k$ reduces us to the case when

$$\text{rank } \beta = \text{rank}(\beta|X_k) = k$$

It is then clear that $X = X_k \oplus R$ is a direct sum decomposition, and part b) of the lemma follows.

Lemma 2.9: Let σ be a representation of $\tilde{P}_m(W)$, and consider $\sigma|\tilde{P}_k(W_k) = \sigma'$. Then

a) $N_k(W_k)$-rank $(\sigma') = \min(k, N_m(W)\text{-rank } (\sigma))$

b) If σ is of pure $N_m(W)$-rank, then σ' is of pure $N_k(W_k)$-rank.

c) If the $N_m(X)$-rank of σ is no more than k, and σ is supported on a single $Ad^*P_m(W)$-orbit in $\hat{N}_m(W)$, then σ' is supported on a single $Ad^*P_k(W_k)$-orbit in $\hat{N}_k(W_k)$.

Proof: This is an immediate consequence of the preceding two lemmas.

We may now prove our first main result concerning the $N_m(W)$-spectrum of representations of $\tilde{S}p(W)$.

Theorem 2.10: Let σ be a unitary representation of $\tilde{S}p(W)$ on a Hilbert space H. Let π_σ be the $N_m(W)$-spectrum of σ. For each $Ad^*P_m(W)$-orbit $O_\beta \subseteq \hat{N}_m$, set

(2.5) $$H_\beta = \pi_\sigma(O_\beta)H$$

so that the spectral measure of N_m acting on H_β is concentrated on H_β. Then for rank $\beta < m$, the subspace H_β is invariant under $\sigma(\tilde{S}p)$.

Proof: We will prove this by induction on $\dim W = 2m$. For $\dim W = 2$, we are dealing with $SL_2(F)$. The only possible ranks are 1 and 0, and the only orbit of rank 0 is the one-point orbit consisting of the origin. The corresponding subspace, H_0, consists of the N_1-fixed vectors. In this case, the theorem follows from [HM] which says that H_0 in fact consists of the fixed vectors for all of $SL_2(F)$.

From now on, we take $m > 1$. It is clear that the H_β are all invariant under $\tilde{P}_m$. Since $\tilde{P}_m$ is a maximal subgroup of $\tilde{S}p$, to prove the result it will suffice to show that for β as specified, there are elements of $\tilde{S}p$, not in $\tilde{P}_m$, which preserve H_β. In particular, consider the parabolic $P_1(W) = P_1$ consisting of transformations in Sp which preserve X_1, the line through e_1. Since $m > 1$, $P_1 \neq P_m$. Hence if we can show H_β is invariant under $\tilde{P}_1$ when rank $\beta < m$, the theorem will follow.

To begin, consider the space $\mathcal{H}_0$ of N_m-fixed vectors Again by [HM] we know that $\mathcal{H}_0$ consists of fixed vectors for all of $\widetilde{Sp}$. Hence the theorem is true for $\mathcal{H}_0$, and we may as well throw $\mathcal{H}_0$ away. Thus from now on we will assume $\mathcal{H}_0 = \{0\}$.

We review the structure of the group P_1. Recall $W_1^\perp$ is the subspace of W orthogonal to the plane spanned by e_1 and f_1. Write $W_1^\perp = W'$. We have the decomposition

$$P_1 = Sp(W') \cdot F^\times \cdot N_1$$

where $N_1 = N_1(W)$ is the unipotent radical of P_1. Furthermore

$$N_1 \simeq H(W')$$

in such fashion that the action by conjugation of $Sp(W')$ and $F^\times$ become the actions described in §1. We let $\tilde{P}_1$, etc., denote the inverse images of these groups in $\widetilde{Sp}$, except that, since N_1 may be lifted in unique fashion to a subgroup of $\widetilde{Sp}$, we will let N_1 denote the lifted group also.

We must clarify one technical point concerning these lifted groups. Since the kernel of the projection map $\widetilde{Sp} \to Sp$ is Z_2, (except when $F = \mathbb{C}$, which we will not explicitly take into account) the same will be true for any of these groups. In particular, we have exact sequences

$$1 \to \mathbb{Z}_2 \xrightarrow{j_1} \widetilde{Sp}(W') \xrightarrow{q_1} Sp(W') \to 1$$

$$1 \to \mathbb{Z}_2 \xrightarrow{j_2} \tilde{F}^\times \xrightarrow{q_2} F^\times \longrightarrow 1$$

$$1 \to \mathbb{Z}_2 \xrightarrow{j_3} (Sp(W')\cdot F^\times)^\sim \xrightarrow{q_3} Sp(W')\cdot F^\times \longrightarrow 1$$

The subgroup $Sp(W')\cdot F^\times \subseteq Sp(W)$ is actually a direct product $Sp(W') \times F^\times$.

Hence we may combine the first two sequences above and map them to the third.

$$\begin{array}{ccccccccc}
1 \to (\mathbb{Z}_2 \times \mathbb{Z}_2) & \xrightarrow{j_1 \times j_2} & \tilde{Sp}(W') \times \tilde{F}^{\times} & \xrightarrow{q_1 \times q_2} & Sp(W')\cdot F^{\times} & \to 1 \\
\downarrow \text{id} \times \text{id} & & \downarrow & & \text{id} \, \| & \\
1 \to \mathbb{Z}_2 & \xrightarrow{j_3} & \tilde{Sp}(W') \cdot \tilde{F}^{\times} & \xrightarrow{q_3} & Sp(W')\cdot F^{\times} & \to 1
\end{array}$$

From this diagram, we see that the kernel of the middle vertical map is the diagonal subgroup $\Delta(\mathbb{Z}_2 \times \mathbb{Z}_2)$. In particular, for later use, we note that a representation ρ of $\tilde{Sp}(W') \times \tilde{F}^{\times}$ will factor to define a representation of $(Sp(W')\cdot F^{\times})^{\sim}$ if and only if $\rho|\ker j_1$ and $\rho|\ker j_2$ define precisely the same representation of $\mathbb{Z}_2$.

Return to consideration of σ. Let ZN_1 denote the center of N_1. Consider the restriction $\sigma|ZN_1$. Since the fixed vectors of N_m have been eliminated, we know from lemma 2.6 (or again from [HM]) that there are no fixed vectors for ZN_1 in H. Thus the only representations of $N_1 \simeq H(W')$ occurring in $\sigma|N_1$ are the representations ρ_t provided for by the Stone-von Neumann Theorem, and described by equations (1.4). Thus we may describe the representations of $\tilde{P}_1$ by expanding on the analysis there. According to formula (1.10), the isotropy group of ρ_t under $Ad^* F^{\times}$ is $\{\pm 1\}^{\sim}$. We may extend ρ_t in 4 possible ways to the group $N_1' = \{\pm 1\}^{\sim} \cdot N_1$. Let us denote the 4 extensions by ρ_t^i for $i = 1,2,3,4$. The ρ_t^i may be obtained from one another by tensoring with characters of $\{\pm 1\}^{\sim}$. Thus we have

$$\rho_t^i = \varphi^i \otimes \rho_t^1$$

where φ_i is a character of $\{\pm 1\}^{\sim}$.

From Mackey's general theory [My] we know that the induced representations

(2.6) $$\tau_t^i = \mathrm{ind}_{N_1'}^{\tilde{F}^\times \cdot N_1} \rho_t^i$$

of $\tilde{F}^\times \cdot N_1$ are irreducible, and constitute all irreducible representations of $\tilde{F}^\times \cdot N_1$ which are non-trivial on ZN_1. Thus we have the description

$$(\tilde{F}^\times \cdot N_1)^\wedge = (\tilde{F}^\times \cdot W')^\wedge \ \cup ((F^\times/F^{\times 2}) \times \{\widetilde{\pm 1}\}^\wedge)$$

We may extend each representation τ_t^i to $(\tilde{Sp}(W') \times \tilde{F}^\times) \cdot N_1$ by means of the oscillator representation of $\tilde{Sp}(W')$. Since $\tilde{Sp}(W')$ is a perfect group, these extensions are unique. We will continue to denote them by τ_t^i. If we extend the characters φ^i of $\{\widetilde{\pm 1}\}$ to characters φ'^i of $\tilde{F}^\times$, then from the compatibility of induction and tensor product we see that

$$\tau_t^i = \varphi'^i \otimes \tau_t^1$$

Precisely 2 of the 4 characters of $\{\widetilde{\pm 1}\}$ will be trivial on the kernel of the projection $\{\widetilde{\pm 1}\} \to \{\pm 1\}$, and 2 will not. Hence from the compatibility criterion noted above, precisely 2 of the τ_t^i will factor from $(\tilde{Sp}(W') \times \tilde{F}^\times) \cdot N_1$ to yield representations of $\tilde{P}_1$. We may assume these are τ_t^1 and τ_t^2. Note then that we can write

$$\tau_t^2 \simeq \varphi \otimes \tau_t^1$$

where φ is a character of $F^\times$ (more precisely, a character of $\tilde{F}^\times$ which factors to $F^\times$) which is non-trivial on $\{\pm 1\}$. (The character φ will factor to $F^\times$ from $\tilde{F}^\times$ because τ_t^2 must also satisfy the compatibility criterion.)

Summarizing, the τ_t^i, $i = 1,2$, are obtained by starting with ρ_t, extending suitably to N_1', inducing to $\tilde{F}^\times \cdot N_1$, then extending again to $\tilde{P}_1$. Alternatively, we can start with ρ_t, extend via the oscillator representation to $\tilde{Sp}(W')\cdot N_1$, extend again, in the 2 possible compatible ways, to $(\{\pm 1\} \cdot Sp(W'))^{\sim} \cdot N_1$, then finally inducing up to $\tilde{P}_1$. In any case, we can see that

$$(2.7) \qquad \tau_t^i | \tilde{Sp}(W') \simeq \infty\omega_t^{(m-1)}$$

where we have labeled the oscillator representation by $m-1 = (\frac{1}{2}) \dim W_1^\perp$, to indicate with what space it is associated.

In fact, with hindsight, we may note the representations τ_t^i are already familiar to us. Indeed, the oscillator representation ω_t^m of $\tilde{Sp}(W)$ decomposes into two irreducible components ω_t^{m+} and ω_t^{m-}. It is not difficult to verify that these representations remain irreducible, and are inequivalent, under restriction to $\tilde{P}_1$. It is also easy to see they have the appropriate restriction to N_1, so that in some order $\omega_t^{m+}|\tilde{P}_1$ and $\omega_t^{m-}|\tilde{P}_1$ are equivalent to the τ_t^i.

Continuing with Mackey's theory we know that any representation σ of $\tilde{P}_1$ which contains no fixed vectors for ZN_1 has the form

$$(2.8) \qquad \sigma \simeq \sum_{i,t} \nu_t^i \otimes \tau_t^i \qquad i=1,2;\ t \in F^\times/F^{\times 2}$$

where the ν_t^i are appropriate representations of $\tilde{Sp}(W')\cdot\tilde{F}^\times$. Although it is not essential, because only $\nu_t^i|\tilde{Sp}(W')$ will be of concern to us, we note that the ν_t^i may be taken of the following form. Let φ^o be a character of $\tilde{F}^\times$ which does not factor to $F^\times$. Then we may write

$$\nu_t^i \simeq \mu_t^i \oplus \left(\mu_t^{oi} \otimes \varphi^o\right)$$

where μ_t^i is a representation of $Sp(W')$, viewed as a quotient of $(Sp(W')\cdot F^{\times})^{\sim}$, and μ_t^{oi} is a representation of $\tilde{S}p$ on which $\ker j_1$ acts by minus the identity, so that the representation $\mu_t^{oi} \otimes \varphi^o$ of $\tilde{S}p(W') \times \tilde{F}^{\times}$ satisfies the compatibility condition and so factors to $(Sp(W')\cdot F^{\times})^{\sim}$.

In particular, the representation σ of $\tilde{S}p(W)$ with which we are concerned can, on restriction to $\tilde{P}_1$, be decomposed in the manner of formula (2.8). Let τ_t^i be realized on a Hilbert space $\mathcal{Y}_t^i$, and let ν_t^i be realized on a Hilbert space $\mathcal{H}_t^i$. Corresponding to (2.8) we have the decomposition

$$\mathcal{H} \simeq \sum_{i,t} \mathcal{H}_t^i \otimes \mathcal{Y}_t^i \qquad i=1,2;\ t \in F^{\times}/F^{\times 2} \tag{2.9}$$

of the space $\mathcal{H}$ of σ.

Set $N_m(W) \cap Sp(W') = N_{m-1}(W') = N'_{m-1}$. We may further refine the decomposition (2.9) by considering the N'_{m-1} spectra of the ν_t^i. For each $Ad^*P'_{m-1}$-orbit $\mathcal{O}_{\beta'}$ in $\hat{N}'_{m-1}$, define the subspace $\mathcal{H}_{t\beta'}^i$ of $\mathcal{H}_t^i$ in analogy with (2.5). Then $\mathcal{H}_t^i$ is the sum of the $\mathcal{H}_{t\beta'}^i$, so that from (2.9) we get

$$\mathcal{H} \simeq \sum_{i,t,\beta'} \mathcal{H}_{t\beta'}^i \otimes \mathcal{Y}_t^i \tag{2.10}$$

The group $N_m(W)$ is a subgroup of $\tilde{P}_1$, so we may consider the N_m-spectra of the representations ν_t^i and τ_t^i. Denote by Q the normalizer $P_1 \cap P_m$ of N_m in P_1. When N_m is identified to $S^2(X_m)$, the action by conjugation on N_m maps Q not to all of $GL(X_m)$, but to the parabolic subgroup Q_1 of $GL(X_m)$ preserving the line X_1. Thus a given $Ad^*\tilde{P}_m$ orbit in $\hat{N}_m$ will decompose into several $Ad^*\tilde{Q}$ orbits Precisely, if $\beta \in S^{2*}(X)$, we may write

(2.11) $$\mathcal{O}_\beta = \cup\, \mathcal{O}_{\beta,t} \cup \mathcal{O}_{\beta,0} \cup \mathcal{O}_{\beta,00} \qquad t \in F^\times/F^{\times 2}$$

where the $\mathcal{O}_{\beta,t}$ are Q_1 orbits in the $GL(X_m)$ orbit $\mathcal{O}_\beta$ and are describable as follows. For $t \neq 0$,

$$\mathcal{O}_{\beta,t} = \{\beta' \in S^{2*}(X)\colon \beta' \in \mathcal{O}_\beta \text{, and } \beta'(e_1,e_1) = s^2 t, \text{ for some } s \in F^\times\}$$

Some of the $\mathcal{O}_{\beta,t}$ may be empty if t is not represented by β. Further

$$\mathcal{O}_{\beta,0} = \{\beta' \in S^{2*}(X)\colon \beta' \in \mathcal{O}_\beta \text{ and } \beta'(e_1,e_1) = 0, \text{ but } \beta'(e_1,e_i) \neq 0 \text{ for some } i\}$$

$$\mathcal{O}_{\beta,00} = \{\beta' \in S^{2*}(X);\ \beta' \in \mathcal{O}_\beta \text{ and } \beta'(e_1,e_i) = 0 \text{ for all } i \le m\}$$

The $\mathcal{O}_{\beta,t}$ for $t \neq 0$ are open in $\mathcal{O}_\beta$, while the union $\mathcal{O}_{\beta,0} \cup \mathcal{O}_{\beta,00}$ is a closed subvariety of $\mathcal{O}_\beta$ of positive codimension.

From the equivalence, noted above, of the representations τ_t^i with the restrictions to $\tilde{P}_1$ of the components of ω_t^m, and from example 2.3, we see that the N_m-spectrum of τ_t^i is concentrated on s single $GL(X)$-orbit in $S^{2*}(X)$, the orbit of the form β_s in the notation of example 2.3, with $s = -(\frac{1}{2})t$. Evidently for $t \neq 0$, the only non-empty Q_1-orbit $\mathcal{O}_{\beta_s,t}$ contained in $\mathcal{O}_{\beta_s}$ is $\mathcal{O}_{\beta_s,s}$. Thus we may regard τ_t^i as being supported on $\mathcal{O}_{\beta_s,s}$.

Consider the subspace $\mathcal{H}^i_{t\beta'}$ for $\beta' \in \hat{N}'_{m-1} \simeq S^{2*}(X \cap W')$. We may extend the form β' on $X \cap W'$ to a form $p(\beta')$ on all of X by letting X_1 be in the radical of $p(\beta')$. Then clearly if $\mathcal{O}_{\beta'}$ is the $GL(X \cap W')$ orbit of β', we have

$$p(\mathcal{O}_{\beta'}) = \mathcal{O}_{p(\beta'),00}$$

Equally clearly we see that the N_m-spectrum of the action of Q_1 on $\mathcal{H}^i_{t\beta'}$ is the Ad^*Q_1 orbit $\mathcal{O}_{p(\beta'),00} \subseteq \hat{N}_m$.

Take $\beta' \in S^{2*}(X \cap W')$, and $t \in F^\times$. Define a form $\gamma(\beta',t) = \gamma$ on X by

$$\gamma(x,x') = \beta'(x,x') \qquad \gamma(e_1,e_1) = t \qquad x,x' \in X \cap W'$$

$$\gamma(x,e_1) = 0$$

Then it is not hard to convince yourself that the sum of the Q_1 orbits $\mathcal{O}_{p(\beta'),00}$ and $\mathcal{O}_{\beta_s,s}$ in $S^{2*}(X)$ is

$$\mathcal{O}_{p(\beta'),00} + \mathcal{O}_{\beta_s,s} = \mathcal{O}_{\gamma(\beta',s),s}$$

Reasoning exactly as in lemma 2.4, we can therefore conclude that the spectrum of the N_m action on $\mathcal{H}^i_{t\beta'} \otimes \mathcal{Y}^i_t$ is concentrated on the $\mathrm{Ad}^*\tilde{Q}$-orbit $\mathcal{O}_{\gamma(\beta',s),s}$. Taking into account the decomposition (2.11), we see by comparing N_m-spectra that, with the $\mathcal{H}_\beta$ from formula (2.5), we may write

$$\mathcal{H}_\beta \simeq \sum_{i,t,\beta'} \mathcal{H}^i_{t\beta'} \otimes \mathcal{Y}^i_t \qquad \gamma(\beta', -(\tfrac{1}{2})t) \simeq \beta \tag{2.12}$$

It is obvious that

$$\operatorname{rank} \gamma(\beta', s) = \operatorname{rank} \beta' + 1$$

Hence if rank $\beta < m$, then rank $\beta' < m-1$. Thus by induction we may assume that $\mathcal{H}^i_{t\beta'}$ is invariant under all of $\widetilde{Sp}(W')$. But then formula (2.12) exhibits $\mathcal{H}_\beta$ as a $\tilde{P}_1$-module. As we noted above, this establishes the theorem.

Corollary 2.11: If $\sigma \in (\tilde{S}p(W))^\wedge$, then σ has pure N_m-rank. If N_m-rank $(\sigma) < m$, then σ is concentrated on a single $Ad^*\tilde{P}_m$ orbit in $\hat{N}_m$.

Proof: Suppose the N_m-rank of σ is ℓ. If an orbit $\mathcal{O}_\beta \subseteq \hat{N}_m$ of rank less than ℓ occurs in σ, then in particular, rank $\beta < m$. Thus theorem 2.10 says that the spectral projection corresponding to $\mathcal{O}_\beta$ yields a non-trivial $\tilde{S}p(W)$ subrepresentation of σ, contradicting the irreducibility of σ. If $\ell < m$, and two orbits $\mathcal{O}_{\beta_1}$ and $\mathcal{O}_{\beta_2}$ occurred in σ, then both rank $\beta_1 < m$ and rank $\beta_2 < m$. Hence theorem 2.10 yields two non-trivial, mutually orthogonal hence proper, subrepresentations of σ, again contradicting irreducibility.

Corollary 2.12: If σ is a representation of $\tilde{S}p_{2m}$ of pure N_m-rank $\ell > 1$, then

$$\sigma | \tilde{S}p_{2(m-1)} \simeq \sum_t \nu_t \otimes \omega_t^{m-1} \qquad t \in F^\times/F^{\times 2} \tag{2.13}$$

where the ν_t are representations of pure N_{m-1}-rank $\ell - 1$, and the ω_t^{m-1} are oscillator representations of $\tilde{S}p_{2(m-1)}$.

Proof: This is immediate from the decomposition (2.12) and formula (2.7).

Corollary 2.13: If σ is a representation of $\tilde{S}p_{2m}$ of pure rank ℓ, and $k \leq \ell$, then $\sigma | \tilde{S}p_{2(m-k)}$ is a (finite) sum of representations of the form

$$\nu_\beta \otimes \omega_\beta^{m-k}$$

where ω_β^{m-k} is the Weil representation of $\tilde{S}p_{2(m-k)}$ associated to the form β of rank k, and ν_β is a representation of $\tilde{S}p_{2(m-k)}$ of pure N_{m-k}-rank $\ell - k$.

Proof: This follows by induction from Corollary 2.12 and the formula (1.15).

Corollary 2.14: If σ is a representation of $\tilde{Sp}_{2m}(F)$ of pure N_m-rank $\ell < m$, then σ factors to $Sp_{2m}(F)$ if and only if ℓ is even, except when $F = \mathbb{C}$.

Proof: If $\ell = 0$, then by [HM] σ is some multiple of the trivial representation. Hence consider $\ell \geq 1$. Then σ factors to Sp_{2m} if and only if $\sigma | \tilde{Sp}_{2(m-1)}$ factors to $Sp_{2(m-1)}$. Look at the decomposition (2.13). By induction we can assume the result is true for the representations ν_t of $\tilde{Sp}_{2(m-1)}$. We know that the oscillator representations ω_t^{m-1} are of rank 1 and do not factor to $\tilde{Sp}_{2(m-1)}$. Considering the restrictions of ν_t and ω_t^{m-1} to the kernel of the projection from $\tilde{Sp}_{2(m-1)}$ to $Sp_{2(m-1)}$, the result follows.

Corollary 2.15: $\hat{Sp}_{2m}$ consists of representations of even pure N_m-rank, or N_m-rank m.

Proof: This is immediate from corollaries 2.13 and 2.11.

For $\ell \leq m$, let $(\tilde{Sp}(W))^\wedge_\ell$ denote the subset of $(\tilde{Sp}(W))^\wedge$ consisting of representations of pure N_m-rank ℓ. For a form $\beta \in S^{2*}(X)$, let $(\tilde{Sp}(W))^\wedge_\beta$ denote the subset of $(\tilde{Sp}(W))^\wedge$ consisting of representations whose N_m-spectrum is concentrated on the Ad^*P_m orbit $\mathcal{O}_\beta \subseteq \hat{N}_m$. Corollary 2.11 tells us we have a disjoint union

$$(2.14) \qquad (\tilde{Sp}(W))^\wedge = (\tilde{Sp}(W))^\wedge_m \,\dot{\cup}\, \Big(\bigcup_{\operatorname{rank}\beta < m} (\tilde{Sp}(W))^\wedge_\beta \Big)$$

The oscillator representations are examples of rank 1 representations. Lemma 2.6 together with formula (1.8) tells us that the Weil representation of $\tilde{Sp}$ associated to the form β decomposes into representations belonging to $\widetilde{Sp}^{\wedge}_{\beta'}$, where $\beta' = -(\frac{1}{2})\beta$. (This slight discrepancy is an artifact of our conventions and could be eliminated. See example 2.3.) In particular, none of the $\widetilde{Sp}^{\wedge}_{\beta}$ are empty. We will study them in more detail in §4. We finish this section with an observation about the rank of the most familiar type of representation, tempered representations.

In order for a representation of an abelian group to be quasi-equivalent to a subrepresentation of the regular representation, its spectral measure must be absolutely continuous with respect to Haar measure on the Pontrjagin dual. Since the canonical measure classes on the $Ad^*P_m(X)$ orbits in $\hat{N}_m(X)$ are absolutely continuous with respect to Haar measure only for the open orbits, which are of rank m, we have

Proposition 2.16: A representation σ of $\widetilde{Sp}$ is $N_m(W)$-regular, in the sense defined at the beginning of this section, if and only if it is of pure N_m-rank m.

Proposition 2.17: All irreducible tempered representations of $\widetilde{Sp}_{2m}$ are of pure N_m-rank m.

Proof: By the preceding proposition, it will suffice to prove tempered representations are N_m-regular. In fact, we will show something much more general.

Proposition 2.18: If G is a reductive group and $N \subseteq G$ is a unipotent subgroup, and $\rho \in \hat{G}$ is tempered, then ρ is N-regular.

It will be convenient to postpone the proof of this until §7.

3: N_o-rank and regularity

In this section we digress slightly from our development of the properties of N_m-rank to put it in a more general setting. Let $P \subseteq Sp(W)$ be an arbitrary parabolic, and let N denote the unipotent radical of P. We will relate the notion of N-regularity to N_m-rank.

We begin by describing the possible parabolics P. As in §1, let X_k be the span of the standard basis vectors e_i for $i \leq k$. Let $0 < k_1 < k_2 < \dots < k_\ell \leq n$ be a sequence of integers. We call the nested sequence of subspaces

$$F = \{X_{k_1}, X_{k_2}, \dots, X_{k_\ell}\}$$

a <u>standard flag</u>. By $P(F,W) = P(F)$, we mean the subgroup of Sp that preserves all the subspaces of F. Each parabolic subgroup is conjugate to $P(F)$ for a unique standard flag F. The unipotent radical $N(F)$ of $P(F)$ consists of transformations which act as the identity on each quotient $X_{k_i}/X_{k_{i-1}}$ and on $X_{k\ell}^{\perp}/X_{k\ell}$. The maximal parabolic subgroups are obviously those $P(F)$ for which $F = \{X_k\}$ is a singleton.

We abbreviate $P(\{X_k\}) = P_k$. We review the structure of P_k for $k < m$. We have the decomposition

$$P_k \simeq Sp(W_k^{\perp}) \cdot GL(X_k) \cdot N_k \tag{3.1}$$

Here as above W_k is the span of $\{e_i, f_i\}$ for $i \leq k$, and $W_k^{\perp}$ is its orthogonal complement in W. We have also abbreviated $N(\{X_k\}, W) = N_k(W) = N_k$. The unipotent radical N_k is two-step nilpotent. It fits in an exact sequence

$$1 \to ZN_k \to N_k \to Hom(W_k^{\perp}, X_k) \to 1 \tag{3.2}$$

Here ZN_k denotes the center of N_k. It is naturally contained in $Sp(W_k)$, and in fact we have

$$(3.3) \qquad ZN_k = N_k(W) \cap Sp(W_k) = N_k(W_k) \simeq S^2(X_k)$$

The final isomorphism is described by formula (1.7). The quotient homomorphism from $N_k(W)$ to $Hom(W_k^{\perp}, X_k)$ is realized by observing that if $x \in N_k(W)$, then $x-1$ defines a map from $W_k^{\perp}$ to X_k.

To finish the description of $N_k(W) = N_k$ we describe the commutator of two of its elements. This will essentially be an $S^2(X_k)$-valued bilinear form on $Hom(W_k^{\perp}, X_k)$. Given $T \in Hom(W_k^{\perp}, X_k)$, we may consider its adjoint $T^* \in Hom((X_k)^*, (W_k^{\perp})^*)$. But by virtue of the restriction of the form $< , >$, the space $W_k^{\perp}$ is isomorphic to its dual $(W_k)^*$. Explicitly, define

$$\alpha : W_k^{\perp} \longrightarrow (W_k^{\perp})^*$$

by

$$(3.4) \qquad \alpha(w_1)(w_2) = <w_2, w_1> \qquad w_1 \in W_k^{\perp}$$

Then $\alpha^{-1} \circ T^*$ maps X_k^* to $W_k^{\perp}$. Hence given two maps $T, S \in Hom(W_k^{\perp}, X_k)$ we may form

$$T \circ \alpha^{-1} \circ S^* \in Hom(X_k^*, X_k) \simeq X_k \otimes X_k$$

We compute

$$(T \circ \alpha^{-1} \circ S^*)^* = S \circ \alpha^{-1*} \circ T^* = - S \circ \alpha^{-1} \circ T^*$$

since $\alpha^* = - \alpha$. Therefore

$$(3.5) \qquad T \circ \alpha^{-1} \circ S^* - S \circ \alpha^{-1} \circ T^*$$

is self-adjoint, and so may be regarded as an element of $S^2(X_k)$. The bilinear form (3.5) is the commutator form on $\mathrm{Hom}(W_k^{\perp}, X_k)$.

Just as in the formula (2.1) we may identify the Pontrjagin dual $(\mathcal{Z}N_k)^{\wedge}$ with $S^{2*}(X_k)$. The action of Ad^*P_k on $(\mathcal{Z}N_k)^{\wedge}$ then becomes the natural action of $GL(X_k)$ on $S^{2*}(X_k)$. (The subgroup $Sp(W_k^{\perp})$ of P_k centralizes $\mathcal{Z}N_k$ and so acts trivially on $(\mathcal{Z}N_k)^{\wedge}$.) Thus the Ad^*P_k orbits in $(\mathcal{Z}N_k)^{\wedge}$ correspond to the isomorphism classes of symmetric bilinear forms on X_k. Choose such a form β and consider it in its role as linear functional on $S^2(X_k) \simeq \mathcal{Z}N_k$. Let N_β denote the quotient of $N(X_k)$ obtained by dividing $\mathcal{Z}N_k$ by the kernel of β. Providing $\beta \neq 0$, the group N_β will still be two-step nilpotent. On $\mathrm{Hom}(W_k^{\perp}, X_k)$, the commutator form of N_β is clearly

$$\langle S,T\rangle_\beta = \beta((T \circ \alpha^{-1} \circ S^* - S \circ \alpha^{-1} \circ T^*)) \tag{3.6}$$

Let $\mathcal{R}_\beta$ denote the radical of the form $\langle\ ,\ \rangle_\beta$, and let $\mathcal{W}$ denote a complement to $\mathcal{R}_\beta$ in $\mathrm{Hom}(W_k^{\perp}, X_k)$, so that

$$\mathrm{Hom}(W_k^{\perp}, X_k) \simeq \mathcal{W} \oplus \mathcal{R}_\beta$$

and $\langle\ ,\ \rangle_\beta$ is non-degenerate on $\mathcal{W}$. Observe that the inverse image of $\mathcal{W}$ in N_β is isomorphic to the Heisenberg group $H(\mathcal{W})$, while the subspace $\mathcal{R}_\beta$ may be lifted to a subgroup, also denoted $\mathcal{R}_\beta$, of N_β. Thus we have

$$N_\beta \simeq H(\mathcal{W}) \oplus \mathcal{R}_\beta \tag{3.7}$$

Let χ_β be the character of $\mathcal{Z}N_k$ attached to β as in formula (2.1). Then there is a unique representation ρ_β of $H(\mathcal{W})$ with central character χ_β. It then follows easily that a general representation of

N_k which is a multiple of χ_β on $Z\,N_k$ has the form

$$(3.8) \qquad \rho_\beta \otimes \psi \qquad \psi \in (R_\beta)^\wedge$$

Here we have made the convention that ρ_β is extended to N_β by letting it be trivial on R_β, and, similarly, ψ is extended to N_β by letting it be trivial on $H(W)$.

It remains to determine the space R_β. Let R_β be the radical of β considered as a symmetric bilinear form on X_k. Then we have

$$(3.9) \qquad R_\beta = \mathrm{Hom}(W_k^\perp, R_\beta)$$

To see this, it is convenient to regard β as a self-adjoint map

$$L_\beta : X_k \to X_k^*$$

by the rule

$$L_\beta(x)(x') = \beta(x,x') \qquad x,x' \in X_k$$

Then $R_\beta = \ker L_\beta$. Given maps $S, T \in \mathrm{Hom}(W_k^\perp, X_k)$, the composite map $T \circ \alpha^{-1} \circ S^*$ goes from X_k^* to X_k. Thus the further composite $T \circ \alpha^{-1} \circ S^* \circ L_\beta$ takes X_k to itself. We may rewrite the formula (3.6) as

$$\langle S,T\rangle_\beta = \mathrm{tr}((T \circ \alpha^{-1} \circ S^* - S \circ \alpha^{-1} \circ T^*)L_\beta)$$

where tr is the usual trace functional on $\mathrm{End}(X_k)$. Suppose S belongs to $\mathrm{Hom}(W_k^\perp, R_\beta)$. Then $\alpha^{-1} \circ S^* \in \mathrm{Hom}(X_k^*, W_k^\perp)$ annihilates $R_\beta^\perp$, the annihilator of R_β in X_k^*. Since the map L_β is self-adjoint, it has image $R_\beta^\perp$. Hence the map $\alpha^{-1} \circ S^* \circ L_\beta$ is zero, while the map

$S \circ \alpha^{-1} \circ T^* \circ L_\beta$ has kernel containing R_β and image contained in R_β and must therefore have trace zero. Thus we see for such S, the number $\langle S,T\rangle_\beta$ is zero independent of T; this is to say $S \in \mathcal{R}_\beta$. The left hand side of equation (3.9) therefore contains the right hand side. The verification of the reverse inclusion is left to the reader.

For β of rank k, which of course is typical, formula (3.9) tells us that $\mathcal{R}_\beta = \{0\}$, so that N_β is itself a Heisenberg group. Hence for β of rank k there will be a unique representation ρ_β of N_k such that ρ_β is a multiple of χ_β on $Z N_k$. We know also [MW] that these ρ_β are square integrable modulo $Z N_k$. The regular representation of $Z N_k$, or any subrepresentation of it, will assign zero spectral measure to the subvariety of $(ZN_k)^\wedge$ consisting of forms of rank less than k. Together, these facts give us one direction of the following result.

Lemma 3.1: A representation σ of $\widetilde{Sp}(W)$ is N_k-regular if and only if it is $Z N_k$-regular.

Proof: As noted, the remarks above show that $Z N_k$-regularity implies N_k-regularity. The other implication is quite general. Let $G \supseteq H_1 \supseteq H_2$ be any locally compact group and two subgroups. Then if a representation σ of G is H_1-regular, it is also H_2-regular. This is because the regular representation of H_1, restricted to H_2, is a multiple of the regular representation of H_2.

Since $Z N_k(W) = N_k(W_k)$ is to $Sp(W_k)$ as $N_m(W)$ is to $Sp(W)$, we have a second result as a consequence of lemma 3.1, lemma 2.9, and proposition 2.14.

Proposition 3.2: A representation σ of $\widetilde{Sp}$ of pure N_m-rank ℓ is N_k-regular if and only if $\ell \geq k$.

Remark: Since Corollary 2.11 allows us to decompose a representation of $\tilde{S}p$ into a direct sum of representations of pure rank, proposition 3.2 is effectively a criterion for a general representation of $\tilde{S}p$ to be N_k-regular.

Now consider a general parabolic subgroup $P(F)$ defined by some standard flag F. Let X_k be the largest subspace of F. The main result of this section gives a criterion for $N(F)$-regularity in terms of rank.

Proposition 3.3: A representation σ of $\tilde{S}p$ of pure N_m-rank ℓ is $N(F)$-regular if and only if $\ell \geq k$.

Proof: We note that $P(F) \subseteq P(X_k)$, and $N_k \subseteq N(F)$. Also ZN_k is a normal, though not central, subgroup of $N(F)$, and N_k is also normal in $N(F)$. Proposition 3.3 will follow from proposition 3.2 and the following lemma.

Lemma 3.4: A representation σ of $\tilde{S}p$ is $N(F)$-regular if and only if it is N_k-regular.

Proof: This result has evidently only to do with the relation between $N(F)$ and N_k. It could be reformulated: a representation of $N(F)$ which is N_k-regular is equivalent to a subrepresentation of the regular representation. We will prove this.

Select $\beta \in S^{2*}(X_k)$. Let β have rank k. Let ρ_β be the unique representation of N_k with central character χ_β. I claim that for β outside a proper subvariety of $S^{2*}(X_k)$, the induced representation

$$\mathrm{ind}_{N_k}^{N(F)} \rho_\beta = \sigma_\beta$$

is irreducible. By Mackey's theory [My], to establish irreducibility of σ_β it is enough to show that $Ad^* m(\rho_\beta) \neq \rho_\beta$ for $m \in N(F)$ but not in N_k.

The decomposition (3.1) of P_k gives us

$$N(F) \simeq N' \cdot N_k$$

where $N' = N(F) \cap GL(X_k)$. The action of $Ad^* N(F)$ on $Z N_k^\wedge$ factors to an action of N'. It is clear from the construction of ρ_β that we have

$$Ad^* m(\rho_\beta) = \rho_{m(\beta)} \qquad m \in N'$$

where $m(\beta)$ indicates the transform of β by m in the standard action of $GL(X_k)$ on $S^{2*}(X_k)$. Hence to prove irreducibility of σ_β we must simply show that N' acts freely on the complement of a proper closed subvariety of $S^{2*}(X_k)$. Since the condition that the isotropy group in N' of $\beta \in S^{2*}(X_k)$ be non-trivial is an algebraic condition, it will suffice to exhibit one β for which said isotropy group is trivial.

In the standard basis $\{e_i\}_{i=1}^k$ of X_k the elements of N' appear as upper triangular matrices with 1's on the diagonal. It is then easy to check that for any inner product β on X_k for which the e_i form an orthogonal basis, the isotropy group in N' of β is trivial. This establishes the claim of generic irreducibility of σ_β.

Now consider a representation σ of $N(F)$. Suppose σ is N_k-regular. Then $\sigma | N_k$ will decompose as a direct integral over the representations ρ_β, for β of rank k in $S^{2*}(X_k)$, and the spectral measure of this decomposition will be absolutely continuous with respect to Haar measure on $S^{2*}(X_k)$. The spectral measure moreover allows N' as automorphism group, so that it is of constant multiplicity along N' orbits. It follows that $\sigma | N_k$ is quasi-equivalent to

$$(\mathrm{ind}_{N_k}^{N(F)}(\sigma|N_k))|N_k$$

But since σ_β is irreducible for almost all β, we see that any N_k-regular representation σ of $N(F)$ is determined by its restriction to N_k. Therefore σ is quasi-equivalent to

$$\mathrm{ind}_{N_k}^{N(F)}(\sigma|N_k)$$

and this representation is obviously $N(F)$-regular. Indeed, if $G \supseteq H$ is a group and a subgroup, then it is easy to see that

$$\mathrm{ind}_H^G \ L^2(H) \simeq L^2(G)$$

It follows that any subrepresentation of the regular representation of H induces a G-regular representation of G. This concludes lemma 3.4.

Remark: In fact, the above analysis, in conjunction with lemma 2.9 allows us to determine the $N(F)$ spectrum of any representation σ of $\tilde{Sp}$ of pure rank $\geq k$ from its N_m-spectrum.

4: Description of $(\tilde{Sp})^\wedge_\beta$.

Fix a symmetric bilinear form β on X. Let $(\tilde{Sp})^\wedge_\beta$ be the subset of $(\tilde{Sp})^\wedge$ defined at the end of §2. The business of this section is to provide a description of $(\tilde{Sp})^\wedge_\beta$.

Let the parabolic $P_k \subseteq Sp$ be as defined in (3.1).

Theorem 4.1: Let σ be a representation of $\tilde{Sp}$ of pure N_m-rank $\ell < m$. Then if $k \geq \ell$, the (weakly closed) algebra generated by $\sigma(\tilde{P}_k)$ is equal to the algebra generated by all of $\sigma(\tilde{Sp})$. In particular σ is irreducible if and only if $\sigma|\tilde{P}_k$ is irreducible. And if σ_1 and σ_2 are of pure N_m-rank ℓ and irreducible, then σ_1 and σ_2 are equivalent if and only if $\sigma_1|\tilde{P}_k$ and $\sigma_2|\tilde{P}_k$ are equivalent.

Proof: As in theorem 2.10, the proof is by induction. For $\ell = 0$, the theorem is true, since from [HM] we know that $(\tilde{Sp})^\wedge_0$ consists of the trivial representation alone.

Assume therefore that $\ell \geq 1$, in which case $m \geq 2$. Consider the restriction of σ to $\tilde{P}_1$. According to the argument of theorem 2.10, we have for $\sigma|\tilde{P}_1$ the decomposition (2.8), which we reproduce:

$$\sigma|\tilde{P}_1 \simeq \sum_{i,t} \nu^i_t \otimes \tau^i_t \qquad i=1,2;\ t \in F^\times/F^{\times 2}$$

Here the ν^i_t are representations of $\tilde{Sp}(W_1^\perp)\cdot\tilde{F}^\times$ on which $\tilde{F}^\times$ acts by scalars. (This follows from the description of the ν^i_t following formula (2.8), and corollary 2.13, since the ν^i_t are of pure rank.) As representations of $\tilde{Sp}(W_1^\perp)$ the ν^i_t are of pure $N_{m-1}(W_1^\perp)$-rank $\ell - 1$.

Consider first the case $\ell = 1$ and $k = 1$. Then the ν^i_t are of rank 0, so that they are simply multiples of the trivial representation. Hence we actually have

$$\sigma|\tilde{P}_1 \simeq \sum_{i,t} n_t^i \tau_t^i$$

where the n_t^i are non-negative integers. To prove the theorem in this case amounts to showing that $\sigma(\tilde{P}_1)$ generates the same algebra as $\sigma(\tilde{S}p)$. Consider $\tilde{S}p(W_1) \subseteq \tilde{S}p(W)$. We note that $Sp(W_1) \cap P_1(W) = P_1(W_1)$. Suppose that $\sigma(\tilde{P}_1(W_1))$ generates the same algebra as $\sigma(\tilde{S}p(W_1))$. Then $\sigma(\tilde{P}_1(W))$ generates the same algebra as $\sigma(\tilde{P}_1(W) \cup \tilde{S}p(W_1))$. But $\tilde{P}_1(W) \cup \tilde{S}p(W_1)$ generates the whole group $\tilde{S}p(W)$.

Thus this case of the theorem will follow if we show $\sigma(\tilde{P}_1(W_1)) = \sigma(\tilde{S}p(W_1))$. From formula (2.7), we know that $\sigma|\tilde{S}p(W_1^{\perp})$ is a sum of oscillator representations. Since $\tilde{S}p(W_1)$ is conjugate in $\tilde{S}p(W)$ to a subgroup of $\tilde{S}p(W_1^{\perp})$, formula (1.15) tells us that $\sigma|\tilde{S}p(W_1)$ is also a sum of oscillator representations. We know the oscillator representations of $\tilde{S}p(W_1)$ form a finite set $\{\omega_t\}$ parametrized by $F^{\times}/F^{\times 2}$. We also know (it is a very special case of theorem 1.1) that each ω_t decomposes into 2 irreducible components ω_t^+ and ω_t^-. In the realization of ω_t on $L^2(Y)$ as given by formula (1.8), the space ω_t^+ can be taken as the even functions: functions f such that $f(-y) = f(y)$. And ω_t^- will be the odd functions: functions such that $f(-y) = -f(y)$. The representations ω_t^+ and ω_t^- are seen by inspection to be inequivalent on $\{\widetilde{\pm 1}\}$ where here $\{\pm 1\}$ indicates the identity transformation of W and its negative. On the other hand, $\omega_t^{\pm}$ and $\omega_s^{\pm}$ for $s^{-1}t$ a non-square have distinct $N_1(W_1)$ spectra. Since $\{\widetilde{\pm 1}\} \cdot N_1(W_1) \subseteq \tilde{P}_1(W_1)$ we see that the $\omega_t^{\pm}$ are pairwise inequivalent on $\tilde{P}_1(W_1)$. Furthermore, it is fairly easy to see that each $\omega_t^{\pm}$ is irreducible on $\tilde{P}_1(W_1)$. Indeed, ω_t has $N_1(W_1)$-spectrum of multiplicity

2 supported on $\mathcal{O}_{\beta_s}$ where $s = -(\frac{1}{2})t$, and this is divided among the $\omega_t^{\pm}$, each of which contains $\mathcal{O}_{\beta_s}$ with multiplicity 1. With $N_1(W_1)$-spectrum of multiplicity one on one $\mathrm{Ad}^*\tilde{P}_1(W_1)$ orbit, irreducibility of $\omega_t^{\pm}$ on $\tilde{P}_1(W_1)$ is immediate.

Remark: The arguments just recited may be regarded as the reverse of the arguments in the proof of theorem 2.10 leading to the construction of the τ_t^i. Also, they justify the comments made there about the connection between the $\omega_t^{\pm}$ and the τ_t^i.

We now know that the $\omega_t^{\pm}$ are irreducible and pairwise inequivalent as representations of $\tilde{P}_1(W_1)$. Since only they occur in $\sigma|\tilde{S}p(W_1)$ in the case at hand, the desired fact, that $\sigma|(\tilde{P}_1(W_1))$ generates the same algebra as $\sigma(\tilde{S}p(W_1))$ (namely the sum of the full matrix algebras on the spaces of the $\omega_t^{\pm}$ which occur in σ). This proves the theorem when $\ell = 1$, $k = 1$.

In all other cases we will have $k > 1$, and thus $k-1 \geq 1$ as well as $k-1 \geq \ell -1$. By induction we may assume the theorem is true for $k-1$, $\ell -1$ and $m-1$. That is, $\nu_t^i(\tilde{P}_{k-1}(W_1^{\perp}))$ will generate the same algebra as $\nu_t^i(\tilde{S}p(W_1^{\perp}))$. But the representations τ_t^i of $\tilde{P}_1(W)$ were obtained from representations of the subgroup $\tilde{F}^{\times} \cdot N_1(W)$ of $\tilde{P}_1(W)$ simply by extension. In particular, the τ_t^i are already irreducible and pairwise distinct on $\tilde{F}^{\times} \cdot N_1(W)$. Hence for $g \in \tilde{S}p(W_1^{\perp})$, the operators $1 \otimes \tau_t^i(g)$, where 1 here is the identity on the space of ν_t^i, already lie in the algebra generated by $\sigma(\tilde{F}^{\times} \cdot N_1(W_1^{\perp}))$. In particular, this is true for $g \in \tilde{P}_{k-1}(W_1^{\perp})$. Hence the operators $\nu_t^i(g) \otimes 1$, where 1 here is the identity on the space of τ_t^i, are in the algebra generated by $\sigma(\tilde{P}_{k-1}(W_1^{\perp}))$ and $\sigma(\tilde{F}^{\times} \cdot N_1(W))$. By the previous case of the theorem, we conclude that the operators $\nu_t^i(g) \otimes 1$ for all $g \in \tilde{S}p(W_1^{\perp})$ are in the algebra generated by $\sigma(\tilde{P}_{k-1}(W_1^{\perp}) \cdot \tilde{F}^{\times} \cdot N_1(W))$. Hence finally the operators

$$\sigma(g) = \Sigma\, \nu_t^i(g) \otimes \tau_t^i(g) = \Sigma\, (\nu_t^i(g) \otimes 1)(1 \otimes \tau_t^i(g)),$$

for all $g \in \tilde{S}p(W_1^{\perp})$ are in this algebra. Thus we have shown $\sigma(\tilde{P}_1)$ and $\sigma(\tilde{P}_{k-1}(W_1^{\perp}) \cdot \tilde{F}^{\times} \cdot N_1(W))$ generate the same algebra. But now observe that

$$\tilde{P}_{k-1}(W_1^{\perp}) \cdot \tilde{F}^{\times} \cdot N_1(W) = \tilde{P}_k(W) \cap \tilde{P}_1(W)$$

Hence $\sigma(\tilde{P}_k(W) \cap \tilde{P}_1(W))$ generates the same algebra as $\sigma(\tilde{P}_1(W))$. Hence $\sigma(\tilde{P}_k(W))$ generates the same algebra as $\sigma(\tilde{P}_1(W) \cup \tilde{P}_k(W))$. Since $k > 1$, the union $\tilde{P}_1(W) \cup \tilde{P}_k(W)$ generates all of $\tilde{S}p(W)$. The theorem follows.

Theorem 4.1 immediately implies that restriction induces an injection

$$(\tilde{S}p(W))^{\wedge}_{\ell} \to \tilde{P}_k(W)^{\wedge}$$

when $k \geq \ell$. In fact, a much sharper result holds, as we shall now show.

We focus attention on $(\tilde{S}p)^{\wedge}_{\beta}$ for a symmetric form β of rank $\ell < m$. Consider $\sigma \in (\tilde{S}p)^{\wedge}_{\beta}$. By definition, the N_m-spectrum of σ is concentrated on the $\mathrm{Ad}^* P_m$-orbit $\mathcal{O}_{\beta} \subseteq \hat{N}_m$. We may assume that the form

$$\gamma = \beta | X_{\ell} \tag{4.1}$$

is non-degenerate on X_{ℓ} and so defines an inner product on X_{ℓ}.

We know from theorem 4.1 that the restriction of σ to the maximal parabolic subgroup $P_\ell(W)$ is irreducible. We want to investigate the structure of $\sigma|P_\ell$. Let $Z\,N_\ell$ be the center of the unipotent radical $N_\ell(W)$ of $P_\ell(W)$. We know that $Z\,N_\ell = N_\ell(W) \cap Sp(W_\ell) = N_\ell(W_\ell)$. Hence lemma 2.9 tells us that the $Z\,N_\ell$-spectrum of σ is concentrated on the $Ad^* P_\ell(W)$ orbit $\mathcal{O}_\gamma \subseteq (ZN_\ell)^\wedge$, with γ as in equation (4.1). If π_σ denotes the $Z\,N_\ell$-spectrum of σ, then π_σ satisfies a transformation law analogous to equation (2.2). But equation (2.2) says π_σ is a system of imprimitivity in the sense of Mackey, for $\tilde{P}_\ell(W)$ and σ, based on $\mathcal{O}_\gamma$. Let $J \subseteq \tilde{P}_\ell(W)$ be the isotropy group of γ under $Ad^* P_\ell$. Then Mackey's Imprimitivity Theorem says that $\sigma|\tilde{P}_\ell(W)$ is realizable as a representation induced from some representation τ of J. In symbols

$$\sigma|\tilde{P}_\ell(W) \simeq \mathrm{ind}_J^{\tilde{P}_\ell}\, \tau \tag{4.2}$$

We will make (4.2) more precise by giving sharper descriptions of J and τ.

We have the decomposition (3.1) of $P_\ell(W)$. It lifts to a decomposition of $\tilde{P}_\ell(W)$. The subgroup $\tilde{Sp}(W_\ell^\perp)\cdot N_\ell(W)$ of $\tilde{P}_\ell$ centralizes $Z\,N_\ell$, so it will belong to J. Clearly the subgroup of $GL(X_\ell)$ that leaves the inner product γ fixed is just O_γ, the isometry group of γ. Hence we have

$$J = \tilde{Sp}(W_\ell\ ) \cdot \tilde{O}_\gamma \cdot N_\ell(W) \tag{4.3}$$

Consider the representation τ of J. Mackey's theory implies that $\tau|Z\,N_\ell$ is a multiple of the character χ_γ, as defined in formula (2.1). Since γ is non-degenerate on X_ℓ, the arguments leading to formula (3.8) tell us that $\tau|N_\ell(W)$ is a multiple of the unique representation ρ_γ of N_ℓ with central character χ_γ. Recall that ρ_γ actually factors to the group

(4.4) $$N_\gamma \simeq H(\mathrm{Hom}(W_\ell, X_\ell)),$$

the Heisenberg group attached to $\mathrm{Hom}(W_\ell, X_\ell)$, with symplectic structure given by formula (3.6).

The action by conjugation of $Sp(W_\ell)\cdot O_\gamma$ on N_ℓ factors to an action on N_γ. It is evident that this action maps $Sp(W_\ell)\cdot O_\gamma$ into $Sp(\mathrm{Hom}(W_\ell, X_\ell), <\ ,\ >_\gamma)$ as a dual pair. Let us abbreviate $Sp(\mathrm{Hom}(W_\ell, X_\ell), <\ ,>_\gamma) = Sp_\gamma$. We know that the representation ρ_γ of N_γ extends to a representation, still to be denoted ρ_γ, of the semi-direct product $\widetilde{Sp}_\gamma \cdot N_\gamma$. Obviously we want to convert ρ_γ into a representation of J. It is fairly clear how to do so, but we encounter the same technicalities as we did in theorem 2.10. We summarize the details.

The formula (1.18) says that $\widetilde{Sp}(W_\ell)$ may be mapped into $\rho_\gamma(\widetilde{Sp}_\gamma)$ by taking an appropriate ℓ-fold tensor product of oscillator representations. Pulling back by ρ_γ maps $\widetilde{Sp}(W_\ell)$ into $\widetilde{Sp}_\gamma$. Since $\widetilde{Sp}(W_\ell)$ is its own commutator subgroup, this mapping may be characterized as the unique lift of the natural embedding of $Sp(W_\ell)$ into Sp_γ. Similarly, O_γ is contained in $Sp(W_\ell)$, and the action by conjugation of O_γ on $N_\ell/Z\,N_\ell$ may be regarded as the $(m-\ell)$ fold direct sum of the natural action of O_γ on W_ℓ. Thus formula (1.15) tells us $\tilde{O}_\gamma$ maps into $\rho_\gamma(\widetilde{Sp}_\gamma)$ by an $(n-\ell)$-fold tensor product of the restriction of the oscillator representation of $\widetilde{Sp}(W_\ell)$.

Thus we can map both $\widetilde{Sp}(W_\ell)$ and $\tilde{O}_\gamma$ into $\widetilde{Sp}_\gamma$, so we can extend the representation ρ_γ to a representation, still denoted ρ_γ, of the semi-direct product

$$(\widetilde{Sp}(W_\ell) \times \tilde{O}_\gamma) \ltimes N_\gamma$$

In order for ρ_γ to factor to a representation of J, we must have a compatibility condition as in theorem 2.10. Let $\mathbb{Z}_2$ denote the common kernel of the projection maps $\widetilde{Sp}(W_\ell^\perp) \to Sp(W_\ell^\perp)$ and $\tilde{O}_\gamma \to O_\gamma$. This group $\mathbb{Z}_2$ is mapped into $\widetilde{Sp}_\gamma$ both as a subgroup of $\widetilde{Sp}(W_\ell^\perp)$ and as a subgroup of $\tilde{O}_\gamma$. In order for ρ_γ to factor to J, these two mappings of $\mathbb{Z}_2$ must coincide. Let ε be the non-identity element of $\mathbb{Z}_2$. By the tensor product descriptions of the mappings to $\widetilde{Sp}_\gamma$, we know that via the $\widetilde{Sp}(W_\ell^\perp)$ mapping we get $\rho_\gamma(\varepsilon) = (-1)^\ell$, while via the $\tilde{O}_\gamma$ mapping we get $\rho_\gamma(\varepsilon) = (-1)^{m-\ell}$. Therefore ρ_γ will factor to J if m is even, and will not factor if m is odd.

When m is odd, we may alter the mapping of $\tilde{O}_\gamma$ to $\widetilde{Sp}_\gamma$ by a character of $\tilde{O}_\gamma$ of order 2, non-trivial on $\mathbb{Z}_2$. This modification will then produce a representation of $(\widetilde{Sp}(W_\ell^\perp) \times \tilde{O}_\gamma)\cdot N_\gamma$ satisfying the compatibility condition. Hence in any case we can produce an extension, now denoted ρ'_γ, to indicate the possibility that some modification of the "natural extension" may have been necessary, of ρ_γ from N_γ to J. The representation ρ'_γ is defined up to modification by a character of order 2 of O_γ.

Having constructed the representation ρ'_γ, we continue according to Mackey's Theory. It tells us that any representation τ of J which is a multiple of χ_γ on $Z\,N_\ell$ has the form

$$\tau \simeq \tau_1 \otimes \rho'_\gamma \tag{4.5}$$

where τ_1 is a representation of $\widetilde{Sp}(W_\ell^\perp)\cdot\tilde{O}_\gamma$. The representation τ will be irreducible if and only if τ_1 is. The factorization (4.5) applies to the representation τ of equation (4.2). But for that τ, the representation τ_1 must be trivial on $\widetilde{Sp}(W_\ell^\perp)$. This may be seen in several ways. For example, we note that $N_m \subseteq J$. In fact

$$(4.6) \qquad N_m = N_{m-\ell}(W_\ell^\perp) \cdot (N_\ell(W) \cap N_m(W))$$

An analysis precisely analogous to that leading to formula (2.12) yields the conclusion that the N_m-rank of τ in (4.5) is ℓ plus the $N_{m-\ell}(W_\ell^\perp)$-rank of τ_1. Hence for our τ from (4.2), the $N_{m-\ell}(W_\ell^\perp)$-rank of τ_1 must be zero, whence $\tau_1|Sp(W_\ell^\perp)$ is trivial, as stated.

Hence, for our τ, the representation τ_1 factors to $\tilde{O}_\gamma$. But the compatibility condition then requires that σ_1 factor further to O_1. Thus we have a sequence of mappings

$$(4.7) \qquad \begin{aligned} \sigma \in (\tilde{S}p(W))^\wedge_\beta &\to \sigma|\tilde{P}_\ell(W) \in (\tilde{P}_\ell(W))^\wedge_\beta \\ \simeq \operatorname{ind}_J^{\tilde{P}_\ell} \tau &\to \tau \in \hat{J} \simeq \tau_1 \otimes \rho'_\gamma \\ &\to \tau_1 \in \hat{O}_\gamma \end{aligned}$$

Combining all these gives us a map

$$(4.8) \qquad \begin{aligned} \Delta : (\tilde{S}p(W))^\wedge_\beta &\to \hat{O}_\gamma \\ \sigma &\to \tau_1 \end{aligned}$$

Theorem 4.2: The map Δ of (4.7) defines an injection from $(\tilde{S}p(W))^\wedge_\beta$ into $\hat{O}_\gamma$.

Proof: This follows directly from theorem 4.1 and Mackey's general results.

To actually compute $\Delta(\sigma)$ for $\sigma \in (\tilde{S}p)^\wedge_\beta$ according to the above recipe would involve going through all the mappings of sequence (4.7). Of these, the factorization (4.5) is the most difficult step. It is in general no very straightforward matter to factor a representation into a tensor product. However, the special structure of the Heisenberg group allows us to compute $\tau_1 = \Delta(\sigma)$ in a relatively efficient manner.

Recall the description (4.4) of N_γ. We may write

$$W_\ell^\perp = (X_m \cap W_\ell^\perp) \oplus (Y_m \cap W_\ell^\perp)$$

The summands will be the spans of e_i for $\ell + 1 \le i \le m$ and f_i for $\ell + 1 \le i \le m$, respectively. This decomposition induces another:

$$\mathrm{Hom}(W_\ell^\perp, X_\ell) \simeq \mathrm{Hom}(X_m \cap W_\ell^\perp, X_\ell) \oplus \mathrm{Hom}(Y_m \cap W_\ell^\perp, X_\ell).$$

It is not hard to check that $\mathrm{Hom}(X_m \cap W_\ell^\perp, X_\ell)$ is the image in N_ℓ/ZN_ℓ of $N_m \cap N_\ell$. Also, the inverse image in N_γ of $\mathrm{Hom}(X_m \cap W_\ell^\perp, X_\ell)$ is a maximal abelian subgroup of N_γ. (Note that in fact $N_m \cap N_\ell$ is abelian in N_ℓ, so its image in N_γ is <u>a fortiori</u> abelian). Finally note that $N_m \cap N_\ell$ is normalized by O_γ and by $P_{m-\ell}(W_\ell^\perp)$, the stabilizer of $X_m \cap W_\ell^\perp$ in $Sp(W_\ell^\perp)$. Of course $Z\,N_\ell \subseteq N_m \cap N_\ell$ is also normalized by these groups. In $N_m \cap N_\ell$ there is a unique complement to $Z\,N_\ell$ which is also normalized by these groups. Call this complement U_ℓ. Then we have

(4.9) $$N_m \cap N_\ell = U_\ell \oplus Z\,N_\ell$$

with both U_ℓ and $Z\,N_\ell$ normalized by O_γ and by $P_{m-\ell}(W_\ell^\perp)$.

The following method for computing τ_1 could be formulated purely in terms of Hilbert space considerations by further use of Mackey's theory, but the alternate formulation chosen seems slightly more convenient in the present context. Let $\mathcal{Y}$ be the Hilbert space on which the representation ρ'_γ of equation (4.5) is defined, and let $\mathcal{H}_1$ be the space of the other factor τ_1 on the right hand side of (4.5). Then the space of the tensor product τ is of course $\mathcal{H} \simeq \mathcal{H}_1 \otimes \mathcal{Y}$. The space $\mathcal{Y}$ is a module for J, but it is already irreducible under the action of N_ℓ acting through its quotient N_γ. Let $\mathcal{Y}^\infty \subseteq \mathcal{Y}$ denote the subspace of smooth vectors for N_γ. Then $\mathcal{Y}^\infty$ has a natural nuclear locally convex topology. If $\mathcal{Y}$ is taken to be $L^2(N_\ell/(N_\ell \cap N_m))$ with action analogous to formulas (1.4), then $\mathcal{Y}^\infty$ is the Schwartz-Bruhat space $\mathcal{S}(N_\ell/(N_\ell \cap N_m))$. In any case the space of smooth vectors for N_γ in $\mathcal{H}$ is clearly

$$\mathcal{H}^\infty = \mathcal{H}_1 \otimes \mathcal{Y}^\infty \tag{4.10}$$

The tensor product is well-defined as a topological vector space since $\mathcal{Y}^\infty$ is nuclear.

Let χ'_γ denote the extension of the character χ_γ from $Z\,N_\ell$ to $N_\ell \cap N_m$ which is trivial on the complement U_ℓ of equation (4.9). A basic fact about the Heisenberg group [H1],[Cr] says there is a unique linear functional λ on $\mathcal{Y}^\infty$ such that

$$\lambda(\rho_\gamma(n)y) = \chi'_\gamma(n)\lambda(y) \qquad n \in N_\ell \cap N_m,\ y \in \mathcal{Y}^\infty \tag{4.11}$$

The character χ'_γ is clearly invariant under the action $\mathrm{Ad}^*(\tilde{P}_{m-\ell}(W_\ell)\cdot\tilde{O}_\gamma)$ on $(N_\ell \cap N_m)^\wedge$. Therefore, since it is unique, the functional λ of equation (4.11) is an eigenfunctional for $\tilde{P}_{m-\ell}(W_\ell)\cdot\tilde{O}_\gamma$. Let ψ be the associated (quasi-) character of $\tilde{O}_\gamma$. Then we see easily that the action of $\tilde{O}_\gamma$ on $\mathcal{H}^\infty/(\mathcal{H}_1 \otimes \ker\lambda) \simeq \mathcal{H}_1$ is just $\rho_1 \otimes \psi$.

We may further observe that, since $N_{m-\ell}(W_\ell)$ is in the commutator subgroup of $\tilde{P}_{m-\ell}(W_\ell^\perp)$, it will actually leave the functional λ invariant. From equation (4.6) we therefore see that λ is an eigenfunctional for all of N_m, and we may write

$$\lambda(\rho'_\gamma(n)y) = \chi_\beta(n)\lambda(y) \qquad n \in N_m,\ y \in \mathcal{Y}^\infty \tag{4.12}$$

where χ_β is the character of N_m which extends χ'_γ and is trivial on $N_{m-\ell}(W_\ell^\perp)$. (Recall γ is related to the original form β by equation (4.1).) We may retrospectively assume that the radical of β is exactly $X_m \cap W_\ell^\perp$. Then our use of χ_β in (4.12) is consistent with our earlier definition (2.1).)

Let us put

$$J' = \tilde{P}_{m-\ell}(W_\ell^\perp)\cdot\tilde{O}_\gamma\cdot N_\ell = \tilde{GL}(X_m \cap W_\ell^\perp)\cdot\tilde{O}_\gamma\cdot(N_m\cdot N_\ell). \tag{4.13}$$

Then we may take the following point of toward the representation $\rho'_\gamma|J'$: that it is the extension to J' of the representation

$$\rho'_\gamma \simeq \mathrm{ind}_{N_m}^{N_m\cdot N_\ell}\ \chi_\beta$$

of $N_m\cdot N_\ell$. The space of smooth vectors for $N_m\cdot N_\ell$ is just the same space of smooth vectors as for N_ℓ above. In summary, we have the following result.

Proposition 4.3: The representation of O_γ on the space

$$\mathcal{H}^\infty/(\mathcal{H}_1 \otimes \ker\lambda),$$

where λ is the N_m-eigenfunctional on the space $\mathcal{Y}^\infty$ defined by equation (4.11), is equivalent to the representation $\tau_1\otimes\psi$ where $\tau_1 = \Delta(\sigma)$ is as in equation (4.5) and ψ is the eigencharacter of $\tilde{O}_\gamma$ defined by λ .

Remarks: a) The analysis leading to proposition 4.3 can be extended to conclude that $\tau_1\otimes\psi$ is equivalent, as admissible representation, to the action of O_1 on the space of functionals on σ^∞, the smooth subrepresentation of σ , which are eigenfunctionals for N_m with eigencharacter χ_β . However, the technicalities leading to this result would lead us too far astray.

b) It is obviously of interest to know the image of the map Δ of formula (4.8). A fairly straightforward argument based on the study of the Weil representation of $\widetilde{Sp}(W)$ attached to O_γ shows that the tempered spectrum of O_γ is certainly in the image of Δ. However, there is evidence that Δ is surjective. See for example the Onofri example discussed in §5.

c) Both $(\widetilde{Sp}(W))^\wedge_\beta$ and $\hat{O}_\gamma$ have the structure of topological spaces, defined by the Fell topology [F1]. (This will be discussed in some detail in §6 and §7). It is obvious that Δ is continuous with respect to these topologies. The techniques that show that Δ hits the whole tempered spectrum also show that Δ has a closed image and is a homeomorphism onto its image.

5: Examples

We offer some examples illustrating theorem 4.2. Our discussion will not include all details.

The simplest situation is offered by the rank 1 representations. For $\dim W = 2m \geq 4$, theorem 4.2 implies that the only N_m-rank 1 representations of $\tilde{S}p(W)$ are the two components ω_t^+ and ω_t^- of the oscillator representations themselves. These correspond respectively to the trivial and the signum characters of $O_1 \simeq \{\pm 1\}$. Altogether, then, taking the different oscillator representations into account, we have

$$(5.1) \qquad (\tilde{S}p(W))_1^\wedge \simeq \{\pm 1\}^\wedge \times \{F^\times / F^{\times 2}\}$$

In particular, this is a finite set.

Next consider rank 2 representations. Here we are dealing with representations of Sp rather than of $\tilde{S}p$, and they are the smallest rank representations of Sp (except for the trivial representation). According to theorem 4.2 they are classified by representations of 2-dimensional orthogonal groups for $\dim W \geq 6$. But theorem 1.1 constructs representations of $Sp(W)$ via the Weil representations associated to 2-dimensional orthogonal groups for $\dim W \geq 4$. All unitary representations of the O_2's are matched to unitary representations of Sp; and for $\dim W \geq 6$ one can see by using proposition 4.3 and developing formulas (1.8) that $\tau \in \hat{O}_2$ is matched to $\Delta^{-1}(\tau) \in \hat{S}p$. Thus all rank 2 representations of Sp occur in this way.

Among binary quadratic forms, all but the hyperbolic plane are anisotropic, so that their isometry groups are compact. Thus the representations of Sp coming from these forms make up a discrete set. The Weil representations of Sp corresponding to these O_2's have been studied by Asmuth [Am]. For $\dim W = 4$ and for the signum character of O_2, one gets when F, the base field, is non-archimedean a supercuspidal representation of Sp, analogous to Srinivasan's representation θ_{10} for Sp over finite fields [Sn]. For $\dim W \geq 6$, the range of validity of theorem 4.1 for rank 2 representations, one gets no supercuspidal representations. Indeed, proposition 2.18 prevents representations of N_m-rank less than m from even being tempered, let alone cuspidal. We note that Asmuth identifies the components of rank 2 Weil representations for $\dim W \geq 6$ as being very small constituents of certain induced representations.

For the hyperbolic plane (the split 2-dimensional form), whose isometry group we label $O_{1,1}$, we get a continuous series of representations instead of a discrete set. We will analyze this case in somewhat more detail. The group $O_{1,1}$ has the semidirect product structure

$$O_{1,1} \simeq \mathbb{Z}_2 \ltimes SO_{1,1} \tag{5.2}$$

$$SO_{1,1} \simeq F^\times$$

where $\mathbb{Z}_2$ acts on $F^\times$ by sending t to t^{-1}. Thus for each character ψ of $F^\times$ that is not of order 2, there is a unique 2-dimensional representation

$$\mu_\psi \simeq \operatorname{ind}_{F^\times}^{O_{1,1}} \psi \tag{5.3}$$

of $O_{1,1}$. The restriction of μ_ψ to $F^\times$ is $\psi \oplus \psi^{-1}$, and this characterizes μ_ψ. For $\psi \in \hat{F}^\times$ of order 2, there are extensions ψ^+ and ψ^- of ψ to linear characters of $O_{1,1}$. Here one has

$$(5.4) \qquad \operatorname{ind}_{F^\times}^{O_{1,1}} \psi = \psi^+ \oplus \psi^- \qquad \psi = \psi^{-1} \in F^\times$$

Consider the Weil representation of $Sp(W)$ associated to $O_{1,1}$. This may be realized on $L^2(Y \oplus Y)$ with action given by formulas (1.8) on one factor and the complex conjugate formulas on the other factor. Alternatively it can be realized on $L^2(W)$ with $Sp(W)$ acting via its linear action on W. It is a direct integral of the representations in $(Sp(W))^\wedge_{(1,1)}$. The individual representations may be described as follows. The action of $SO_{1,1} \simeq F^\times$ on $L^2(W)$ is

$$(5.5) \qquad \omega(t)f(w) = |t|^{-n}f(t^{-1}w) \qquad t \in F^\times,\ w \in W,\ f \in L^2(W)$$

Fix a character ψ of $F^\times$ and consider the space $\mathcal{Y}^\infty(\psi)$ of smooth functions f on $W - \{0\}$ such that $\omega(t)f = \psi(t)f$, or in other words

$$f(tw) = |t|^{-n}\,\psi^{-1}(w)f(w) \qquad t \in F^\times,\ w \in W$$

Consider the evaluation map

$$(5.6) \qquad \eta : \mathcal{Y}^\infty(\psi) \to C^\infty(Sp)$$

defined by

$$\eta(f)(g) = f(g^{-1}(e_1)) \qquad f \in \mathcal{Y}^\infty(\psi),\ g \in Sp$$

where e_1 is as before the first standard basis vector for W. On the parabolic P_1 of Sp stabilizing X_1, the line through e_1, define a rational character α_1 by

$$(5.7) \qquad p(e_1) = \alpha_1(p)e_1 \qquad p \in P_1$$

It is easy to compute that

$$\eta(f)(pg) = |\alpha_1(p)|^n \psi(\alpha_1(p))f(g)$$

Since the projective space $\mathbb{P}(W)$ is isomorphic to $P_1\backslash Sp$, it is easy to conclude that η defines an isomorphism between $\mathcal{Y}^\infty(\psi)$ and the smooth vectors in the induced representation

$$\mathrm{ind}_{P_1}^{Sp} \psi \circ \alpha_1 = \sigma_\psi$$

Here we are using the convention of normalized induction, which sends unitary representations to unitary representations.

The action of the $\mathbb{Z}_2$-factor of $O_{1,1}$ on $L^2(W)$ is by a "symplectic Fourier transform". We will not write down the exact formula. It is evident, however, that $\omega(\mathbb{Z}_2)$ must interchange $\mathcal{Y}^\infty(\psi)$ and $\mathcal{Y}^\infty(\psi^{-1})$. Therefore, when $\psi \neq \psi^{-1}$ the action of $O_{1,1} \times Sp$ on $\mathcal{Y}^\infty(\psi) \oplus \mathcal{Y}^\infty(\psi^{-1})$ is equivalent to $\mu_\psi \otimes \sigma_\psi$. However, if $\psi = \psi^{-1}$, then $\omega(\mathbb{Z}_2)$ provides an extra endomorphism of $\mathcal{Y}^\infty(\psi)$, which therefore breaks up further into eigenspaces $\mathcal{Y}^\infty(\psi^+)$ and $\mathcal{Y}^\infty(\psi^-)$. Since Sp must respect this decomposition, we see σ_ψ must in this case be reducible into 2 inequivalent parts. Theorem 4.2 prevents any further decomposition of the σ_ψ. In summary:

Proposition 5.1: All irreducible representations of $Sp(W)$, $\dim W \geq 6$, which have split rank 2 N_m-spectrum are constituents of $\sigma_\psi = \mathrm{ind}_{P_1}^{Sp} \psi \circ \alpha_1$ for some $\psi \in \widehat{F}^\times$. If $\psi \neq \psi^{-1}$, then σ_ψ is irreducible, and $\Delta(\sigma_\psi) = \mu_\psi$. If $\psi = \psi^{-1}$, then σ_ψ decomposes into 2 inequivalent constituents, and these correspond in some order to the extensions ψ^+ and ψ^- of ψ to $O_{1,1}$.

For a more direct approach to the σ_ψ, see [Fr] [Gs].

As a third example, we mention holomorphic representations of $Sp_{2m}(\mathbb{R})$. For our purposes, a holomorphic representation may be defined as a representation whose N_m-spectrum is supported on positive semidefinite classes of bilinear forms. Theorems 1.1 and 4.2 yield the following result.

Proposition 5.2: The holomorphic unitary representations of $\tilde{S}p_{2m}(\mathbb{R})$ of rank $\ell < m$ are in natural bijection with $\hat{O}_\ell$. They are realized as summands of the Weil representation of $\tilde{S}p_{2m}(\mathbb{R})$ corresponding to O_ℓ. These representations account for all holomorphic representations of $\tilde{S}p_{2m}(\mathbb{R})$ which are not N_m-regular.

Our final example is rather different. In the first place, it is a representation of $O_{n,2}(\mathbb{R})$ instead of Sp. Secondly, it has rank 2 rather than rank less than 2. (The ranks of representations of orthogonal groups are always even integers, and the only unitary representation of rank 0 is the trivial representation.) Therefore it may not be apparent from our current vantage point that this example really illustrates the phenomenon under discussion. However it does, and very nicely. On that account, and because of its intrinsic interest, we present it.

The example is constructed by Onofri [On]. It is a unitary representation ρ of $SO_{2m,2}(\mathbb{R})$ on $L^2(S^{2m-1})$, where S^k is the k-sphere in Euclidean (k+1)-space. Actually Onofri constructed a representation for $O_{n,2}$ for all n, not only even n, but it seems that for n odd, his representations must be representations of a covering group of $SO_{n,2}$. The maximal connected compact subgroup of $SO_{2m,2}$ is $SO_{2m} \times SO_2$. The action of $\rho(SO_{2m})$ is the natural action, and by the classical theory of spherical harmonics breaks up into a direct sum

$$L^2(S^{2m-1}) \simeq \sum_{j=0}^{\infty} H_j$$

of irreducible representations, where H_j is the action of SO_{2m} on the harmonic polynomials on $\mathbb{R}^{2m}$ of degree j. The H_j are the eigenspaces for $\rho(SO_2)$, which acts on H_j by χ^{j+m-1}, where χ is the basic character of SO_2. The subgroup $SO_{2m,1}$ acts irreducibly via ρ, and ρ defines a principal series representation for $SO_{2m,1}$. (The sphere S^{2m-1} is a homogeneous space for $SO_{2m,1}$ acting by conformal transformations, and the principal series may be realized as "multiplier representations", i.e., actions on homogeneous vector bundles (in our case, line bundles) on S^{2m-1}.)

It is clear from this description that ρ is a very small representation of $SO_{2m,2}$. It has only a one-parameter family of K types, as opposed to the typical 4-parameter family. It has functional dimension 2m-1, as opposed to the typical 4m-3. Moreover, it is the only known example of such a small representation of this group.

How are we to understand this ρ? Onofri constructed it as a quantization of the Kepler problem*, inspired by the philosophy of "geometric quantization". Here we point out its relation to the theory of reductive dual pairs and the oscillator representation. We remark generally that geometric quantization and the oscillator representation are pretty clearly compatible, indeed deeply connected, but the relation is not well understood.

---

*It has been suggested therefore [Z1] to call it the <u>planetary</u> <u>representation</u>.

Consider the dual pair $(O_{2m,2}(\mathbb{R}), SL_2(\mathbb{R}))$. This is a stable pair in the sense of [H2], and so we get a pairing, as in Theorem 1.1, between the tempered dual of $SL_2(\mathbb{R})$ and certain representations of $O_{2m,2}(\mathbb{R})$. These representations of $O_{2m,2}$ have a 2-parameter K-spectrum and functional dimension 2n, so they are not Onofri's ρ. However, if one takes the distributional approach, one can look for other representations of $O_{2m,2}$ corresponding to non-tempered representations of SL_2. The least-tempered unitary representation of SL_2, and the only one which is smaller than average is the trivial representation. And indeed, when we look at the representation of $O_{2m,2}$ which is paired in the sense of [H3] with the trivial representation we find it is the representation of $O_{2m,2}$ induced from Onofri's ρ on $SO_{2m,2}$. It's restriction to $SO_{2m,2}$ is just $\rho \oplus \rho^*$.

This is an encouraging fact. One thing it suggests is that the map Δ of Theorem 4.2 may be surjective. This would be very interesting to know. However, at the moment while one can see that the representation of $O_{2m,2}$ corresponding to $1 \in \hat{SL}_2$ will be uniformly bounded, no _a priori_ proof of unitarity now exists. Unitarity now comes only from Onofri's construction.

6: Asymptotics of matrix coefficients, generalities.

In this section we switch our attention to a rather different topic, namely, the asymptotic behavior of matrix coefficients of representations. The topic is also somewhat technical. But it has been of basic importance in the work of Harish-Chandra, and has had a number of surprising and important applications elsewhere [Kn], [BW], [Z2]. We have already had occasion to use the weak results of [HM] in Theorem 2.10. We will see in §8 that these seemingly general and unrelated considerations are in fact closely connected with the theory of N_m-rank for representations of Sp.

We begin very generally. Let G be a locally compact group, and let ρ be a unitary representation of G on a Hilbert space $\mathcal{H}$. Given $x,y \in \mathcal{H}$, the function

$$\varphi^{\rho}_{x,y}(g) = \varphi_{x,y}(g) = (x, \rho(g)y) \qquad g \in G \tag{6.1}$$

where (,) here indicates the inner product on $\mathcal{H}$, is called the (left) <u>matrix coefficient</u> of ρ with respect to x and y. Let L and R denote the left and right action of G on functions on G:

$$Lg(f)(g') = f(g^{-1}g') \qquad Rg(f)(g') = f(g'g) \quad g,g' \in G \tag{6.2}$$

It is trivial to verify that

$$Lg(\varphi_{x,y}) = \varphi_{\rho(g)x,y} \qquad Rg(\varphi_{x,y}) = \varphi_{x,\rho(g)y} \tag{6.3}$$

Hence for fixed y the map $x \to \varphi_{x,y}$ intertwines ρ with the (not unitary) left regular action L. Similarly, for fixed x, the map $y \to \varphi_{x,y}$ intertwines ρ with the right action R. The functions $\varphi_{x,y}$ are trivially bounded: one has the obvious estimate

(6.4) $$|\varphi_{x,y}(g)| \le \|x\| \, \|y\|$$

where $\| \ \|$ indicates the norm in H. Much of this section will be devoted to improving estimate (6.4) in certain cases. Here let us simply note that (6.4) implies we may convolve $\varphi_{x,y}$ with L^1 functions. Suppose G is unimodular and $f \in L^1(G)$. Then it follows directly from (6.3) that, for $f \in L^1(G)$ and $x,y \in H$,

(6.5) $$f * \varphi_{x,y} = \varphi_{\rho(f)x,y} \qquad \varphi_{x,y} * f^* = \varphi_{x,\rho(f)y}$$

Here $*$ denotes convolution on G, and

(6.6) $$f^*(g) = \overline{f(g^{-1})} \ .$$

where $\overline{}$ denotes complex conjugation.

We are interested in the asymptotic behavior of $\varphi_{x,y}(g)$ as $g \to \infty$ in G. If G is a simple algebraic group such as Sp and ρ does not contain the trivial representation, then we know from [HM] that the matrix coefficients of ρ all vanish at ∞. Here we shall look more closely at the rate of decay of $\varphi_{x,y}$. In this connection we entertain several definitions.

(6.7) a) ρ is said to be <u>strongly mixing</u> if for all $x,y \in H$, the matrix coefficient $\varphi_{x,y}$ vanishes at ∞ on G.

b) ρ is <u>absolutely continuous</u> if ρ is quasi-equivalent to a subrepresentation of the regular representation on $L^2(G)$.

c) ρ is <u>strongly L^p</u> if there is a dense subspace of H such that for x,y in this subspace, $\varphi_{x,y} \in L^p(G)$.

For our last definition we need additional structure. Let $K \subseteq G$ be a compact subgroup, and let $\hat{K}$ be the unitary dual of K. We may decompose the space H of ρ into an orthogonal direct sum

$$\mathcal{H} = \bigoplus_{\mu \in \hat{K}} \mathcal{H}_\mu$$

such that $\mathcal{H}_\mu$ is $\rho(K)$-invariant, and the action of K on $\mathcal{H}_\mu$ is equivalent to $n\mu$ for some integer n or $+\infty$, called the multiplicity of μ in ρ. The subspace $\mathcal{H}_\mu$ is the <u>μ-isotypic component</u> of ρ or of $\mathcal{H}$.

(6.8) Let $K \subseteq G$ be a compact subgroup. Let Φ be a positive function on G, invariant under left and right translations by K, and such that $\Phi(g) = \Phi(g^{-1})$, and let Ψ be a function on $\hat{K}$. We say ρ is <u>(Φ,Ψ)-bounded</u>, if given $\mu, \nu \in \hat{K}$, and $x \in \mathcal{H}_\mu$, $y \in \mathcal{H}_\nu$, one has the estimate

$$|\varphi_{x,y}(g)| \leq \|x\| \, \|y\| \quad \Psi(\mu)\, \Psi(\nu)\, \Phi(g)$$

We list below some straightforward observations concerning these concepts.

a) Because of estimate (6.4), a representation ρ will be strongly mixing if there is a dense subspace of $\mathcal{H}$ such that $\varphi_{x,y}$ vanishes at ∞ when x and y belong to this dense subspace.

b) If ρ is absolutely continuous, then ρ is strongly mixing. For if $u,v \in L^2(G)$, then

$$(6.9) \qquad \varphi_{u,v}(g') = \int_{G'} u(g)\, \overline{v(g'^{-1}g)}\,dg = u * v^*(g)$$

where — denotes complex conjugation, and v^* is as in (6.6).

An easy argument shows $u * v^*$ vanishes at ∞ for $u,v \in L^2(G)$. Hence $L^2(G)$ is strongly mixing.

Suppose $\rho = \rho_1 \oplus \rho_2$ is a direct sum of two subrepresentations, and $H = H_1 \oplus H_2$ is the corresponding decomposition of H. For $x \in H$, write $x = x_1 + x_2$ with $x_i \in H_i$. Then for $x,y \in H$

(6.11) $$\varphi_{x,y} = \varphi_{x_1,y_1} + \varphi_{x_2,y_2}$$

Therefore a direct sum of strongly mixing representations is strongly mixing. Every absolutely continuous representation embeds into a direct sum of copies of $L^2(G)$, by definition. Hence such representations are also strongly mixing.

c) If ρ is strongly L^p for $p < \infty$, then ρ is strongly mixing. For suppose $\varphi_{x,y} \in L^p$ for $x,y \in H$. If $f \in C_c(G)$, i.e., the function f is continuous of compact support, then $f * \varphi_{x,y} \in C_0(G)$, i.e., vanishes at ∞. Letting f run through a "Dirac sequence" - a sequence of non-negative functions each with total integral equal to 1 and whose supports shrink to the identity - and using formula (6.5) shows $\varphi_{x,y} \in C_0(G)$ for a dense set of x,y in H. By remark a), then, we see ρ is strongly mixing.

d) If ρ is strongly L^2, then ρ is absolutely continuous. This is shown in [HM] and in [Ra]. Let $y \in H$ be such that $\varphi_{x,y} \in L^2(G)$ for a dense subspace of H. Then $\varphi_y : x \to \varphi_{x,y}$ is a densely defined intertwining operator from H to $L^2(G)$, and φ_y is easily verified to have a closable graph. Hence the general Schur's lemma [Ra] produces a unitary embedding of $(\ker \varphi_y)^{\perp}$ into $L^2(G)$. We may splice together a collection of such embeddings coming from a dense collection of y's to embed H in $\infty L^2(G)$, establishing the absolute continuity of ρ.

e) Since matrix coefficients are bounded, we see that if ρ is strongly L^p, then ρ is strongly L^q for all $q \geq p$. We will say ρ is strongly $L^{p+\varepsilon}$ if ρ is strongly L^q for $q > p$.

f) If ρ is (Φ,Ψ)-bounded for some compact subgroup $K \subseteq G$, then Φ is strongly mixing or strongly L^p, according as $\Phi \in C_0(G)$ or $\Phi \in L^p(G)$. This is because the algebraic sum of the K-isotypic subspaces H_μ (the space of K-finite vectors) is dense in H. Also (Φ,Ψ)-boundedness need only be checked for a dense subspace in H, by estimate (6.4).

g) A direct sum of representations satisfying any of conditions (6.7) a),b),c) or (6.8) again satisfies the same condition. This follows from the formula (6.11). The only condition that is not completely obviously maintained under direct sums is (Φ,Ψ)-boundedness. For this, observe that, with notation as in (6.11), if $x \in H_\mu$, then $x_i \in (H_i)_\mu$, and since $\|x\|^2 = \|x_1\|^2 + \|x_2\|^2$, one has

$$\|x_1\| \, \|y_1\| + \|x_2\| \, \|y_2\| \le \|x\| \, \|y\|$$

by the Schwartz inequality. The preservation of (Φ,Ψ)-boundedness under direct sums follows .

h) It is obvious that the property of being strongly mixing, absolutely continuous, or (Φ,Ψ)-bounded is inherited by subrepresentations. However, the property of being strongly L^p is not. For example, the regular representation of G on $L^2(G)$ is strongly L^1, since $C_c(G) \subseteq L^2(G)$ is dense, and $\varphi_{u,v}$ for $u,v \in C_c(G)$ is compactly supported by formula (6.9). But there are subrepresentations of $L^2(G)$ (e.g., non-integrable discrete series) which are not strongly L^1.

i) Given $x,y \in H$, if $\varphi_{x,y}$ belongs to $C_0(G)$ or $L^p(G)$, then the same will be true for $\rho(f_1)x$ and $\rho(f_2)y$ for any $f_i \in L^1(G)$ by formula (6.5). Hence to check strong mixing or strong

L^p-ness of ρ, it is enough to check it for x,y belonging to a subspace of H which generate H as G-module. In particular if ρ is irreducible, then ρ will be strongly mixing or strongly L^p if only one matrix coefficient is in $C_0(G)$ or $L^p(G)$ respectively.

j) There is a slight modification of (Φ,Ψ)-boundedness which comes to essentially the same thing; it is less pleasant to define, but more pleasant to work with. We will say ρ is <u>modified (Φ,Ψ)-bounded</u> if the same estimates on $\varphi_{x,y}$ hold, except that when both x and y belong to a given K-isotypic space, we only require the estimate of definition (6.8) to hold when $x=y$. Of course, this would make no sense for x and y belonging to different K-isotypic spaces. For any x,y we the have easily verified identity

$$4\,\varphi_{x,y} = \varphi_{x+y,x+y} - \varphi_{x-y,x-y} + i\,\varphi_{x+iy,x+iy} - i\,\varphi_{x-iy,x-iy}$$

(here $i = \sqrt{-1}$.) Suppose x and y are unit vectors in the K-isotypic subspace H_μ. Then if we assume modified (Φ,Ψ)-boundedness, we get

$$|4\,\varphi_{x,y}| \le (\|x+y\|^2 + \|x-y\|^2 + \|x+iy\|^2 + \|x-iy\|^2)\ \ \Psi(\mu)^2\ \Phi$$

Using the parallelogram law

$$\|x+y\|^2 + \|x-y\|^2 = 2(\ \|x\|^2 + \|y\|^2)$$

and the assumption $\|x\| = \|y\| = 1$, we get

$$|4\,\varphi_{x,y}| \le 8\ \|x\|\ \|y\|\ \Psi(\mu)^2\ \ \Phi$$

But clearly if (Φ,Ψ)-boundedness holds for unit vectors, it holds for all vectors. Hence we see that modified (Φ,Ψ)-boundedness implies $(\Phi, \sqrt{2}\ \Psi)$-boundedness.

We need some further facts about definitions (6.7) and (6.8) that are more involved than the above, though still not difficult. These concern results on restriction and induction, products of groups and tensor products of representations, and the Fell topology.

Consider two representations ρ_1 and ρ_2 acting on spaces H_1 and H_2. Form the tensor product $\rho_1 \otimes \rho_2$ acting on $H_3 = H_1 \otimes H_2$. The definition of the inner product on H_3 is

$$(x_1 \otimes x_2, y_1 \otimes y_2)_3 = (x_1, y_1)_1 \, (x_2, y_2)_2 \qquad x_i, y_i \in H_i \tag{6.12}$$

A dense subspace of H_3 is spanned by the finite rank tensors

$$x = \sum_{i=1}^{n} x_1^i \otimes x_2^i \qquad x_j^i \in H_j \tag{6.13}$$

If we take x and y of the form (6.13) and use (6.12) to compute the matrix coefficient $\varphi_{x,y}$ of $\rho_1 \otimes \rho_2$, we find

$$\varphi_{x,y} = \Sigma \, \varphi_{x_1^i, y_1^j} \, \varphi_{x_2^i, y_2^j} \tag{6.14}$$

Thus matrix coefficients of $\rho_1 \otimes \rho_2$ formed from finite rank tensors are sums of products of matrix coefficients of ρ_1 and ρ_2.

Proposition 6.1: a) If either ρ_1 or ρ_2 is strongly mixing, then $\rho_1 \otimes \rho_2$ is also.

b) If either ρ_1 or ρ_2 is absolutely continuous, then so is $\rho_1 \otimes \rho_2$.

c) If ρ_1 is strongly L^p and ρ_2 is strongly L^q, then $\rho_1 \otimes \rho_2$ is strongly L^r where

$$\frac{1}{r} = \frac{1}{p} + \frac{1}{q}$$

Remark: There is no result analogous to these for (Φ,Ψ)-boundedness because taking tensor products mixes up the K-types so much. However, we will see in proposition 6.2 that (Φ,Ψ)-boundedness works well in situations where the above concepts do poorly.

Proof: Statement a) is completely obvious from formula (6.14), and statement c) is also, because of the well-known facts governing the products of L^p functions [DS]. For statement b), it will suffice to consider the case when ρ_1 is the (right) regular representation, so that $H_1 = L^2(G)$. The space $L^2(G) \otimes H_2$ can also be regarded as the space $L^2(G; H_2)$ of H_2-valued functions on G with square integrable norm. Then the tensor product action becomes

$$R \otimes \rho_2(g)(f)(g') = \rho_2(g)(f(g'g))$$

Define a map A on $L^2(G;H_2)$ by

$$A(f)(g) = \rho_2(g)f(g) \qquad g \in G,\ f \in L^2(G;H_2)$$

It is clear that A is unitary. We compute

$$\begin{aligned} A(R \otimes \rho_2(g)(f))(g') &= \rho_2(g')(R \otimes \rho_2(f)(g')) = \rho_2(g')(\rho_2(g)(f(g'g))) \\ &= \rho_2(g'g)(f(g'g)) = (Af)(g'g) = Rg(Af)(g') \end{aligned}$$

Thus A defines a unitary equivalence between $R \otimes \rho_2$ and $(\dim \rho_2)R$, proving b).

The behavior of (Φ,Ψ)-boundedness under tensor products is more difficult to explicate. To do so requires some preliminary work.

A topology has been defined on the unitary dual $\hat{G}$ of G by Fell [F1]. If $\{\rho_n\}$ is a sequence of representations in $\hat{G}$, and $\sigma \in \hat{G}$ is fixed then $\{\rho_n\}$ converges to σ, if given a matrix coefficient $\varphi^{\sigma}_{x,y}$ of

σ, there are matrix coefficients $\varphi^{n}_{x_n,y_n}$ of ρ_n which converge to $\varphi^{\sigma}_{x,y}$ uniformly on compact sets. A related concept is that of weak containment. A representation σ is weakly contained in a representation ρ if the matrix coefficients of σ can be uniformly approximated on compacta by matrix coefficients of ρ. In particular, to an arbitrary representation ρ of G, we may assign a set $\text{supp}\ \rho \subseteq \hat{G}$ consisting of all $\sigma \in \hat{G}$ which are weakly contained in ρ. Clearly supp ρ is closed in $\hat{G}$. Further, if G is of type I, as we will assume, then supp ρ is exactly the support of the projection-valued measure on $\hat{G}$ defining ρ up to unitary equivalence.

Lemma 6.2: Let G be a locally compact group with compact subgroup K. Let Φ be a function on G and Ψ a function on $\hat{K}$.

a) The subset of $\hat{G}$ consisting of representations which are modified (Φ,Ψ)-bounded is closed in $\hat{G}$.

b) A representation ρ of G is modified (Φ,Ψ)-bounded if and only if all $\sigma \in \text{supp}\ \rho$ are modified (Φ,Ψ)-bounded.

Proof: Let $\{\rho_n\}$ be a sequence of representations in $\hat{G}$ converging to $\sigma \in \hat{G}$. Suppose the ρ_n are (Φ,Ψ) -bounded. Fix a vector x in the space H of σ. Then we know from [F1] that we can choose vectors x_n in the space H_n of ρ_n such that all the x_n have the same length as x and $\varphi^{\sigma}_{x,x}$ is the uniform-on-compacta limit of the $\varphi^{\rho_n}_{x_n,x_n}$. Select $\mu, \nu \in \hat{K}$, and choose unit vectors $x \in H_\mu$ and $y \in H_\nu$. Then, assuming $\mu \neq \nu$, the vectors x and y will be orthogonal, so $\|x+y\|^2 = 2$. Thus we can find vectors $v_n \in H_n$ such $\|v_n\|^2 = 2$, and $\varphi^{\rho_n}_{v_n,v_n}$ converges uniformly on compacta to $\varphi^{\sigma}_{x+y,x+y}$. Let e_μ, e_ν be the central idempotents in $L^1(K)$ corresponding to μ and ν respectively. Then from formula (6.5) we see

$$e_\mu * \varphi^\sigma_{x+y,x+y} * e_\nu^* = \varphi^\sigma_{x,y}$$

Therefore $\varphi^\sigma_{x,y}$ is the uniform on compacta limit of

$$e_\mu * \varphi^{\rho_n}_{v_n,v_n} * e_\nu^* = \varphi^{\rho_n}_{u_n,w_n}$$

where

$$u_n = \rho(e_\mu)(v_n) \qquad\qquad w_n = \rho(e_\nu)(v_n)$$

Clearly u_n and w_n are orthogonal, and

$$\|u_n\|^2 + \|w_n\|^2 \le \|v_n\|^2 = 2.$$

Hence the product $\|u_n\| \, \|w_n\|$ is at most 1, so that

$$|\varphi^{\rho_n}_{u_n,w_n}(g)| \le \Psi(\mu)\ \Psi(\nu)\ \ \Phi(g)$$

by definition of (Φ,Ψ)-boundedness. In the limit, remembering that $\|x\| = \|y\| = 1$ by choice, we get

$$|\varphi^\sigma_{x,y}| \le \|x\| \ \|y\| \ \ \Psi(\mu)\ \Psi(\nu)\ \Phi$$

Since it is clearly enough to verify the condition for (Φ,Ψ)-boundedness on unit vectors, we see that σ does satisfy that condition, at least when $\mu \neq \nu$. The proof when $\nu = \mu$ is completely straightforward and is omitted.

Essentially the same argument shows that for an arbitrary representation ρ of G, all σ in supp $\rho \subseteq \hat{G}$ are modified (Φ,Ψ)-bounded if ρ is. It remains to prove the converse. Let x belong to the space $\mathcal{H}$ of ρ and consider the matrix coefficient $\varphi^\rho_{x,x}$. The direct integral theory [Nk] implies that $\varphi^{\rho}_{x,x}$ is a uniform-on-compact limit of sums of the form

(6.15) $$\Sigma\, \varphi^{\sigma_m}_{v_m,v_m}$$

where the σ_m range through supp ρ. Comparing values at 1, we conclude that $\|x\|^2$ is a limit of $\Sigma\|v_m\|^2$. We now perform the same trick as above. Select K-types μ and ν and choose unit vectors $x \in H_\mu$, $y \in H_\nu$. Approximate the matrix coefficient $\varphi^{\rho}_{x+y,x+y}$ by sums of the form (6.16). Then $\varphi^{\sigma_m}_{x,y}$ is approximated by sums

$$\Sigma\, \varphi^{\sigma_m}_{u_m,w_m}$$

where

$$u_m = \sigma_m(e_\mu)v_m \qquad w_m = \sigma_m(e_\nu)v_m \,.$$

Here as above e_μ and e_ν are the central idempotents for μ. and ν. Then we again have

$$\|u_m\|^2 + \|w_m\|^2 \;\le\; \|v_m\|^2$$

since u_m and w_m are the projections of v_m onto μ - and ν - isotypic subspaces of the space of σ_m. Therefore

$$\Sigma\, \|u_m\|\,\|v_m\| \;\le\; \frac{1}{2}\,\Sigma\,\|v_m\|^2$$

Therefore, by (Φ,Ψ)-boundedness of the $\sigma \in$ supp ρ, one has

$$\left|\Sigma\, \varphi^{\sigma_m}_{u_m,w_m}(g)\right| \;\le\; \frac{1}{2}\,(\,\Sigma\,\|v_m\|^2)\,\Psi(\mu)\,\Psi(\nu)\,\Phi(g)$$

In the limit, the modified (Φ,Ψ)-boundedness of ρ follows.

Again let G be a locally compact group with compact subgroup K. Following Fell [F3] once more, we say K is a uniformly large subgroup of G if there is a function M on $\hat{K}$, such that for any $\sigma \in \hat{G}$, the multiplicity of $\mu \in \hat{K}$ in ρ is at most $M(\mu)$. We call M a multiplicity bound for K in G. To have uniformly large compact subgroups is a very strong condition for a group to satisfy. Such groups are in particular type I.

Proposition 6.3: Let G_1 and G_2 be two locally compact groups with uniformly large compact subgroups K_1 and K_2. Let M_i be the multiplicity bound for K_i in G_i. Suppose ρ is a representation of $G_1 \times G_2$ such that $\rho|G_i$ is modified (Φ_i, Ψ_i)-bounded for suitable functions Φ_i on G_i and Ψ_i on $\hat{K}_i$. Then ρ is modified $(\Phi_1 \times \Phi_2, \Psi)$-bounded as a representation of $G_1 \times G_2$, where for $\mu_1 \otimes \mu_2 \in (K_1 \times K_2)^\wedge$, we define

$$\Psi(\mu_1 \otimes \mu_2) = 2 \prod_{i=1}^{2} \Psi_i(\mu_i)(M_i(\mu_i)\dim \mu_i)^{1/4} \tag{6.16}$$

Remark: A result like this completely fails for the strong mixing, absolute continuity, or strong L^p properties of representations. For example consider the joint left and right action of $\mathbb{R}$ on $L^2(\mathbb{R})$, taken as a representation of $\mathbb{R} \times \mathbb{R}$. This action is strongly L^1 on each factor, but the diagonal subgroup acts trivially. Hence the matrix coefficients of this action are constant on cosets of the diagonal subgroup, so the representation of $\mathbb{R} \times \mathbb{R}$ is not even strongly mixing.

Proof: By lemma 6.2 it is enough to consider the case when ρ is irreducible for $G_1 \times G_2$. Since the G_i, having uniformly large compact subgroups, are of type I we may factor ρ into a tensor product:

$$\rho \simeq \rho_1 \otimes \rho_2 \qquad\qquad \rho_i \in \hat{G}_i$$

Select K_i types μ_i and ν_i. Then

$$\mu = \mu_1 \otimes \mu_2 \qquad \nu = \nu_1 \otimes \nu_2$$

are typical elements of $(K_1 \times K_2)^\wedge$. If ρ_i is realized on H_i, then ρ is realized on $H_1 \otimes H_2 = H$, and

$$H_\mu = (H_1)_{\mu_1} \otimes (H_2)_{\mu_2},$$

and likewise for H_ν.

A typical element of H_μ can be represented in the form

$$x = \Sigma\, \alpha_i \otimes \beta_i$$

where $\alpha_i \in (H_1)_{\mu_1}$ and $\beta_i \in (H_2)_{\mu_2}$. The number of summands is at most

$$\min(\dim(H_i)_{\mu_i}) \leq \min(M_i(\mu_i) \dim \mu_i)$$

Furthermore, the spectral theorem tells us that we may so select the summands so that the α_i are mutually orthogonal in $(H_1)_{\mu_1}$ and the β_i are mutually orthogonal in $(H_2)_{\mu_2}$. We then have the relation

$$\|x\|^2 = \Sigma\, \|\alpha_i\|^2 \|\beta_i\|^2 \tag{6.17}$$

where each norm is taken in the appropriate space. A typical element y of H_ν can be written in analogous fashion:

$$y = \Sigma\, \gamma_j \otimes \delta_j$$

With this notation, we may compute

$$\varphi_{x,y}(g_1,g_2) = \sum_{i,j} \varphi_{\alpha_i,\gamma_j}(g_1)\ \varphi_{\beta_i,\delta_j}(g_2)$$

Since ρ_1 and ρ_2 are assumed modified (Φ_i,Ψ_i)-bounded, we get the estimate

(6.18) $$|\varphi_{x,y}(g_1,g_2)| \le 4(\sum_{i,j} \|\alpha_i\|\|\beta_i\|\|\gamma_j\|\|\delta_j\|)\Psi_1(\mu_1)\Psi_1(\nu_1)\Psi_2(\mu_2)\Psi_2(\nu_2)\Phi_1(g_1)\Phi_2(g_2)$$

Here the factor 4 comes from remark j) above. From the equation (6.17) and the Schwartz inequality, we can estimate

$$\Sigma\ \|\alpha_i\|\ \|\beta_i\| \le \|x\| \min_{k=1,2} (M_k(\mu_k) \dim \mu_k)^{1/2}$$

Plugging this into (6.18) gives the relation for $(\Phi_1 \times \Phi_2,\Psi)$-boundedness, with Ψ as specified in formula (6.16). Actually, it gives a slightly sharper estimate, since in (6.16) we have replaced the minimum of $M_k(\mu_k)\dim \mu_k$ by the geometric mean, to make the formula more symmetric.

It seems appropriate in this general discussion to make some observations about the behavior of the properties of definition (6.7) under induction and restriction.

Proposition 6.4: Let G be a unimodular locally comapct group and let H be a unimodular closed subgroup.

a) If ρ is strongly mixing, absolutely continuous, or strongly L^p as a representation of G, then $\rho|H$ has the same properties relative to H.

b) If σ is a representation of H that is strongly mixing, or absolutely continuous, then the representation

$$\rho = \operatorname{ind}_H^G \sigma$$

has the same properties relative to G.

Proof: It is obvious that if ρ is strongly mixing, then $\rho|H$ is also. For absolute continuity, it suffices to consider the case when

$\rho \simeq L^2(G)$; but then it is clear that $\rho|H$ is a multiple of $L^2(H)$, and in particular is absolutely continuous. Finally suppose ρ is strongly L^p and let $\varphi_{x,y}$ be an L^p matrix coefficient of ρ. Let m denote a typical element in a nice set of coset representatives for the quotient space $H\backslash G$, and let dm denote the G-onvariant measure on $H\backslash G$. Then

$$\infty > \int_G |\varphi_{x,y}(g)|^p dg = \int_{H\backslash G} (\int_H |\varphi_{x,y}(hm)|^p dh) dm$$

$$= \int_{H\backslash G} (\int_H |\varphi_{x,\rho(m)y}(h)|^p dh) dm$$

Since the integrand is everywhere positive and the total integral is finite, the inner integral must be finite for almost all m by Fubini. Since a set whose complement has measure zero is everywhere dense, we see that $\varphi_{x,\rho(m)y}$ belongs to $L^p(H)$ for m arbitrarily close to the identity in G. The set of such pairs x and $\rho(m)y$ are clearly dense in the space of ρ. This concludes part a) of the proposition.

Consider now a representation ρ of G induced from a representation σ of H. Because the representation induced from a direct sum is the sum of the representations induced from the summands, to deduce absolute continuity of ρ from that of σ, it will suffice to take $\sigma \simeq L^2(H)$. But then obviously $\rho \simeq L^2(G)$, which is absolutely continuous.

To prove the other assertions of part b) of the proposition, we must compute some matrix coefficients of ρ. We will suppose we can find a compact set $C \subseteq G$ such that the map $(h,c) \to hc$) defines an injection of $H \times C$ onto a neighborhood of the identity in G. For all groups with which we shall deal, the existence of such a local cross-section to the H cosets will be obvious. Given x in the space of σ, define $\tilde{x}$

in the space of ρ by

$$\tilde{x}(hc) = \sigma(h)x \qquad h \in H,\ c \in C$$

$$\tilde{x}(g) = 0 \qquad g \notin HC$$

The set of such functions, for variable x and C, clearly generate the space of ρ as G-module. Let us compute

$$\varphi_{\tilde{x},\tilde{y}}(g) = \int_{H\backslash G} (\tilde{x}(m),\tilde{y}(mg))dm$$

$$= \int_C (\tilde{x},\ \tilde{y}(cg)\ dc$$

where dc is the measure on C pulled back from its image in $H\backslash G$. If $\tilde{y}(cg) \neq 0$, the we may write

$$cg = hc' \qquad h \in H,\ c' \in C$$

Thus

$$\varphi_{x,y}(g) = \int_{C\cap HCg^{-1}} (x,\sigma(cgc'^{-1})y)dc = \int_{C\cap HCg^{-1}} \varphi_{x,y}(cgc'^{-1})dc$$

If $g \to \infty$ in G, clearly $c\,g\,c'^{-1} \to \infty$ in G also, and so those cgc'^{-1} which are in H go to ∞ in H. Hence if $\varphi_{x,y}$ vanishes at ∞ on H, so will $\varphi_{\tilde{x},\tilde{y}}$ vanish at ∞ on G. In other words, if σ is strongly mixing, so is ρ.

We conclude this section with a trick of Cowling [Cg] showing how an estimate for matrix coefficients of K-invariants vectors can be parleyed into an estimate for more general vectors.

Theorem 6.5: (Cowling) Let G be a locally compact group with compact subgroup K. Let ρ be a representation of G, and let $\rho\otimes\rho^*$

be the tensor product of ρ with its contragredient. Suppose that if x and y are two K-fixed vectors in the space of $\rho \otimes \rho^*$, then

$$(6.18) \qquad |\varphi_{x,y}| \le \|x\| \, \|y\| \, \Phi$$

for some K-bi-invariant function Φ on G. Then ρ is $(\Phi^{1/2}, 2 \dim \mu)$-bounded.

Proof: Let ρ be realized on the Hilbert space H. Then $\rho \otimes \rho^*$ may be realized on the space $H.S.$ of Hilbert-Schmidt operators on H, via the action

$$\rho \otimes \rho^*(g)(T) = \rho(g) \, T \, \rho(g)^{-1} \qquad g \in G, \; T \in H.S.$$

The inner product on $H.S.$ is given by

$$(S,T) = tr(ST^*) \qquad S,T \in H.S.$$

where tr is the standard trace functional.

Given x,y in H, we can form the dyad $E_{x,y} \in H$, by the recipe

$$E_{x,y}(z) = (z,y)x \qquad z \in H .$$

It is then easy to check that

$$E_{x,y}E_{u,v} = (u,y)E_{x,v} \qquad E^*_{x,y} = E_{y,x}$$

and

$$T \, E_{x,y} = E_{Tx,y} \qquad E_{x,y} \, T = E_{x,T^*y}$$

$$tr(TE_{x,y}) = (Tx,y)$$

for an operator T on H. In particular, for $x,y,u,v \in H$, we have

$$\begin{aligned}\varphi_{E_{x,y},E_{u,v}}(g) &= tr(E_{x,y}(\rho(g)\, E_{u,v}\, \rho(g)^{-1})^*) \\ &= tr(E_{x,y}\, \rho(g)\, E_{v,u}\, \rho(g)^{-1}) \\ &= tr(E_{\rho(g)^{-1}x,y}\, E_{\rho(g)v,u}) \\ &= (\rho(g)v,y)\ tr\,(E_{\rho(g)^{-1}x,u}) \\ &= (x,\ \rho(g)u)\overline{(y,\ \rho(g)v)} \\ &= \varphi_{x,u}\overline{\varphi_{y,v}}\end{aligned}$$

Take a vector x in H. Let $x_1, \ldots, x_d$ be an orthonormal basis for the linear span of the K-orbit of x in H. We may assume $x_1 = \|x\|^{-1} x$. Then $S = \Sigma\, E_{x_i,x_i}$ is orthogonal projection onto the span of $\rho(K)x$, and so is invariant under $\rho \otimes \rho^*(K)$. Therefore inequality (6.18) applies to $\varphi_{S,S}$. On the other hand, we see $(S,S) = d$, the dimension of the span of the K-orbit $\rho(K)(x)$. Also

$$\varphi_{S,S} = \sum_{i,j} |\varphi_{x_i,x_j}|^2 \geq |\varphi_{x_1,x_1}|^2 = \|x\|^{-4}\ |\varphi_{x,x}|^2$$

Therefore we conclude

$$(6.19)\qquad |\varphi_{x,x}| \leq d^{1/2}\ \|x\|^2\ \Phi^{1/2}$$

Choose representations μ and $\nu \in \hat{K}$. Choose vectors $x \in H_\mu$ and $y \in H_\nu$. Observe that $L^1(K)$ acts on the span of $\rho(K)(x)$ by means of a matrix algebra of rank $\dim \mu$. Hence the dimension of the span of $\rho(K)(x)$ is at most $(\dim \mu)^2$. Similarly the span of $\rho(K)(x+y)$ has dimension at most $(\dim \mu)^2 + (\dim \nu)^2$. Hence (6.19) specializes to of F. Let $|\alpha|$ denote the absolute value of α. Then $|\alpha|$ is a homomorphism from A to $\mathbb{R}^{+x}$, the positive real numbers. For $s \in \mathbb{R}^{+x}$, set

$$A_s^+ = \{a \in A: |\alpha|(a) \geq s \quad \text{for all} \quad \alpha \in \Sigma^+\}$$

Evidently A_s^+ is a subsemigroup in A when $s \geq 1$. We call A_1^+ the positive Weyl chamber in A. We will suppose we have the decompositions of G into

$$G = KB \qquad \text{(Iwasawa decomposition)} \tag{7.1}$$

$$G = KA_1^+K \qquad \text{(Cartan decomposition)}$$

The Iwasawa realization is always achievable with appropriate choice of K. The Cartan decomposition is not. However, it is achievable for groups over $\mathbb{R}$ or $\mathbb{C}$ (Lie groups), and for many groups over non-Archimedean fields, in particular for Sp. In general, it almost holds, and our arguments can be modified to cope with the general case. However, we will assume the Cartan decomposition in the form (7.1) for simplicity.

Let Ξ be the basic zonal spherical function for K defined by Harish-Chandra [HC1], [HC3]. Precisely, Ξ is the matrix coefficient of the K-fixed vector in the representation of G induced from the trivial representation of B. More explicitly, Ξ is produced as follows. Let δ_B denote the modular function of B, so that if $d_\ell b$ is a left-invariant Haar measure on B, one has

$$\delta_B(b')\ d_\ell(bb') = d_\ell b$$

Then $d_r b = \delta_B(b)\ d_\ell(b)$ is a right-invariant Haar measure on B. Let ψ be a unitary character of B. Consider the space $\mathcal{H}_\psi^\infty$ of smooth

$$(6.20) \qquad |\varphi_{x+y,x+y}| \le ((\dim\mu)^2 + (\dim\nu)^2)^{1/2}(\|x\|^2 + \|y\|^2)\ \Phi^{1/2}$$

If we use the polarization identity of remark j) above, we can conclude

$$(6.21) \qquad |\varphi_{x,y}| \le ((\dim\mu)^2 + (\dim\nu)^2)^{1/2}(\|x\|^2 + \|y\|^2)\ \Phi^{1/2}$$

Since $\dim\mu \ge 1$ we have

$$2\dim\mu\dim\nu \ge ((\dim\mu)^2 + (\dim\nu)^2)^{1/2}$$

Hence, if we choose x and y so that $\|x\| = \|y\|$, estimate (6.21) says

$$(6.22) \qquad |\varphi_{x,y}| \le 4(\dim\mu)(\dim\nu)\ \|x\|\ \|y\|\ \Phi^{1/2}$$

This is precisely the estimate for $(\Phi^{1/2}, 2\dim\mu)$-boundedness of ρ. But it clearly suffices to prove such an estimate when $\|x\| = \|y\|$, for we may achieve this by simply multiplying x or y by a scalar factor. Thus theorem 6.5 is proved.

7: Asymptotics of matrix coefficients for semisimple groups

In this section, we use the general concepts of §6 to study matrix coefficients of representations of $\widetilde{Sp}$. For abelian groups the asymptotic properties of matrix coefficients of representations are relatively delicate analytic properties. For example for abelian G, $L^2(G)$ is resolved into a direct integral of characters, each of which individually is only L^∞. However, some things are known which suggest the situation is rather different for semisimple groups. For example, Harish-Chandra's theory of the Plancherel formula for semisimple groups shows that for semisimple G, the regular representation on $L^2(G)$ is resolved into representations which are strongly $L^{2+\varepsilon}$ (c.f. theorem 7.1). At the other end of the spectrum, there is Kazhdan's result [Kn] that if G has split rank at least 2, then the identity representation of G is isolated in $\hat{G}$. These facts suggest that for semisimple groups the asymptotic properties of matrix coefficients reflect something relatively robust about the representations from which they come, and are related to the topology of the unitary dual $\hat{G}$. The goal here is to study this phenomenon systematically, especially in the exemplary case of symplectic groups.

Our first result in this direction is valid for general semisimple groups. Let G be a semisimple group over the local field F, and let K be a maximal compact subgroup of G. We will assume K is "good" in the following sense. Let $B \subseteq G$ be a minimal parabolic subgroup. Let $A \subseteq B$ be a maximal split torus, and let $N \subseteq B$ be the unipotent radical of B. The action $\mathrm{Ad}\, A$ of A on N by conjugation gives rise to a collection Σ^+ of positive roots of A. Each root α is a rational character of A, a homomorphism from A to $F^\times$, the multiplicative group functions f on G satisfying

$$f(bg) = \delta_B^{1/2}\,\psi(b)f(g) \qquad b \in B,\ g \in G.$$

Then G acts on H_ψ^∞ by right translations. The inner product

$$(f_1,f_2) = \int_K f_1(k)\,\overline{f_2(k)}\,dk$$

defines a G-invariant inner product on H_ψ^∞. The completion H_ψ of H_ψ^∞ in the associated Hilbert space norm is the space of the unitary representation $\mathrm{ind}_B^G\,\psi$, the (normalized) induced representation from ψ on B. For the moment we will abbreviate it to ind ψ. The representations ind ψ are collectively termed the unitary principal series.

The ind ψ which contain a K-fixed vector are called the spherical principal series. These will consist of the ind ψ such that ψ is trivial on $B \cap K$. They will contain a unique K-invariant function $f_\psi^o = f^o$. It will be given by the formula

$$f_\psi^o(bk) = \delta_B^{1/2}(b)\,\psi(b) \qquad b \in B,\ k \in K \tag{7.2}$$

If we then compute the matrix coefficient φ_{oo}^ψ of ind ψ with respect to f^o, we find

$$\begin{aligned} \varphi_{oo}^\psi(k_1 g\, k_2) &= \varphi_{oo}(g) \qquad g \in G,\ k_i \in K \\ \varphi_{oo}^\psi(b) &= \int_K f_\psi^o(k)\,\overline{R_b(f_\psi^o)(k)}\,dk \\ &= \int_K \overline{f_\psi^o(kb)}\,dk \end{aligned} \tag{7.3}$$

Harish-Chandra's function Ξ is given by

$$\Xi = \varphi_{oo}^1 \tag{7.4}$$

where 1 here denotes the trivial representation of B. It is then clear from (7.3) that

$$(7.5) \qquad |\varphi^{\psi}_{oo}(g)| \leq \Xi(g) \qquad g \in G$$

Harish-Chandra [HC1][S1]has proven some basic facts about the asymptotic behavior of Ξ. We will recall them. The modular function δ_B is related to the positive roots of the torus $A \subseteq B$ by

$$(7.6) \qquad \delta_B = \prod_{\alpha \in \Sigma^+} |\alpha|^{m(\alpha)}$$

where $m(\alpha)$ is a positive integer, the "multiplicity" of α.

By virtue of the Cartan decomposition (7.1), the function Ξ is determined by its restriction to A_1^+. Harish-Chandra has shown that

$$(7.7) \qquad c_1\delta_B^{-1/2}(a) \leq \Xi(a) \leq c_2(\varepsilon)\,\delta_B^{-1/2+\varepsilon}(a) \qquad a \in A_1^+$$

for some positive constants c_1 and $c_2(\varepsilon)$, for any $\varepsilon > 0$. If we write Haar measure in terms of the Cartan decomposition, then we have [Hn][Wr]

$$(7.8) \qquad dg = \Delta(a)\, dk_1 da\, dk_2$$

where $\Delta(a)$ is a positive function on A_1^+ satisfying

$$(7.9) \qquad d_1(t)\delta_B(a) \leq \Delta(a) \leq d_2\,\delta_B(a) \qquad a \in A_t^+$$

for some constant d_2, and constant $d_1(t)$ which is positive for $t > 1$.

It follows from formulas (7.7) to (7.9) that the representations ind ψ are strongly $L^{2+\varepsilon}$. It will follow from our first result that all representations in the support in $\hat{G}$ of the regular representation are strongly $L^{2+\varepsilon}$. Estimates like that of Theorem 7.1 are found frequently in the work of Harish-Chandra [HC2] and work based on his [Ar], [TV], [V]. However the simple dependence of estimate (7.10) on the auxiliary parameters, e.g., the different tempered irreducible representations and the K-types, is essential to us and is not readily dug out of that literature. Also the methods of theorem 7.1 are different from those of Harish-Chandra.

Theorem 7.1: Let G be semisimple and $K \subseteq G$ the maximal compact subgroup specified above. Let ρ be an absolutely continuous representation of G. Then ρ is $(\Xi, (\dim \mu)^2)$-bounded. In particular, ρ is strongly $L^{2+\varepsilon}$.

Proof: Since (Φ,Ψ)-boundedness is inherited by subrepresentations and preserved under taking direct sums, it will suffice to prove the theorem when $\rho \simeq L^2(G)$. From formula (6.9) we see that this amounts to showing that if u and v are in $L^2(G)$, and u transforms under left translations by K by a multiple of an irreducible representation μ of K, and v transforms by another $\nu \in \hat{K}$, then

$$(7.10) \qquad |u * v^*| \leq \|u\|_2 \, \|v\|_2 \dim \mu^2 \dim \nu^2 \, \Xi ,$$

where $\|u\|$ indicates the L^2-norm of u, and similarly for v.

We will establish inequality (7.10) in three steps. We will first prove it for K-biinvariant functions. Then we will establish a weaker analogue of (7.10) for functions which also transform under right translations by K according to a multiple of an irreducible representation. Finally we will reduce estimate (7.10) to this weaker version.

The case of (7.10) when u and v are K-bi-invariant follows directly from the Plancherel formula of Harish-Chandra [HC1][HC2] (for Lie groups) and MacDonald [McD] (for p-adic groups) for K-bi-invariant functions.

A simplified proof of Harish-Chandra's theorem is given in [Rg]. These theorems say that the representation $\mathrm{ind}_K^G 1$ of G decomposes into a direct integral over the unitary spherical principal series. The K-bi-invariant functions in $L^2(G)$ form exactly the space of K-fixed vectors in $\mathrm{ind}_K^G 1$, so a function $u * v^*$, with u and v K-bi-invariant in $L^2(G)$, is just a matrix coefficient of $\mathrm{ind}_K^G 1$ with respect to 2 K-fixed vectors. Therefore the Plancherel Theorem says

$$u * v^* = \int p(u)\, \overline{p(v)}\; \varphi_{oo}^{\psi}\, d\,\mu(\psi) \tag{7.11}$$

where $d\mu(\psi)$ is Plancherel measure and $p(u)$ and $p(v)$ are the "spherical transforms" of u and v. One has

$$\int |p(u)\, \overline{p(v)}|\quad d\mu(\psi) \leq \|u\|\, \|v\| \tag{7.12}$$

where $\|u\|$ is the L^2 norm of $u \in L^2(G)$. Equation (7.11) and (7.12) combine with estimate (7.4) to yield estimate (7.10) when u and v are K-bi-invariant.

Next, suppose u transforms to the left under K according some multiple of an irreducible representation μ, and transforms to the right under K by a multiple of some other representation μ'. Similarly suppose v transforms to the right and to the left under K by multiples of ν and $\nu' \in \hat{K}$.

Consider the restriction of u to a given (K,K) double coset KgK. Via the mapping $(k_1,k_2) \to k_1 g\, k_2$, $k_i \in K$, we may pull u back to a function u' on $K \times K$. By our assumptions about u, we know that as a function on $K \times K$, u' will belong to the minimal ideal associated to the representation $\mu \otimes \mu'$ of $K \times K$. Suppose $\mu \otimes \mu' = \mu''$ is realized on a space $\mathcal{J}$. Then there is an operator T on $\mathcal{J}$ such that

(7.13) $$u'(x) = \operatorname{tr}(\mu''(x)T) \qquad x \in K \times K$$

where tr denotes the usual trace function on End(J). Let $\| \|_\infty$ and $\| \|_2$ denote as usual the supremum and L^2 norms for functions on $K \times K$. (Here it is understood that the measure of $K \times K$ is normalized to be 1.) Let $\| \|_{2,J}$ denote the Hilbert-Schmidt norm on End(J). The formula (7.13) will be recognized as defining $u'(x)$ by taking the Hilbert-Schmidt inner product of $\mu''(x)$ with T^*, the adjoint of T. By the Schwartz inequality, we have

(7.14) $$|u'(x)| \leq \|\mu''(x)\|_{2,J} \|T\|_{2,J} = (\dim \mu'')^{1/2} \|T\|_{2,J}$$

On the other hand, the Schur Orthogonality relations tell us

(7.15) $$\|u\|_2 = (\dim \mu'')^{-1/2} \|T\|_{2,J}$$

Combining (7.14) and (7.15) yields

(7.16) $$\|u'\|_\infty \leq (\dim \mu'') \|u\|_2$$

Return to consideration of the functions u and v on G. Define

(7.17) $$\tilde{u}(g) = \max \{|u(k_1 g\, k_2)| : k_i \in K\}$$

and define $\tilde{v}$ similarly. It is clear from its definition that $|u(g)| \leq \tilde{u}(g)$, and that $\tilde{u}(g)$ is K-bi-invariant. From the integration formula (7.8) we find

(7.18) $$\begin{aligned} \int_G \tilde{u}^2\, dg &= \int_{K \times K \times A_1^+} \Delta(a)\, \tilde{u}(k_1 g\, k_2)^2\, da\, dk_1 dk_2 \\ &\leq (\dim \mu)^2 (\dim \mu')^2 \int_{K \times K \times A_1^+} \Delta(a) |u'(k_1 g k_2)|^2 da\, dk_1 dk_2 \\ &= (\dim \mu)^2 (\dim \mu')^2 \int_G |u|^2\, dg \end{aligned}$$

Analogous estimates apply to v. Therefore using estimate (7.10) for K-bi-invariant functions and estimate (7.18) gives us

$$|u * v^*(g)| \leq \tilde{u} * \tilde{v}^*(g) \leq \|\tilde{u}\|_2 \, \|\tilde{v}\|_2 \, \Xi(g) \tag{7.19}$$
$$\leq ((\dim \mu)(\dim \nu)(\dim \mu')(\dim \nu')) \, \|u\|_2 \, \|v\|_2 \, \Xi\,(g)$$

(Actually, $u * v^* = 0$ if $\mu' \neq \nu'$; but that is not important here).

Finally consider $u, v \in L^2(G)$, such that u transforms to the left under K by a multiple of $\mu \in \hat{K}$, and v transforms by a multiple of $\nu \in \hat{K}$. By an obvious approximation argument, to prove inequality (7.10) it is enough to prove it when u has compact support. Let $H_1 \subseteq L^2(G)$ be the closed span of left translates of u, and let v_1 be the projection of v into H_1. Then $\|v_1\|_2 \leq \|v\|_2$, and $u * v_1^* = u * v^*$. Hence we may as well assume $v \in H_1$. Let $L^2(G; \mu^*)$ denote the subspace of $L^2(G)$ consisting of functions which transform to the right under K by a multiple of μ^*, the representation contragredient to μ. Define

$$T: H_1 \to L^2(G; \mu^*)$$

by

$$T(w) = w * u^* \qquad\qquad w \in H_1$$

Since u has compact support, it is in $L^2(G)$, so T is a bounded operator. By inspection of the formula for $w * u^*$, we see that the kernel of T is the space of functions orthogonal to all left translates of u. By definition of H_1, we see T is injective. The general Schur's lemma [Ra] therefore gives us an isometric embedding

$$S: H_1 \to L^2(G; \mu^*)$$

which intertwines the left action of G on these two spaces. (This is essentially an instance of Frobenius Reciprocity.)

Because of the interpretation of $u * v^*$ as a matrix coefficient, we will have

$$u * v^* = (S(u)) * (S(v))^*$$

But $S(u)$ and $S(v)$ transform to the right according to a multiple of μ^*. Therefore the estimate (7.19) is applicable. Remembering that S is isometric we get

$$|u * v^*| \le \|u\|_2 \, \|v\|_2 \, (\dim \mu)^3 \dim \nu \ \Xi$$

But we may assume that $\dim \mu \le \dim \nu$. Then replacing $(\dim \mu)^3 \dim \nu$ by $(\dim \mu)^2 (\dim \nu)^2$ for purposes of symmetry, we obtain estimate (7.10) in general. This concludes Theorem 7.1.

We can immediately parlay theorem 7.1 into an estimate for matrix coefficients of strongly L^p representations, for any $p < \infty$.

Corollary 7.2: Let ρ be a strongly L^p representation of G, and suppose $p \le 2m$ for some integer m. Then ρ is $(\Xi^{1/m}, (\dim \mu)^2)$ bounded.

Proof: If ρ is strongly L^p with $p \le 2m$, then the m-fold tensor product $(\otimes \rho)^m$ of ρ is strongly L^2, hence absolutely continuous, by remark d) of §6 and proposition 6.1. Therefore Theorem 7.1 applies to $(\otimes \rho)^m$. Let H be the space of ρ. For representations $\mu, \nu \in \hat{K}$, select vectors $x \in H_\mu$ and $y \in H_\nu$. Let x' and y' denote the m-th tensor powers of x and y. Then

(7.20) $$\varphi_{x',y'} = (\varphi_{x,y})^m$$

We may decompose the m-th tensor power of μ into irreducible components:

$$(\otimes\,\mu)^m \simeq \Sigma\; a_i\, \mu_i$$

for appropriate $\mu_i \in \hat{K}$ and multiplicities a_i. Let x_i' denote the projection of x' into the μ_i -th isotypic component of $(\otimes\,\mu)^m$. Decompose $(\otimes\,\nu)^m$ similarly into a sum of $\nu_j \in \hat{K}$, and let y_j' be the component of y' in the ν_j-isotypic subspace of $(\otimes\, H_\nu)^m$. Then inequality (7.10) gives us

(7.21) $$\begin{aligned} |\varphi_{x',y'}| &= \Big|\sum_{i,j} \varphi_{x_i',y_j'}\Big| \le \sum_{i,j} |\varphi_{x_i',y_j'}| \\ &\le \sum_{i,j} \big(\|x_i'\|\, \|y_j'\|\, (\dim\mu_i)^2 (\dim \nu_j)^2\big)\, \Xi \\ &= \Big(\sum_i \|x_i'\|\, (\dim \mu_i)^2\Big)\Big(\sum_j \|y_j'\|\, (\dim \nu_j)^2\Big)\, \Xi\,. \end{aligned}$$

The Schwartz inequality gives

$$\Sigma\, \|x_i'\|\;\; (\dim \mu_i)^2 \le (\Sigma\, \|x_i\|^2)^{1/2} (\Sigma\, (\dim \mu_i{}^4))^{1/2}$$

But

$$\Sigma\, \|x_i'\|^2 = \|x'\|^2 = \|x\|^{2m}$$

since the x_i' are orthogonal. Furthermore

$$\Sigma\, \dim \mu_i \le \dim(\otimes^m\, \mu) = (\dim \mu)^m$$

Therefore

$$(7.22) \qquad \Sigma \, \|x_i'\| \, (\dim \mu_i)^2 \leq (\|x\| \, (\dim \mu)^2)^m$$

Similar estimates hold for y. Combining (7.20), (7.21) and (7.22) yields the estimate defining $(\Xi^{1/m}, (\dim \mu)^2)$-boundedness, so the corollary is proved.

Combining corollary 7.2 with Harish-Chandra's estimate (7.7) and estimate (7.8), and applying lemma 6.2 we obtain the following result.

Corollary 7.3: For any p, let $(\hat{G})^p$ denote the subset of $\hat{G}$ consisting of representations which are strongly L^p. Then for any integer $m \geq 1$, the closure of $(\hat{G})^{2m}$ in $\hat{G}$ is contained in $(\hat{G})^{2m+\varepsilon} = \bigcap_{q > 2m} (\hat{G})^q$.

Remarks: a) These corollaries illustrate a dramatic difference between semisimple harmonic analysis and abelian, or more generally, amenable harmonic analysis. One can also show (c.f. Theorem 8.4) that there is a $p < \infty$ such that $\hat{G} - (\hat{G})^p$ consists of the trivial representation alone, providing a strong and quantitative version of Kazhdan's Theorem [Kn], and further emphasizing the distinctive nature of semisimple harmonic analysis.

b) A serious weakness of these corollaries is that they provide sharp estimates only for even integral p. In particular they provide no distinctions between $L^{2+\varepsilon}$ and L^4, a very serious lack of resolution. It is possible by various <u>ad hoc</u> tricks to improve the situation for symplectic groups. However, it is natural to wonder whether, for any $p \geq 2$, if a representation of G is strongly L^p, is it then $(\Xi^{2/p}, (\dim \mu)^2)$-bounded ?

We conclude by filling in the proof of proposition 2.18. If $\rho \in \hat{G}$ is tempered, then the K-finite matrix coefficients of ρ are bounded by some constant times Ξ , according to theorem 7.1, or the estimates of Harish-Chandra. But Harish-Chandra has shown [HC1], [S1] that the restriction of Ξ to the maximal unipotent subgroup N of G is in $L^{1+\varepsilon}(G)$. Thus $\rho|N$ is strongly $L^{1+\varepsilon}$, hence in particular absolutely continuous; or in other words, ρ is N-regular.

Remark: After this was written, I realized the argument for theorem 7.1 could be simplified, and the result improved to $(\Xi, \dim\mu)$-boundedness. Furthermore, Cowling showed me that the estimate for left K-invariant functions follows directly from elementary considerations, and does not need the Plancherel formula. Thus theorem 7.1 can be strengthened and given a much simpler, essentially elementary proof.

8: Asymptotics of matrix coefficients and rank for Sp.

We finally apply the results of the foregoing sections to symplectic groups. Let W be our symplectic vector space and let $\{e_i, f_i\}$ be the standard symplectic basis of formula (1.1). Recall X_j is the span of the e_i for $i \le j$. The group B preserving the maximal flag consisting of all the X_i is a minimal parabolic subgroup of $Sp(W)$. The diagonal subgroup A of B is a split Cartan subgroup. Define rational characters α_i of A by

(8.1) $$T(e_i) = \alpha_i(T)e_i \qquad Tf_i = \alpha_i(T)^{-1}f_i \qquad T \in A.$$

These characters form a basis for the lattice of all rational characters of A. The set Σ^+ of, positive roots of A with respect to B are the characters

(8.2) $$\alpha_i\,\alpha_j^{-1} \qquad i < j$$
$$\alpha_i\,\alpha_j \qquad i \le j$$

Thus the modular function δ_B of B is

(8.3) $$\delta_B = \prod_{i=1}^{m} |\alpha_i|^{2(m+1-i)}$$

Here as before $2m = \dim W$. We also note

(8.4) $$A_1^+ = \{T \in A: |\alpha_i(T)| \ge |\alpha_{i+1}(T)| \ge 1\}.$$

If the base field F is $\mathbb{R}$ or $\mathbb{C}$, let K be the maximal compact subgroup preserving the (Hermitian) inner product for which the e_i and f_i are orthonormal. For F non-Archimedean, let K be the maximal compact subgroup preserving the lattice generated by the e_i and f_i.

These choices for K satisfy the conditions (7.1).

We begin by studying the asymptotics of the matrix coefficients of the oscillator representations ω_t. Then we establish a relation between asymptotics and rank.

Consider the oscillator representation ω_t. It is realized on $L^2(Y)$, according to formulas (1.8). Take u and v in $S(Y)$, the Schwartz space of Y. We can identify Y with F^n by introducing coordinates with respect to the basis vectors f_i. Then we compute, for $T \in A_1^+$,

$$
\begin{aligned}
(8.5)\qquad \varphi_{u,v}(T) &= \int_Y u(y)\,\overline{\omega_t(T)(v)(y)}\,dy \\
&= \Pi\,|\alpha_i(T)|^{1/2} \int_{F^n} u(y_1,\dots,y_n)\overline{v(\alpha_1(T)y_1,\dots,\alpha_n(T)y_n)}\,dy_1\dots dy_n \\
&= \Pi|\alpha_i(T)|^{-1/2} \int u(\alpha_1(T)^{-1}y_1,\dots,\alpha_n(T)^{-1}y_n)\overline{v(y_1,\dots,y_n)}\,dy_1\dots dy_n \\
&\le \Pi\,|\alpha_i(T)|^{-1/2}\ \|u\|_\infty\ \|v\|_1
\end{aligned}
$$

Furthermore, if u is constant in a neighborhood of 0 in Y, and v is supported in this neighborhood, which we will assume invariant by $(A_1^+)^{-1}$, and $u(0) = \|u\|_\infty$, and $v \ge 0$, then inequality (8.5) is actually an equality. And in any case we have an asymptotic formula

$$
\varphi_{u,v}(T) \sim \Pi|\alpha_i(T)|^{-1/2}\, u(0)\, \overline{\int_Y v(y)dy}
$$

Comparing these facts with the integration formula (7.8) and the formula (8.3) with δ_B, we can come to the following conclusion (c.f. [HM], proposition 6.4).

Proposition 8.1: The oscillator representations ω_t are strongly $L^{4m+\varepsilon}$, but are not strongly L^4.

Remark: For appropriate t, there will be a vector u_o in $L^2(Y)$

which is an eigenvector for $\omega_t(\tilde{K})$. When F is non-archimedean, this vector u_o can be arranged to be the characteristic function of the lattice spanned by the f_i's. For this vector, one has

$$|\varphi_{u_o,u_o}(k_1Tk_2)| = \Pi\ |\alpha_i(T)|^{-1/2}$$

When $F = \mathbb{R}$, the vector u_o can be made to be the Gaussian function

$$e^{-(\pi/2)\,\Sigma\, y_i^2}$$

Then one can compute that

$$|\varphi_{u_o,u_o}(k_1Tk_2)| = \Pi\ |\alpha_i(T)|^{1/2}(1 + |\alpha_i(T)|^2)^{-1/2}$$

Proposition 8.1 shows that the ω_t just miss having the decay necessary for a sharp application of corollary 7.2. Nevertheless, we can get the estimate corollary 7.2 fails to yield here by another means. In fact, it is not hard to do by analyzing $\omega_t \otimes \omega_t^*$. However, the following argument will give us a good, though not the best possible, result, and is what we need for later developments.

Theorem 8.2: If ρ is an m-fold tensor product of oscillator representations of $\tilde{Sp}_{2m}$ (i.e., a Weil representation associated to a quadratic form of degree m), then ρ is $(\Xi^{1/2}, 2\dim\mu)$-bounded.

Proof: By theorems 6.5 and 7.1, it will suffice to show that the K-fixed vectors in $\rho \otimes \rho^*$ generate an absolutely continuous representation. We will only give the proof for p-adic groups (i.e., for non-Archimedean base fields). The proof for $F = \mathbb{R}$ or $\mathbb{C}$ is similar in spirit but more technically involved.

We know from, e.g., [H1] II §3, that $\rho \otimes \rho^*$ factors to a representation of Sp_{2m}, and can be realized on $L^2((F^{2m})^m)$, with Sp_{2m} acting via its diagonal linear action on $(F^{2m})^m = V$. With K as specified above for F non-archimedean, the open K orbits are dense in V. Hence the characteristic functions of the open K orbits form an orthogonal basis for $L^2(V)^K$, the K-fixed vectors in $L^2(V)$. The K orbits are described in [H1] I, §11. Let R denote the ring of integers of F. Then each K-orbit is determined by the R module it spans. This R module will clearly be invariant by K. If we think of V as $F^m \otimes F^{2m}$, and let L_0 be the R-module in F^{2m} generated by the standard basis of e_i's and f_i's, then a K-invariant module in V has the form $\Lambda \otimes L_0$, where Λ is any R module in F^m. The characteristic functions of these lattices thus also form a spanning set for $L^2(V)^K$, although not an orthogonal one. Fix a K-invariant lattice $\Lambda \otimes L_0$. Let x denote the characteristic function of $\Lambda_0 \otimes L_0$, and let H denote the closed subspace of $L^2(V)$ generated by x under the action of Sp. Suppose we can find a function $y \in L^2(X)^K$ such that the map $z \to \varphi_{z,y}$, for $z \in H$, embeds H into $L^2(Sp)$. Then clearly we will have shown that the Sp module generated by $L^2(X)^K$ is absolutely continuous, and the theorem would be proved. Moreover let us observe that $GL_m(F)$ acts on V via its action on F^m, and commutes with Sp. From the form of the K-invariant lattices, we see that $GL_m(F)$ permutes them transitively. Hence it will suffice to deal with a single lattice.

Naturally we will choose the standard lattice $L_0^m \simeq R^m \otimes L_0$ to work with. Let x be the characteristic function of L_0^m. Observe that $L^2(V)$ is the m-fold tensor product of $L^2(F^{2m})$ with itself, and that x is the m-fold tensor product of the characteristic function of L_0 with itself. We will also construct our function y as a tensor product of m different functions in $L^2(F^{2m})^K$. Let u denote the characteristic function of $L_0 \subseteq F^{2m}$. Let π be a prime element of R, and let $q^{-1} = |\pi|$ be the absolute value of π. Thus $\#(R/\pi R) = q$ is the cardinality of the residue class field of R. Let v be the characteristic function of πL_0. Normalize Haar measure on F^{2m} so that L_0 has measure 1. Then

$$\|u\| = 1 \qquad \|v\|^2 = (u,v) = q^{-2m}$$

where $\| \ \|$ is the norm in $L^2(F^{2m})$ and $(\ ,\ )$ is the inner product.

Let A be our standard Cartan subgroup of Sp, defined at the beginning of this section, and let α_i be the rational characters of formula (8.1). Let ρ_1 denote the action of Sp on $L^2(F^{2m})$. Then for $T \in A$, the function $\rho_1(T)(u)$ is the characteristic function of the lattice TL_0, spanned by $\{\alpha_i(T)e_i, \alpha_i(T)^{-1}f_i\}$. The inner product $(u, \rho_1(T)u)$ is the volume of the intersection $L_0 \cap TL_0$. This volume is not hard to compute. The result is

$$\text{(8.6)} \qquad \varphi_{u,u}(T) = \prod_{i=1}^{m} |\alpha_i(T)|^{-1} \qquad T \in A_1^+$$

In the same way we find

$$\text{(8.7)} \qquad \varphi_{u,v} = q^{-m} \prod_{i=1}^{m} |\alpha_i(T)|^{-1} \min(1, |\pi\, \alpha_i(T)|)$$

$$= q^{-m}\, \varphi_{u,u}(T) \prod_{i=1}^{m} \min(1, |\pi\, \alpha_i(T)|)$$

The quantity

$$\min(1, |\pi\, \alpha_i(T)|) \qquad T \in A_1^+$$

is equal to 1 or to q^{-1}, according to whether $|\alpha_i(T)| > 1$ or $|\alpha_i(T)| = 1$. Therefcre the quotient $\varphi_{u,u}\varphi_{u,v}^{-1}$ takes on only the values q^{m+j} for $0 \le j \le m$; and it takes on the value q^{2m} only on K. Therefore, if we set $z_j = u - q^{m+j}v$, one of the functions

$$\varphi^{(j)} = \varphi_{u,z_j} \qquad 0 \le j < m$$

vanishes at any point of Sp_{2m}-K. Hence the product of the $\varphi^{(j)}$ vanishes everywhere but on K. But if

$$y = \bigotimes_{j=0}^{m-1} z_j \in L^2(V)$$

then the product of the $\varphi^{(j)}$ is just the matrix coefficient $\varphi_{x,y}$ of $\rho \otimes \rho^*$. Since $\varphi_{x,y}$ is just a multiple of the characteristic function of K, we see that y has the desired properties. This proves theorem 8.2.

Recall that $W_\ell \subseteq W$ is the subspace spanned by the e_i and f_i for $i \le \ell$. We want to study the relation between (Φ,Ψ)-boundedness on Sp(W) and on $Sp(W_\ell)$. We will take as compact subgroup of W_ℓ the intersection $K \cap Sp(W_\ell) = K(W_\ell)$ where K is the standard maximal compact subgroup of Sp(W) specified above. Let $\Xi(W_\ell)$ be Harish-Chandra's spherical function for $Sp(W_\ell)$ with respect to $K(W_\ell)$.

Proposition 8.3: Let ρ be a representation of $\tilde{S}p(W)$, and suppose that $\rho|\tilde{S}p(W_\ell)$ is $(\Xi(W_\ell)^\beta, \Psi(W_\ell))$-bounded for $\beta = \frac{1}{s}$, the reciprocal of some integer s, and some function $\Psi(W_\ell)$ on $K(W_\ell)^\wedge$. Then for some $\varepsilon > 0$, the representation ρ is itself $(\Xi^{\gamma+\varepsilon},\Psi_\varepsilon)$-bounded, for some function Ψ_ε on $\hat{K}$, where

$$\gamma = (s([\tfrac{m}{\ell}] + 1))^{-1} \tag{8.8}$$

where $[x]$ denotes the largest integer less than x. In particular if $\rho|\tilde{S}p(W_k)$ is strongly L^{2s}, then ρ is strongly L^{2q} where $q\gamma = 1$.

Proof: For convenience in the proof we suppress all $\sim$'s, as in $\tilde{S}p$, etc. Write $m = b\ell + r$ for non-negative integers b and r, with $r < \ell$. We will decompose W into a direct sum of b subspaces V_i of dimension 2ℓ, and another space of dimension $2r$. Each of the V_i and U will be spanned by certain of the pairs e_j, f_j belonging to the standard basis, but we will not specify until later which pairs e_j, f_j belong to which spaces. In any case the decomposition $W \simeq (\bigoplus_i V_i) \oplus U$ induces an embedding of

$$(\prod_i Sp(V_i)) \times Sp(U) \simeq Sp_{2\ell}^{b} \times Sp_{2r}$$

into $Sp(W)$. If we set $B(V_i) = B \cap Sp(V_i)$, and $A(V_i) = A \cap Sp(V_i)$, then $B(V_i)$ is a minimal parabolic subgroup of $Sp(V_i)$ and $A(V_i)$ is a split Cartan subgroup of $B(V_i)$. Also

$$A(V_i)_1^+ = A_1^+ \cap A(V_i)$$

Similar notations and remarks apply to U. We have

$$A_1^+ \subseteq (\prod_i A(V_i)_1^+) \times A(U)_1^+ \tag{8.9}$$

Set $K(V_i) = K \cap SP(V_i)$ and do likewise for U. Then the $K(V_i)$ and $K(U)$ satisfy conditions (7.1). Note that each $Sp(V_i)$ is conjugate in $Sp(W)$ to $SP(W_\ell)$. In fact, the conjugation may be accomplished by a permutation of the e_j's and the f_j's, and in such a way that $B(V_i)$ is taken to $B(W_\ell)$, and similarly for $A(V_i)$ and $K(V_i)$. Also the group $Sp(U)$ is conjugate in similar fashion to the subgroup $Sp(W_r)$ of $Sp(W_\ell)$ by a conjugation with analogous properties.

Let $\Xi(V_i)$ denote Harish-Chandra's spherical function for $Sp(V_i)$ with respect to $K(V_i)$. Define $\Xi(U)$ similarly. By the conjugacy properties of the $Sp(V_i)$, we see that the hypotheses of the proposition imply that $\rho|Sp(V_i)$ is $(\Xi(V_i)^\beta, \Psi(V_i))$-bounded, where $\Psi(V_i)$ is the function on $K(V_i)^\wedge$ obtained from the function $\Psi(W_\ell)$ on $K(W_\ell)^\wedge$ by conjugation. It will also hold that $\rho|Sp(U)$ is $(\Xi(U)^\beta, \Psi(U))$-bounded for some function $\Psi(U)$. It will be convenient to delay slightly the derivation of this estimate.

It is known [Wr], [Bn], [Ly] that the subgroup K of $Sp(W)$ is uniformly large in the sense of §6. Therefore proposition 6.3 tells us that the restriction $\rho|((\prod_i Sp(V_i)) \times Sp(U)$ is (Φ,Ψ)-bounded, where

$$\Phi = (\prod_i \Xi(V_i))^\beta \, \Xi(U)^\beta \tag{8.10}$$

and Ψ is whatever it turns out to be.

Define a function Φ^1 on $Sp(W)$ by the recipe

$$\Phi^1(k_1 T k_2) = \Phi(T) \qquad k_i \in K, \quad T \in A_1^+ \tag{8.11}$$

This definition makes sense by virtue of inclusion (8.9). I claim that ρ is (Φ^1, Ψ^1)-bounded for an appropriate function Ψ^1 on $\hat{K}$. Indeed, select $\mu \in \hat{K}$, and let x belong to the μ-isotypic component of the space of ρ. The restriction of μ to $(\prod_i K(V_i)) \times K(U)$ will decompose into a sum of finitely many irreducible representations μ_i of the smaller group. Let x_i be the μ_i-component of x. Similarly, select another K-type ν, and a vector y in the ν-isotypic component. Let ν decompose into representations ν_j on restriction to $(\prod_i K(V_i)) \times K(U)$, and let y_j be the ν_j-component of y. Then for $T \in A_1^+$ we have

$$(8.12)\qquad |\varphi_{x,y}(T)| = |\sum_{i,j} \varphi_{x_i,y_j}(T)| \le \sum_{i,j} |\varphi_{x_i,y_j}(T)|$$

$$\le \sum_{i,j} \|x_i\| \, \|y_j\| \quad \Psi(\mu_i)\,\Psi(\nu_j)\,\Phi(T)$$

$$= (\sum_i \|x_i\| \quad \Psi(\mu_i))\,(\sum_j \|y_j\| \quad \Psi(\nu_j)) \quad \Phi(T)$$

$$\le (\sum \|x_i\|^2)^{1/2} (\sum \Psi(\mu_i)^2)^{1/2} (\sum \|y_j\|^2)^{1/2} (\sum \Psi(\nu_j)^2)^{1/2} \quad \Phi(T)$$

$$= \|x\| \, \|y\| \, (\sum \Psi(\mu_i)^2)^{1/2} \, (\sum \Psi(\nu_j)^2)^{1/2} \quad \Phi(T)$$

The last step follows because the x_i are mutually orthogonal and sum to x, and similarly for the y_j. Now observe that for $k_i \in K$,

$$\varphi_{x,y}(k_1\,Tk_2) = \varphi_{\rho(k_1)^{-1}x,\ \rho(k_2)y}(T)$$

Since $\rho(k_1)x$ still is in the μ-isotypic component of ρ, and has the same norm as x, and similatly for y, we get estimate (8.12) for $\varphi_{x,y}(k_1Tk_2)$ as well as for $\varphi_{x,y}(T)$. But this is precisely the estimate necessary to establish (Φ^1, Ψ^1)-boundedness, with

$$\Psi^1(\mu) = (\sum \Psi(\mu_i)^2)^{1/2} \qquad \mu \in \hat{K}$$

To finish proving the proposition, it remains to specify how the e_j and f_j are distributed among the V_i and U, and then to relate the function Φ^1 to the function Ξ. The idea is to perform the distribution to maximize the compatibility between Φ^1 and Ξ. Our recipe is

$$(8.13)\qquad V_i = \text{span}\,\{e_j, f_j : j = bk + i \ \text{ for } 0 \le k \le \ell - r, \text{ and } j = m - (b+1)k + i \ \text{ for } \ell \le k \le r\}$$

$$U = \text{span}\,\{e_j, f_j : j = m - (b+1)k \ \text{ for } 0 \le k < r\} \ .$$

Recall the inequalities (7.7) relating the function Ξ to the modular function δ_B of B. Recall also formula (8.3) explicitly describing δ_B for Sp(W). These formulas apply <u>mutatis mutandis</u> to the $Sp(V_i)$ and to Sp(U). Combining them we see that for any $\varepsilon > 0$, there is a constant $c(\varepsilon)$ such that on A_1^+,

$$(8.14)\qquad (\Phi^1)^s \le c(\varepsilon)\ \delta_B^{\varepsilon}\ \Big[\Big(\prod_{k=0}^{\ell-r-1}\prod_{j=1}^{b}|\alpha_{bk+j}|^{\ell-k}\Big)\Big(\prod_{k=-r}^{\ell-1}\prod_{j=1}^{b+1}|\alpha_{(b+1)k+r-\ell+j}|^{\ell-k}\Big)\Big]^{-1}$$

(Here recall $s\beta = 1$.)

At this point we can demonstrate the asymptotic estimate we claimed for $\rho|Sp(U)$. We see from the same formulas used for inequality (8.14) that if a representation ρ of $Sp(W_\ell)$ is $(\Xi(W_\ell)^\beta, \Psi')$-bounded, then the $K(W_\ell)$-finite matrix coefficients, restricted to $A_1^+ \cap Sp(W_r)$, decay faster than

$$\Big(\prod_{j=1}^{r}|\alpha_j|^{\ell+1-r}\Big)^{\beta-\varepsilon}$$

for any $\varepsilon > 0$. Thus $\rho|Sp(W_r)$ is strongly $L^{p+\varepsilon}$ where

$$p = \frac{2r}{\ell\beta} < 2m$$

Hence corollary 7.2 provides the desired estimate for $\rho|Sp(W_r)$.

We need to compare the product of the $|\alpha_i|$ in square brackets on the right hand side of (8.14) with the product defining δ_B. We see that the exponent with which $|\alpha_{m-j}|$ contributes to $\delta_B^{1/2}$ is $j+1$. If we write $j+1 = (b+1)k + i$, with $i \le b$, then the exponent with which $|\alpha_{m-j}|$ contributes to the right hand side of (8.14) is at least $k+1$. Thus the exponent of $|\alpha_i|$ in δ_B is never more than $b+1$ times the exponent of $|\alpha_i|$ in the product of (8.14). Moreover, the factor $|\alpha_1|$ appears in the two functions with exponents m and ℓ respectively. Since $(b+1)\ell$ is strictly larger than m, and since $|\alpha_1|$ dominates all the $|\alpha_i|$ on A_1^+, we see that for all sufficiently small $\varepsilon > 0$, there is some constant d such that

$$(8.15) \qquad (\Phi^1)^{s(b+1)} \leq d \quad \Xi^{1+\varepsilon}$$

From estimate (8.15) the statement of the proposition is immediate.

Remark: Inspection of the proof of proposition 8.3 shows that if ℓ divides m exactly, then estimate (8.8) for γ can be improved. Specifically if β is greater than $\frac{1}{s}$ instead of exactly equal to it, then γ can be changed to $\gamma' = (\ell/ms)$. Or if $\beta = \frac{1}{s}$ then we can replace $\gamma + \varepsilon$ by $\gamma' - \varepsilon$, for any $\varepsilon > 0$.

We are now in a position to relate rank to asymptotic decay of matrix coefficients. Recall that $N_m(W)$ is the unipotent radical of the parabolic $P_m(W)$ which preserves X_m, the maximal isotropic subspace spanned by the e_i. We recall the notion of N_m-rank of representations studied in §2. We will prove two main results. One result relates rank to asymptotic decay of matrix coefficients. It says that the larger the rank of a representation, the faster its matrix coefficients tend to decay. The second result, based on the first, relates rank to the topology in $\hat{Sp}$. Unfortunately, our control on asymptotics still leaves much to be desired - the remark b) following corollary 7.3 applies here with a vengeance. Consequently, these final results are far from best possible. However, they do illustrate the phenomenon at question.

Theorem 8.4: Let ρ be a representation of $\tilde{Sp}(W)$ of pure N_m-rank $r \le m$.

a) If $m > 1$, then ρ is strongly L^q where

$$q = 4([\frac{m}{r}] + 1) \tag{8.16}$$

where $[x]$ again denotes the greatest integer not larger than x.

b) Additionally, if $m > 1$, then all N_m-rank 1 representations are strongly $L^{4m+\varepsilon}$. If $m > 2$, then all N_m-rank 2 representations are strongly $L^{2m+\varepsilon}$.

c) If $m > 3$, and the integer ℓ satisfies $\ell \le \frac{m}{3}$, then if $r > 2\ell$, the representation ρ is $\tilde{Sp}(W_\ell)$-regular; that is, the restriction $\rho|\tilde{Sp}(W_\ell)$ is absolutely continuous.

Remark: Parts a) and b) of this theorem together imply that when $m > 1$, all non-trivial irreducible representations of $\tilde{Sp}_{2m}$ are $L^{4m+\varepsilon}$, and when $m > 2$ the only representations which are not L^{4m} (indeed $L^{2m+\varepsilon}$) are the components of the oscillator representations. Also for $m > 2$ every representation of Sp_{2m} is strongly $L^{2m+\varepsilon}$, and if $m > 3$, the only representations which are not strongly L^{2m} are the rank 2 representations described in §5.

Proof: If $m > 1$, then theorem 4.2 implies the only rank 1 representations are the components of oscillator representations, and these are strongly $L^{4m+\varepsilon}$ by proposition 8.1. For rank 2 representations, we may argue similarly. Or, independently of classification, we may reason as follows. If $m > 3$, and ρ is of pure rank 2, then corollary 2.13 implies $\rho|Sp(W_2)$ is a sum of two-fold products of oscillator representations. Thus $\rho|Sp(W_1)$ is $(\Xi(W_2)^{1/2}, 2 \dim \mu)$-bounded. Then a slight adaptation of the argument of proposition 8.3 shows ρ is strongly $L^{2m+\varepsilon}$. With these remarks, we consider part b) of the theorem proven.

Next consider part c) of the theorem. Consider the restriction to $Sp(W_\ell)$ of a representation ρ of $Sp(W)$ of pure rank r. According to corollary 2.13, the restriction $\rho|Sp(W_\ell)$ is a finite sum of representations of $\sigma \otimes \tau$ where τ is a tensor product of oscillator representations, involving $\min(r, m-\ell)$ factors, and σ is a representation of $Sp(W_\ell)$ of rank $\max(0, r+\ell-m)$. Our restriction on ℓ implies $m - \ell \geq 2\ell$. Since $r > 2\ell$ by assumption, we see that either τ is a product of more than 2ℓ oscillator representations, or of exactly 2ℓ oscillator representations, and rank $\sigma > 1$. In the latter case, our restriction on m tells us $\ell > 2$. Hence propositions 8.1 and 6.1 imply τ is absolutely continuous, and proposition 6.1 then applies again and says $\sigma \otimes \tau$ is absolutely continuous. Hence part c) of the theorem is true.

Finally consider part a). Consider the case when $r = \text{rank } \rho$ is at most $\frac{m}{2}$. Then $m - r \geq r$, so that again by corollary 2.13, the restriction $\rho|Sp(W_r)$ is a sum of r-fold tensor products of oscillator representations. Thus theorem 8.2 implies $\rho|Sp(W_r)$ is $(\Xi(W_r)^{1/2}, 2 \dim \mu)$-bounded, and the estimate (8.16) follows from proposition 8.3.

This argument extends also to the case $2r = m+1$. For in this case corollary 2.13 says that $\rho|Sp(W_r)$ is a sum of representations $\sigma \otimes \tau$ where τ is an $(r-1)$-fold tensor product of oscillator representations and σ has rank 1. Hence σ is a sum of components of oscillator representations. Therefore theorem 8.2 and proposition 8.3 again give the result.

The case when $2r \geq m+2$ is easier. Set

$$m' = [\frac{m-1}{2}] + 1$$

Thus $2m' = m$ if m is even, and $2m' = m+1$ if m is odd. Another application of corollary 2.13 says that $\rho|Sp(W_{m'})$ is a sum of representations of the form $\sigma \otimes \tau$ where τ is an $(m-m')$-fold tensor product of oscillator representations, and σ has rank $r-(m-m') \geq 2$. Again applying propositions 6.1 and 8.1, and part b) of this theorem, we conclude that $\rho|Sp(W_{m'})$ is strongly L^4. Then corollary 7.2 and proposition 8.1 give the desired conclusion. This proves part a) of the theorem.

We close with a result that shows that representations of small rank cannot be obtained as limits of complementary series. Compare Duflo's [Df] computation of the unitary dual of $Sp_{2\cdot 2}(\mathbb{C})$. Let β be a symmetric bilinear form on X_m. Let $(\tilde{Sp})^\wedge_\beta$ be the subset of $(\tilde{Sp})^\wedge$ associated to β in §2.

Theorem 8.5: The subset $(\tilde{Sp})^\wedge_\beta$ of $(\tilde{Sp})^\wedge$ is both open and closed if rank $\beta \leq \frac{2m}{3} - 2$.

Proof: It is completely clear that the union of the $(\tilde{Sp})^\wedge_\beta$ for rank β less than some given bound is closed in $(\tilde{Sp})^\wedge$. This follows simply by looking at N_m-spectra. It is also clear that $(\tilde{Sp})^\wedge_\beta$ is relatively open in the union of the $(\tilde{Sp})^\wedge_{\beta'}$ with rank $\beta' \leq$ rank β. Hence to prove the theorem, it will suffice to show that no element $\rho \in (\tilde{Sp})^\wedge_\beta$ is a limit of representations of larger rank. Alternatively, it will suffice to show that if σ is a representation, not necessarily irreducible, of $\tilde{Sp}$, and σ has pure N_m-rank $r >$ rank β, then σ does not contain weakly any representation ρ of N_m-rank equal to rank β.

By consideration of the action of the center of $\tilde{Sp}$ under σ and under ρ, we see (by corollary 2.14) that if σ weakly contains ρ, then the N_m-ranks of σ and of ρ have the same parity, so that the rank of σ is at least 2 more than the rank of ρ. Also, let ω be an oscillator representation. Then if σ weakly contains ρ, the tensor product $\sigma \otimes \omega$ will weakly contain $\rho \otimes \omega$. Hence, by this device, we may assume that the rank of ρ is odd, and that it is at most $\frac{2m}{3} - 1$.

Define an integer ℓ by

$$2\ell = N_m\text{-rank } \rho + 1$$

Then $2\ell \leq \frac{2m}{3}$, so theorem 8.4, part c) is applicable. Since the N_m-rank of σ is greater than 2ℓ, that result implies that $\sigma|Sp(W_\ell)$ is absolutely continuous. Therefore corollary 7.2 implies that any representation in the weak closure of σ must be $(\Xi(W_\ell), (\dim \mu)^2)$-bounded on $Sp(W_\ell)$. But corollary 2.13 says $\rho|Sp(W_\ell)$ is a $(2\ell-1)$-fold tensor product of oscillator representations. Then proposition 4.1 says this representation is only $L^{q+\varepsilon}$ where

$$q = \frac{4\ell}{2\ell-1}$$

Hence $\rho|Sp(W_\ell)$ is not $(\Xi(W_\ell),(\dim \mu)^2)$-bounded, and ρ cannot be in the weak closure of σ. This concludes theorem 8.5, except in the case $F = \mathbb{C}$. A slight refinement of the argument covers that case too. We omit details.

Remark: This result implies for example that holomorphic representations of $\tilde{Sp}_{2n}(\mathbb{R})$ of sufficiently small rank are isolated in $(\tilde{Sp}_{2n}(\mathbb{R}))^\wedge$.

In conclusion, I would like to thank Professor Michael Atiyah and the Mathematical Institute at Oxford University for a very pleasant stay in May-June 1978, during which time some of the ideas developed here germinated. Also, thanks are due to Mrs. Mel DelVecchio for an excellent and expeditious job of typing.

References

[Ar] J. Arthur, Harmonic analysis of the Schwartz space on a reductive Lie group, I and II, mimeographed notes.

[Am] C. Asmuth, Weil Representations of Symplectic p-adic groups, Am. J. Math., 101(1979), 885-908.

[Bn] I. N. Bernstein, All reductive p-adic groups are tame, Fun. Anal. and App. 8 (1974), 91-93.

[BW] A. Borel and N. Wallach, Continuous Cohomology, Discrete Subgroups, and Representations of Reductive Groups, Annals of Math. Studies 94, Princeton University Press, 1980, Princeton, New Jersey.

[Cr] P. Cartier, Quantum mechanical commutation relations and theta functions, Proc. Sym. Pure Math. IX, A.M.S., 1966, Providence, R.I.

[Cg] M. Cowling, Sur l'algèbre de Fourier-Stieltjes d'un group semisimple, to appear.

[Dx] J. Dixmier, Les C^*-algebres et leurs representations, Gauthier-Villars, 1964, Paris.

[Df] M. Duflo, Représentations unitaires irréductibles des groupes simples complexes de rang deux, preprint.

[DS] N. Dunford and J. Schwartz, Linear operators, Interscience 1958-1971, New York.

[Fr] T. Farmer, On the reduction of certain degenerate principal series representations of $Sp(n,\mathbb{C})$, Pac. J. Math. 84, No.2(1979), 291-303.

[F1] J. Fell, The dual spaces of C^*-algebras. T.A.M.S., v.94(1960), 364-403.

[F2] J. Fell, Weak containment and induced representations of groups, Can. J. Math., v. 14(1962), 237-268.

[F3] J. Fell, Non-unitary dual spaces of groups, Acta Math., v. 114 (1965), 267-310.

[Gs] K. Gross, The dual of a parabolic subgroup and a degenerate principal series of $Sp(n,C)$, Am. J. Math. 93 (1971), 398-428.

[HC1] Harish-Chandra, Discrete series for semisimple Lie groups II, Acta Math., v. 116 (1966), 1-111.

[HC2] Harish-Chandra, Harmonic analysis on semisimple Lie groups, B.A.M.S., v. 76 (1970), 529-551.

[HC3] Harish-Chandra, Harmonic analysis on reductive p-adic groups, Proc. Symp. Pure Math. XXVI, A.M.S. 1973, Providence, R.I.

[Hn] S. Helgason, Differential Geometry and Symmetric Spaces, Academic Press, 1962, New York.

[H1] R. Howe, Reductive dual pairs and the oscillator representation, in preparation.

[H2] R. Howe, L^2 duality for stable reductive dual pairs, preprint.

[H3] R. Howe, θ-series and invariant theory, Proc. Symp. Pure Math. XXXIII, Part I, 275-286.

[HM] R. Howe and C. Moore, Asymptotic properties of unitary representations, J. Fun. Anal. 32 (1979), 72-96.

[Kn] D. Kazhdan, Connection of the dual space of a group with the structure of its closed subgroups, Fun. Anal. and App., v. 1 (1967) 63-65.

[Ly] J. Lepowsky, Algebraic results on representations of semisimple Lie groups, T.A.M.S. 176 (1973), 1-44.

[My] G. Mackey, Unitary of group extensions, Acta Math., v. 99 (1958), 265-301.

[McD] I. MacDonald, Spherical functions on groups of p-adic types, Publ. Ramanujan Inst. 2, Univ. Madras, 1971, Madras, India

[MW] C. Moore and J. Wolf, Square integrable representations of nilpotent groups, T.A.M.S. 185 (1973).

[Nk] M. Naimark, Normed Rings, P. Noordhoff 1964, Groningen, Netherlands.

[On] E. Onofri, Dynamical quantization of the Kepler manifold, J. Math. Phys. 17 (1976), 401-408.

[Ra] J. Repka, Tensor products of unitary representations of $SL_2(R)$, Am. J. Math. 100 (1978), 747-774.

[Rg] J. Rosenberg, A quick proof of Harish-Chandra's Plancherel Theorem for spherical functions on a semisimple Lie groups, P.A.M.S. 63 (1971), 143-149.

[Sh] D. Shale, Linear symmetries of free boson fields, T.A.M.S. 103 (1962), 149-167.

[Sl] A. Silberger, Introduction to Harmonic Analysis on Reductive p-adic Groups, Princeton University Press, 1979, Princeton, New Jersey.

[Sn] B. Srinivasan, The characters of the finite symplectic group Sp(4,q) T.A.M.S. 131 (1968), 488-525.

[TV] P. Trombi and V. Varadarajan, Asymptotic behavior of eigenfunctions on a semisimple Lie group, the discrete spectrum, Acta Math. 129(1972), 237-280.

[V] V. Varadarajan, Characters and discrete series for Lie groups, Proc. Symp. Pure Math. XXVI, A.M.S. 1973, Providence, R.I.

[Wr] G. Warner, Harmonic Analysis on Semi-Simple Lie Groups I, II. Grundlehren der Math. Wiss. 188, 189, Springer-Verlag, 1972, Heidelberg, New York.

[Wi1] A. Weil, Sur certains groupes d'opérateurs unitaires, Acta Math. 111, (1964), 143-211.

[Wi2] A. Weil, Basic Number Theory, Grund. der Math. Wiss. 144, Second Edition, Springer-Verlag 1973, Heidelberg, New York.

[Z1] G. Zuckerman - oral communication.

[Z2] G. Zuckerman, Continuous cohomology and unitary representations of real reductive groups, Ann. Math. 107 (1978), 495-516.

CENTRO INTERNAZIONALE MATEMATICO ESTIVO

(C.I.M.E.)

SOME APPLICATIONS OF GELFAND PAIRS IN CLASSICAL ANALYSIS

ADAM KORANYI

SOME APPLICATIONS OF GELFAND PAIRS IN CLASSICAL ANALYSIS

Adam Koranyi
(Washington University)

Introduction

Let G be a unimodular Lie group and K a compact subgroup. As well known, (G,K) is called a Gelfand pair if the convolution algebra of left-and-right K-invariant functions on G is commutative; this is equivalent to saying that every irreducible representation of K occurs at most once in the regular representation of G on $L^2(G/K)$.

The most important classical example is the case where G/K is a Riemannian symmetric space; the fact that (G,K) is a Gelfand pair is then crucial in the representation theory of non-compact semisimple Lie groups. Even in the case of a compact group U, the observation that $(U \times U$, diag $U \times U)$ is a Gelfand pair leads to the most illuminating way to prove the Peter-Weyl theorem.

The other main classical example, as well-known, is the case (G,K) where $G = K \times_s A$ is a semidirect product with an Abelian normal subgroup A; this gives the best framework for the harmonic analysis of radial functions on $\mathbb{R}^n$.

In these lectures I will describe two further examples that have arisen naturally in the context of some questions of classical analysis. In one of these G is a semidirect product of a compact group and a special type of nilpotent group, and the purpose it is used for is the study of certain analogues of radial functions on the nilpotent group. In the other example we will consider some of the most well-studied compact groups, but in a context which involves a slight extension of the notion of a Gelfand pair.

Both examples to be discussed originate from the same source: the theory of singular integrals on certain homogeneous vector bundles developed in [10]. Let us remark in passing that in [10] this theory is not put in the language of vector bundles, but it goes through without any change in this setting. In fact, let B_i $(i = 1,2)$ be homogeneous vector bundles over $X = G/K$, with the fibre over $x \in X$ denoted $E_x^{(i)}$ and the action of $g \in G$ on B_i denoted $\sigma_i(g)$. The linear operators A mapping sections of B_1 to sections of B_2 can be written, at least symbolically, as

$$(Af)(x) = \int_X S(x,y)f(y)dy$$

where $S(x,y)$ is a linear transformation $E_y^1 \to E_x^2$ and dy is a G-invariant measure. (In general, of course, S will be a distribution-valued kernel.) A will commute with the action T of G on sections given by

$$(T_g f)(x) = \sigma(g)f(g^{-1}\cdot x)$$

if and only if

$$S(gx,gy) = \sigma_2(g)S(x,y)\sigma_1(g)^{-1}$$

for all $x,y \in X$, $g \in G$. Introducing the function

$$k(g) = S(g,e)$$

and denoting by o the base point in G/K, A can also be written as

$$(Af)(g\cdot o) = \int_G \sigma_2(u)k(u^{-1}g)\sigma_1(u)^{-1}f(u)du$$

(with du denoting Haar measure). This is exactly the same formula as in [10] except that σ_1 and σ_2 have a more special meaning there. However, the results and proofs remain valid under the present interpretation. So, if G/K has a G-invariant pseudometric satisfying the conditions listed in [10], one has a complete corresponding theory of G-equivariant singular integral operators.

We should also mention that if the homogeneous vector bundles B_i are given in the form $G \times_K V^i$ $(i = 1,2)$, where the V^i are finite-dimensional K-modules, and the sections are identified with functions $v: G \to V$ satisfying

$$v(gk) = k^{-1} \cdot v(g) \qquad (k \in K,\ g \in G)$$

(so the action of G on sections becomes simply left translation), then the G-equivariant operators appear in the form

$$(A^o v)(g \cdot o) = \int_G k^o(u^{-1}g)v(u \cdot o)du$$

with a kernel such that $k^o(g): V^1 \to V^2$ is linear for all $g \in G$ and

$$k^o(k_1^{-1}gk_2) = k_2^{-1} \cdot k^o(g) \cdot k_1$$

for all $k_1, k_2 \in K$; $g \in G$.

The results to be discussed in the present lectures arise from the attempt to use the harmonic analysis of G and K in the study of G-equivariant singular integral operators, i.e., to look at these operators "on the Fourier transform side", where they appear as multiplier operators.

Of course this kind of harmonic analysis is most convenient to use in situations where every irreducible representation of the group occurs at most once. In the present case this means that we are considering vector bundles $G \times_K V$ where the natural representation of K on the space of all L^2-sections contains every irreducible representation at most once. This is the extension of the notion of Gelfand pair referred to above; in the case where $V = C$ and the action of K on V is trivial, it coincides with the usual notion of a Gelfand pair. In our examples it will even be true, although we will not make use of this fact, that for <u>every</u> irreducible K-module V, all irreducible representations of K occur at most once in $G \times_K V$. One could call (G,K) a "strong Gelfand pair" in such a case.

§1. Vector-valued functions on spheres

In our first example we consider $L = L^2(S^{n-1}, \mathbb{R}^n)$ $(n \geq 5)$, the L^2-space of $\mathbb{R}^n$-valued functions on the unit sphere of $\mathbb{R}^n$. As customary, we will refer to these as vector-valued functions, although it is functorially more correct and more illuminating to think of them as covector-valued functions, i.e., differential forms.

Let $G = SO(n)$, acting on $\mathbb{R}^n$ in the usual way, and let $K \cong SO(n-1)$ be the stabilizer of the point $e_n = (0,\ldots,0,1)$. The elements of L can be regarded as sections of the trivial bundle $S^{n-1} \times \mathbb{R}^n$ with G acting by $\sigma(g)\colon (x',v) \to (g\cdot x', g\cdot v)$. So the action of G on the sections, i.e., on $f \in L$, is given by

$$(T_g f)(x') = g \cdot f(g^{-1} \cdot x') .$$

The subspace $H \subset L$ formed by the boundary values of Riesz systems, i.e. of gradients of harmonic functions in the unit ball, is clearly G-invariant. The orthogonal projection $P\colon L \to H$ was shown in [11] to be a G-equivariant singular integral operator, bounded in every L^p $(p > 1)$; similar results were proved in [10] about the "Riesz transform", i.e., the map R that associates to the normal component of every element of H the corresponding tangential component.

Here we wish to describe the main results of [12] concerning the harmonic analysis of the underlying vector bundle and to mention some of the main applications. Let us denote by $D^{r,o}$ resp. $D^{r,1}$ the irreducible representations of $SO(n)$ whose maximal weight in the usual parametrization [2] is $(r,0,\ldots,0)$ resp. $(r,1,0,\ldots,0)$. Let $\mathcal{H}_r$ denote the space of homogeneous harmonic polynomials of degree r on $\mathbb{R}^n$; as well known, G acts on $\mathcal{H}_r$ by the representation $D^{r,o}$.

First of all, we clearly have the orthogonal sum decomposition

$$L = L_{Tan} \oplus L_{Nor}$$

Here L_{Tan} is the set of "tangential vector fields", i.e., all f such that $f(x')\cdot x' = 0$ for all $x' \in S^{n-1}$ (the dot here denotes the natural inner product on $\mathbb{R}^n$). L_{Nor} is the "normal vector fields", i.e., such that $f(x') = \varphi(x')x'$ with some scalar-valued $\varphi \in L^2(S^{n-1}, \mathbb{R})$. It is then obvious that

$$L_{Nor} = \bigoplus_{r \geq o} L^{r,o}_{Nor}$$

$L^{r,o}_{Nor}$ being the space $\{\varphi(x')x' \mid \varphi \in \mathcal{H}_r\}$, and carrying the representation $D^{r,o}$.

To decompose L_{Tan} into irreducible subspaces we have to observe that it is exactly the homogeneous vector bundle $G \times_K \mathbb{R}^{n-1}$ with the natural action of $K = SO(n - 1)$ on $\mathbb{R}^{n-1}$. So the Frobenius reciprocity theorem implies that the representations of G occurring in L_{Tan} are exactly those whose restriction to K contains the natural representation (of type $D^{1,o}$) of $SO(n - 1)$ on $\mathbb{R}^{n-1}$. By the Branching Theorem (cf. [2]) this gives exactly those representations $(m_1,\ldots,m_p)$ for which

$$m_1 \geq 1 \geq m_2 \geq 0 \geq m_3 \geq \ldots$$

i.e., $D^{r,o}$ and $D^{r,1}$, with multiplicity one for each $r \geq 1$. (It is here that the hypothesis $n \geq 5$ is used; the cases $n \leq 4$ require some modifications.) With an obvious notation we can then write

$$L_{Tan} = \bigoplus_{r \geq 1} (L^{r,o}_{Tan} \oplus L^{r,1}_{Tan}).$$

As for H, since the operator ∇ commutes with the action of G, we clearly have

$$H = \bigoplus_{r \geq 1} H^{r,o}$$

with

$$H^{r,o} = \{(\nabla\varphi) \mid S^{n-1} : \varphi \in \mathcal{H}_r\}.$$

This space carries the representation $D^{r,o}$ and consists of restrictions to S^{n-1} of harmonic polynomials of degree $r - 1$; it is contained in $L^{r,o}_{Nor} \oplus L^{r,o}_{Tan}$ but is itself neither normal nor tangential. One proceeds to get more explicit information by noting that

$$L = \bigoplus_{r\geq o} (\mathcal{H}_r \otimes \mathbb{R}^n)|_{S^{n-1}}$$

and that G acts on the r'th summand by $D^{r,o} \otimes D^{1,o}$. A Clebsch-Gordan type formula says that this is $D^{r,1} \oplus D^{r+1,o} \oplus D^{r-1,o}$. Comparing representations and degrees this shows that

$$(\mathcal{H}_r \otimes \mathbb{R}^n)|_{S^{n-1}} = L^{r,1}_{Tan} \oplus H^{r+1,o} \oplus M^{r-1,o}$$

where $M^{r-1,o}$ is a subspace of type $D^{r-1,o}$; it can also be described more precisely in terms of harmonic polynomials (cf. [12]).

Using some classical facts about $\mathcal{H}_r$, it is now easy to describe the action of the Riesz transform R on the representations: It maps $L^{r,o}_{Nor}$ onto $L^{r,o}_{Tan}$ with the multiplier $(1 + \frac{n-2}{r})^{\frac{1}{2}}$, which shows in particular that it is a compact perturbation of an isometry. Similarly, P has an obvious description in terms of the representations.

One can use these methods to solve some further problems of analysis on S^{n-1}. For example one may ask, in analogy with the case of holomorphic functions on the unit ball in $\mathbb{C}^n$, whether the subspace H of L can be characterized by differential equations. As well known, the Riesz systems $f = (f_1,\ldots,f_n)$ on $\mathbb{R}^n$ are characterized by the analogues of the Cauchy-Riemann equations:

$$D_i f_k = D_k f_i \qquad (1 \leq i,k \leq n)$$

$$\Sigma D_i f_i = 0$$

It is easy to give a meaning to the restriction of these equations to S^{n-1}. They automatically annihilate every normal vector field, and the theorem one can prove is the following [12].

The vector fields on S^{n-1} that are the tangential projections of some element in H are characterized by the restricted Cauchy-Riemann equations (taken in the sense of distributions).

For the proof one shows that the restricted Cauchy-Riemann equations amount to a certain single G-equivariant differential operator annihilating the vector field in question (it is really just the operator of exterior differentiation, if the vector field is reinterpreted as a differential form on S^{n-1}). By equivariance and by the decomposition of L_{Tan} into irreducible subspaces it suffices now to show that this operator does not annihilate any of the $L^{r,1}_{\mathrm{Tan}}$; this can be done by exhibiting a concrete element in the space that is not mapped to zero.

Another very interesting application of the harmonic analysis of L was given recently by M. Reimann [14]. To describe this, we recall that Ahlfors [1] defined an operator S mapping $\mathbb{R}^n$-valued functions $u = (u_1,\dots,u_n)$ defined on the unit ball of $\mathbb{R}^n$ to matrix-valued functions ; the (i,k)-th entry is given by

$$(Su)_{ik} = \frac{1}{2}(D_i u_k + D_k u_i) - \frac{1}{n}(\Sigma D_j u_j)\delta_{ik} .$$

S is an analogue of the $\bar{\partial}$-operator and is of interest for the theory of quasiconformal mappings. Of obvious importance in connection with S is the equation $S^*Su = 0$, where S^* denotes the adjoint of S . (The Riesz systems are particular solutions of this equation.) By working with the irreducible pieces of L it is shown in [14] that, given an arbitrary $f \in L$, there is a unique solution u of $S^*Su = 0$ whose boundary values coincide with f , i.e., the "Dirichlet problem" with L^2 boundary data has a unique solution.

§2. Biradial functions on nilpotent groups.

Let G be a connected real semisimple Lie group of real rank one. Let $G = KAN$ be an Iwasawa decomposition, and let M be the centralizer of A in K . We are going to consider the groups N which occur in this manner. They are of interest since they can be identified with open dense subsets of the boundaries of the rank one symmetric spaces G/K ; the N given by $G = SU(n,1)$ is, by the way, the well-known Heisenberg group.

The Lie algebra of N is of the form

$$\underline{n} = \underline{g}^1 + \underline{g}^2$$

where $\underline{g}^1, \underline{g}^2$ are the A-root spaces corresponding to the roots $\gamma, 2\gamma$. Let p, q denote the dimensions of $\underline{g}^1, \underline{g}^2$ respectively; it is possible that $p = 0$. (This convention may seem strange, but in this way the case where there is only one root fits much more smoothly into the general case.) Since M commutes with A, the adjoint action of M preserves $\underline{g}^1$ and $\underline{g}^2$. It follows that M normalizes N, so MN is a semidirect product.

If S_i denotes the unit sphere in $\underline{g}^i$ (with respect to the usual inner product given by the Cartan involution and the Killing form), then M obviously operates on $S_1 \times S_2$. By Kostant's double transitivity theorem [13], [15] this action is transitive when $q > 1$. It follows that the functions on N which, when written in the form $f(\exp(X + Y))$, $X \in \underline{g}^1$, $Y \in \underline{g}^2$ depend only on the norms $|X|$, $|Y|$, are exactly the functions that are invariant under the action of M. As in [7], we call these functions biradial. (When $q = 1$, the M-invariant functions are the ones that depend on $|X|$, Y. In this case we can still call this slightly larger class biradial, since all that follows applies, with only minimal modifications. The special status of the case $q = 1$ would disappear entirely if instead of M we used its analogue in the entire, not necessarily connected isometry group of the symmetric space.)

Under the map $n \to MnM$ the M-invariant functions on N are identified with the left-and-right M-invariant functions on MN. The main observation of this section is now the following.

(MN,M) is a Gelfand pair. Hence the L^1-algebra $\underline{A}$ of biradial functions on N is commutative.

The proof is immediate: writing $n = \exp(X + Y)$ as before, one has $n^{-1} = \exp(-X - Y)$, and by the double transitivity theorem (if $q > 1$) there exists $m \in M$ such that $mnm^{-1} = n^{-1}$. It follows that $MnM = Mn^{-1}M$ and this implies the desired commutativity. The case $q = 1$ can also be discussed without using the classification of symmetric spaces, along the lines of [15]. If one is willing to use the classification, then one knows that $q = 1$ occurs only when N is the Heisenberg group; the statement (even a stronger statement) follows then from the remarks below.

In general we may ask if there exists a larger class than the biradial functions that form a commutative algebra under convolution. In the special case of the Heisenberg group H_{2n+1} parametrized as (z,t); $z \in \mathbb{C}$, $t \in \mathbb{R}$ with the product

$$(z,t)\cdot(z',t') = (z + z', t + t' + 2\,\mathrm{Im}\,z\cdot\bar{z}') ,$$

a function $f(z,t)$ is called multiradial [4] if it depends only on $|z_1|,\ldots,|z_n|,t$. It is an important fact [4] that these form a commutative algebra. There is a Gelfand pair in the background here too: let the elements $\epsilon = (\epsilon_1,\ldots,\epsilon_n)$ of the n-torus T^n act on H_{2n+1} by the automorphisms $(z,t) \to (\epsilon z, t)$ where $\epsilon z = (\epsilon_1 z_1,\ldots,\epsilon_n z_n)$. Then $T^n H_{2n+1}$ is a semidirect product and $(T^h H_{2n+1}, T^n)$ is a Gelfand pair. In fact, the map α: $(z,t) \to (\bar{z},-t)$ is an automorphism of H_{2n+1} and it is easy to see that the double coset of $\alpha(g)$ is the same as that of g^{-1}, for every $g = (z,t)$. Since α induces an automorphism of the convolution algebra of bi-invariant functions while $g \to g^{-1}$ induces an antiautomorphism, commutativity follows.

A similar larger commutative algebra was exhibited by Cygan [3] in the case of the quaternionic analogue of the Heisenberg group.

Let us return to biradial functions in the general case and describe their Fourier analysis. Since we are dealing with a Gelfand pair, it is well-known [5, Ch. X] that the maximal ideal space of $\underline{A}$ is given by the bounded spherical functions; the spherical functions, in turn, are the joint eigen-functions of the algebra $\underline{D}$ of MN-invariant differential operators without constant term on N.

Let $\{X_i\}$, $\{Y_i\}$ be orthonormal bases of $\underline{g}^1$, $\underline{g}^2$. By double transitivity the algebra of M-invariant symmetric tensors on $\underline{n}$ is generated by ΣX_i^2 and ΣY_i^2. Using the "symmetrization map" [5, Ch. X] one sees easily that $\underline{D}$ is generated by the operators

$$\Delta_1 = \Sigma X_i^2$$

$$\Delta_2 = \Sigma Y_i^2$$

where, this time, the square is taken in the enveloping algebra. (If $q = 1$, $\Delta_2 = Y_i$; we are not going to keep track of the trivial modifications required in this case.)

If u is biradial, let u^o be defined by

$$u(\exp(X + Y)) = u^{o}(|X|^2, |Y|^2)$$

and Δ_i^o by

$$(\Delta_i u)^o = \Delta_i^o u^o .$$

Since Δ_2^o involves only the center $\underline{g}^2$ of the Lie algebra $\underline{n}$, it is simply the radial Laplacian of the Euclidean space $\underline{g}^2$,

$$\Delta_2^o u^o = (4|Y|^2 D_2^2 + 2q D_2)u^o .$$

A simple computation, carried out in [7] gives

$$\Delta_1^o u^o = (4|X|^2 D_1^2 + 2p D_1 + b^2|X|^2 \Delta_2^o)u^o$$

where

$$b = 2^{-\frac{1}{2}}|\gamma|$$

(of course, the length of the root $|\gamma|$ can easily be expressed in terms of p and q).

Now suppose that u is spherical, i.e., that it is a biradial function such that

$$\Delta_i u = \chi_i u \qquad (i = 1,2)$$

with some numbers χ_1, χ_2, and $u(e) = 1$. Since Δ_2^o involves only the second variable, u^o is a product of functions of one variable; the factor depending on Y is a radial eigenfunction of the Euclidean Laplacian. So

$$u = j^{(q)}(\lambda Y)\Phi(|X|^2) , \quad \chi_2 = -\lambda^2$$

where

$$j^{(q)}(Y) = \int_{S_2} e^{i\sigma \cdot Y} d\sigma$$

($d\sigma$ denotes the normalized surface measure; $j^{(q)}$ is, of course, essentially a Bessel function). The equation $\Delta_1 u = \chi_1 u$ now gives an ordinary differential equation for Φ. If $\lambda = 0$, we again get the radial Euclidean Laplacian. If $\lambda \neq 0$ and we set $\Phi(t) = \exp(-\frac{b}{2}\lambda t)v(b\lambda t)$, the equation becomes

$$xv''(x) + (\frac{p}{2} - x) + nv(x) = 0$$

i.e., the confluent hypergeometric equation, with

$$n = -\frac{1}{4}(p + \frac{\chi_1}{b\lambda}) .$$

Every solution which gives a smooth function on N (i.e., which is regular at $x = 0$) gives a spherical function.

However, we are really interested only in the bounded spherical functions. It is not hard to see that we get these exactly when λ is real (then we may take $\lambda \geq 0$, since $j^{(q)}$ is even) and if n is a natural number. The solutions v of our differential equation are then the Laguerre polynomials L_n^α $(\alpha = \frac{p}{2} - 1)$.

So finally we find the following bounded spherical functions:

$$\varphi_\mu(\exp(X + Y)) = j^{(p)}(\mu X)$$

$$\varphi_{\lambda,n}(\exp(X + Y)) = c_{\alpha n}\, j^{(q)}(\lambda Y)e^{-\frac{b}{2}\lambda|X|^2} L_n^\alpha(b\lambda|X|^2)$$

where $u \geq 0$, $\lambda > 0$, $n \in \mathbb{N}$, and $c_{\alpha n}$ is a constant making sure that $\varphi_{\lambda,n}(e) = 1$.

Given a biradial function f on N, we will write $\hat{f}(\lambda,n)$ for $\int f\varphi_{\lambda,n}$. (This is really only part of the Gelfand transform, but it is the only part that will occur in the Plancherel formula; the remaining part $\int f\varphi_\mu$ can anyway be obtained by taking limits of the $\hat{f}(\lambda,n)$.) Since f is biradial, one can also write

$$\hat{f}(\lambda,n) = c_{\alpha,n} \int f(\exp(X + Y)e^{i\lambda\sigma\cdot Y} e^{-\frac{b}{2}\lambda|X|^2} L_n^\alpha(b\lambda|X|^2)dXdY$$

(the integral is independent of $\sigma \in S_2$).

It is now easy to prove the Plancherel formula for the Gelfand transform. Given a biradial function f, we write

$$F(|X|^2,\lambda) = \int f(\exp(X + Y))e^{i\lambda\sigma\cdot Y}\,dY .$$

Our last expression for $\hat{f}$ then takes the form, after introducing polar coordinates in the X-variable,

$$\hat{f}(\lambda,n) = c\int_0^\infty F(\tfrac{x}{b\lambda},\lambda)e^{-\frac{x}{2}}\,x^\alpha\,L_n^\alpha(x)dx$$

where c is a positive constant. (Even though it would present no difficulty, we stop keeping track of the constants and just write c , c' , c'' .) Now, since $\{e^{-x/2}\,x^{\alpha/2}\,L_n^\alpha(x)\}$ is a complete orthonormal system on $(0,\infty)$, we have by Parseval's formula,

$$\sum_n |\hat{f}(\lambda,n)|^2 = c'\,\lambda^{-p}\int_0^\infty |F(\tfrac{x}{b\lambda},\lambda)|^2\,x^\alpha\,dx$$

Multiplying by $\lambda^{\alpha+q}$, integrating in λ , then reintroducing the variable X and using the Plancherel theorem for R^n in the Y-variable, we get

$$\sum_n\int_0^\infty \lambda^{\alpha+q}\,|\hat{f}(\lambda,n)|^2\,d\lambda = c''\|f\|_2^2$$

which is the desired result.

Now we describe a few applications of these results.

(i) We first consider a singular integral operator on N in the sense of [10],

$$(Tf)(x) = \text{v.p.}\int k(xy^{-1})f(y)dy$$

with the principal value taken with respect to the gauge $|\exp\,(X + Y)| = (b|X|^4 + 4|Y|^2)^{\frac{1}{4}}$. Suppose that the kernel k is <u>biradial</u> besides the usual conditions of homogeneity, $k(\exp(t^{1/2}X + tY)) = t^{-q-\frac{p}{2}}\,k(\exp(X + T))$, and of having mean zero on "spheres" with respect to the gauge. Biradial kernels occur naturally: the kernels considered in [10, §6] and in [7], as well as some occurring in the work of Knapp-Stein on intertwining operators [8], are of this type. To prove the continuity of T

in $L^2(N)$, the usual method found by Knapp and Stein [8] and also used in [10] makes use of a lemma of M. Cotlar, and requires a strong smoothness hypothesis on k . If k is biradial, this condition can be relaxed, and the proof based on Fourier analysis. In fact, the problem reduces to showing that $\hat{k}(\lambda,n)$ is bounded. Now a change of variable shows that the homogeneity condition of variable shows that the homogeneity condition on k means that $\hat{k}(\lambda,n)$ is independent of λ , and another computation shows that the mean value condition guarantees that $\hat{k}$ is bounded as a function of n as well.

(ii) Another application is to give a simplified proof of a result of Hulanicki-Ricci [6] about Hermitian hyperbolic space and extended by Cygan [3] to all non-compact symmetric spaces X of rank one. The result says that if F is a bounded harmonic function on X and if $\lim_{n\to\infty} f(n\cdot x)$ exists for some $x \in X$, then it exists, and has the same value, for <u>all</u> $x \in X$ (n here denotes the generic element of the Iwasawa group N) . The proof, in a nutshell, consists of writing F as a Poisson integral, $F(n\cdot x) = f * P_x(n)$, noticing that P_x is biradial, checking that $\hat{P}_x(\lambda,n)$ is nowhere zero, and finally applying the Wiener Tauberian Theorem. The simplification is on one hand that there is no need of using the classification and making case-by-case computations, on the other that one works directly with the spherical functions and does not have to go through an explicit description of the representations of N .

(iii) One can use the present ideas in the study of potential theory on N . An MN-invariant Riemannian metric (actually a family of such metrics) on N can be defined by letting the length of the tangent vector $X + Y$ be $|X|^2 + c^{-1}|Y|^2$, $(c > 0)$. The corresponding Laplace-Beltrami operator is $\Delta_c = \Delta_1 + c\Delta_2$. (The particularly interesting potential theory for the subelliptic operator Δ_1 appears then as a limiting case when $c \to 0$.) Taking the case of a fixed $c > 0$, one can, similarly as in the case of a Riemannian symmetric space, consider weakly harmonic functions (i.e., such that $\Delta_c f = 0$) and strongly harmonic functions (i.e., $\Delta_1 f = \Delta_2 f = 0$) on N : In the analogous situation on a symmetric space a well-known theorem of Furstenberg states that every bounded weakly harmonic function is strongly harmonic. A proof of Furstenberg's theorem given by Guivar'ch applies to the case of N as well (the essential point being again that (MN,M) is a Gelfand pair). Applying the classical Liouville theorem twice, one sees easily that a bounded strongly harmonic function on N must be a constant. So we have obtained the following result: every bounded weakly harmonic function on N is a constant.

References

[1] L. Ahlfors, A singular integral equation connected with quasiconformal mappings in space, Enseignement Math. 24 (1978), 225-236.

[2] H. Boerner, Darstellungen von Gruppen, 2nd ed., Springer 1976.

[3] J. Cygan, A tangential convergence for bounded harmonic functions on a rank one symmetric space, to appear.

[4] D. Geller, Fourier analysis on the Heisenberg group, Proc. Nat'l Acad. Sci., USA 74 (1977) 1328-1331.

[5] S. Helgason, Differential Geometry and Symmetric Spaces, Academic Press, New York 1969.

[6] A. Hulanicki and F. Ricci, A Tauberian theorem and tangential convergence for bounded harmonic functions on balls in C^n, to appear in Inv. Math.

[7] A. Kaplan and R. Putz, Boundary behavior of harmonic forms on a rank one symmetric space, Trans. Amer. Math. Soc. 231 (1977), 369-384.

[8] A. Knapp and E.M. Stein, Intertwining operators for semisimple groups, Ann. of Math. 93 (1971), 489-578.

[9] A. Korányi, Fourier analysis of biradial functions on certain nilpotent groups, to appear.

[10] A. Korányi and S. Vági, Singular integrals on homogeneous spaces and some problems of classical analysis, Ann. Scuola Norm. Sup. Pisa 25 (1971), 576-648.

[11] ________, Cauchy-Szegö integrals for systems of harmonic functions Ann. Scuola Norm. Sup. Pisa 26 (1972), 181-196.

[12] ________, Group-theoretic remarks on Riesz systems in balls, to appear.

[13] B. Kostant, On the existence and irreducibility of certain series of representations, in "Lie groups, Summer school on group representations", Budapest 1971.

[14] M. Reimann, A rotation-invariant differential equation for vector fields, to appear.

[15] G. Schiffmann, Travaux de Kostant sur la série principale, in "Analyse harmonique sur les groupes de Lie", pp. 460-510, Lecture Notes in Mathematics #739, Springer 1979.

CENTRO INTERNAZIONALE MATEMATICO ESTIVO

(C.I.M.E.)

EIGENFUNCTION EXPANSIONS ON SEMISIMPLE LIE GROUPS

V. VARADARAJAN

EIGENFUNCTION EXPANSIONS ON SEMISIMPLE LIE GROUPS

V. Varadarajan
Department of Mathematics
University of California
Los Angeles, CA 90024

1. Representations of the Principal Series. Harish Chandra's Plancherel formula.

1. In these lectures it will be my aim to discuss some aspects of the problem of obtaining an explicit Plancherel formula for a connected real semisimple Lie group with finite center, and the close connection of this problem with the theory of eigenfunction expansions on the group. The central results are those of Harish Chandra, and it is impossible to give anything more than a partial outline of his monumental work that began in the early '50's and has spanned almost three decades.

For a given locally compact group which is separable and unimodular, the fundamental problem is that of decomposing its regular representation into irreducible constituents. If the group is commutative or compact this is quite classical. However, apart from some general existence theorems (see for instance Segal [1]), there is no systematic development of harmonic analysis on general locally compact groups. The category of locally compact groups (even separable and unimodular) is so extensive and the structure of its individual members so varied that it has so far proved impossible to develop analysis on these groups beyond a few general theorems. For Lie groups the situation is much better, and among these the semisimple groups (both real and $\mathfrak{p}$-adic) occupy a central position. We know their structure in great detail and are able to use this knowledge in formulating and solving the questions of harmonic analysis in a significant manner.

Although our interest is essentially only in the semisimple groups we consider a somewhat wider class of groups for a variety of reasons. For example, many theorems in the subject are proved by induction on the dimension of the group via a descent principle that transfers the problem from the given group to a Levi factor of one of its parabolic subgroups; these Levi factors are in general neither semisimple nor connected, even if the ambient group is. Furthermore, in number theoretic applications, the groups whose representations are important are often the real points of a reductive algebraic group defined over $\mathbb{Q}$. These and other reasons suggest that it will be convenient to work with a class of reductive Lie groups which are not necessarily connected. Following Harish Chandra we shall work with groups G with the following properties:

(i) G is reductive (i.e., $\mathfrak{g}$, the Lie algebra of G, is a real reductive Lie algebra)

(ii) $[G : G^0] < \infty$ where G^0 is the connected component of G containing the identity

(iii) If G_1 is the analytic subgroup of G defined by $\mathfrak{g}_1 = [\mathfrak{g}, \mathfrak{g}]$, then G_1 is closed in G and has finite center

(iv) If G_c is the (complex) adjoint group of $\mathfrak{g}_c$ (= the complexification of $\mathfrak{g}$), then $Ad(G) \subset G_c$

These are the groups of the so-called Harish Chandra class $\mathcal{H}$; for a more detailed discussion of their properties, see Varadarajan [1]. Connected semisimple real Lie groups with finite center are in $\mathcal{H}$; if $\mathbf{G}$ is an algebraic group defined over R such that $\mathbf{G}(C)$ is irreducible and reductive, then $\mathbf{G}(R)$ is of class $\mathcal{H}$. From now on we fix a group G of class $\mathcal{H}$ and a Cartan involution θ of G; the fixed point set of θ, which is a maximal compact subgroup of G, will be denoted by K. We shall write $\mathfrak{g}$ (resp. $\mathfrak{k}$) for the Lie algebra of G (resp. K). Complexifications will be indicated by a suffix c. One knows that K meets all connected components of G.

Let $\hat{G}$ be the set of equivalence classes of irreducible unitary representations of G. If $\omega \in \hat{G}$ and π is a representation in the class ω, it is well known that π has a character, namely the distribution $f \mapsto \mathrm{tr}(\pi(f))$ $(f \in C_c^\infty(G))$. This distribution depends only on ω and is written as Θ_ω; moreover Θ_ω determines ω uniquely, and is an invariant (= invariant under all inner automorphisms of G) distribution on G which is an eigendistribution for the algebra $\mathfrak{Z}$ of all bi-invariant differential operators on G. By an <u>explicit Plancherel formula</u> is meant an "expansion" of the Dirac measure δ on G at the identity element as an integral of the Θ_ω:

$$\delta(\cdot) = \int_{\hat{G}} \Theta_\omega(\cdot)\, d\mu(\omega).$$

Here μ is a nonnegative measure on $\hat{G}$; and we require explicit descriptions of Θ_ω for μ-almost all ω as well as of μ. The measure μ is called the <u>Plancherel measure</u> for G.

It is a remarkable fact that if G is nontrivial in the sense that G_1 is not compact, the "support" of the Plancherel measure is not the whole of $\hat{G}$. In fact Harish Chandra discovered that one can introduce a notion of <u>temperedness</u> of distributions on G, and that if $\hat{G}_t$ is the subset of all tempered classes, i.e., of $\omega \in \hat{G}$ for which Θ_ω is tempered, then $\mu(\hat{G}\backslash\hat{G}_t) = 0$. The classes ω in $\hat{G}\backslash\hat{G}_t$ are called <u>exceptional</u>. If G is nontrivial, the trivial representation of G is exceptional or, what is the same thing, Haar measure on G is not a tempered distribution. The meaning and significance of the exceptional representations is one of the outstanding puzzles of the harmonic analysis of semisimple groups.

2. Let us now proceed to an explicit statement of Harish Chandra's Plancherel formula. To do this we need a description of the irreducible representations that will enter the Plancherel formula.

First of all we have the discrete series $\hat{G}_d \subset \hat{G}_t$. A class ω of $\hat{G}$ belongs to $\hat{G}_d$ by definition if and only if the matrix coefficients of the representations of the class are in $L^2(G)$, or equivalently, if and only if the regular representation of $L^2(G)$ has a direct summand that belongs to ω. For $\hat{G}_d$ to be nonempty it is necessary and sufficient that $rk(G) = rk(K)$; examples are $G = SL(2,R)$, $G = Sp(n,R)$, $G = SO(1,2k)$; nonexamples are $G = SO(1,2k+1)$, G = any connected <u>complex</u> semisimple group. If G is compact, $\hat{G} = \hat{G}_d$; in this case, assuming further that G is semisimple and simply connected, if $\mathfrak{b} \subset \mathfrak{k} = \mathfrak{g}$ is a CSA (= Cartan subalgebra) of $\mathfrak{k}$, we have Hermann Weyl's parametrization of $\hat{G}$ by the orbits under the Weyl group $W(\mathfrak{g},\mathfrak{b})$ of the lattice of integral elements in $(-1)^{1/2}\mathfrak{b}^*$. In the general case when G is not compact but $rk(G) = rk(K)$, Harish Chandra's theory of the discrete series allows us to proceed in an almost completely analogous fashion. For instance, let G be a connected real form of a simply connected complex semisimple group with $rk(G) = rk(K)$; let $B \subset K$ be a CSG (= Cartan subgroup) with Lie algebra $\mathfrak{b} \subset \mathfrak{k}$. We write $\mathbb{L}$ for the lattice of integral elements in $(-1)^{1/2}\mathfrak{b}^*$; $\mathbb{L}$ is canonically isomorphic to the character group $\hat{B}$ of B. Let $\mathbb{L}'$ be the set of regular elements of $\mathbb{L}$, i.e., the set of all $\lambda \in \mathbb{L}$ with $\langle\alpha,\lambda\rangle \neq 0$ for each root α of $(\mathfrak{g}_c,\mathfrak{b}_c)$. If $W = W(G,B)$ is the group $\tilde{B}/B$ where $\tilde{B}$ is the normalizer of B in G, W operates on $\mathbb{L}'$. Harish Chandra's theory gives a unique bijection

$$W\backslash\mathbb{L} \xrightarrow{\sim} \hat{G}_d$$

such that if $\lambda \in \mathbb{L}'$ and $\omega = \omega(\lambda)$ the corresponding class of $\hat{G}_d$,

$$\Theta_\omega(\exp H) = (-1)^q \operatorname{sgn} \varpi(\lambda) \frac{\sum_{s\in W} \varepsilon(s) e^{s\lambda(H)}}{\Delta(\exp H)} \qquad (H \in \mathfrak{b},\ \exp H \text{ regular})$$

Here $q = \frac{1}{2}\dim(G/K)$, $\varpi(\lambda) = \Pi_{\alpha\in P}\langle\alpha,\lambda\rangle$ and $\Delta(\exp H) = \Pi_{\alpha\in P}(e^{\alpha(H)/2} - e^{-\alpha(H)/2})$ where P is a fixed positive system of roots of $(\mathfrak{g}_c, \mathfrak{b}_c)$; of course the formula is independent of the choice of P. For an arbitrary G in $\mathcal{H}$ with $\mathrm{rk}(G) = \mathrm{rk}(K)$, the parametrization is a little more subtle; for instance, even for semisimple G, it may happen that G is the real form of a complex group whose character lattice does not contain ρ where ρ is as usual half the sum of positive roots. To see what the formula is in the general case we note that the usual formula

$$\sum_{s\in W} \varepsilon(s) e^{s(\Lambda+\rho)(H)} \Big/ \prod_{\alpha\in P} (e^{\alpha(H)/2} - e^{-\alpha(H)/2})$$

can we rewritten as

$$\sum_{s\in W} \varepsilon(s) e^{s\Lambda(H)}\, e^{(s\rho-\rho)(H)} \Big/ \prod_{\alpha\in P} (1 - e^{-\alpha(H)})$$

which has the advantage that $s\rho - \rho$ is in the root lattice and so $\exp H \mapsto e^{(s\rho-\rho)(H)}$ as well as $\exp H \mapsto \Pi_{\alpha\in P}(1 - e^{-\alpha(H)})$ are functions on B (even $\mathrm{Ad}(B)$). We introduce a CSG $B \subset K$ and the set B^* of irreducible characters of B; $W = W(\bar{B}/B)$ operates on B^*. If $d(b^*)$ is the dimension of the representation of B corresponding to b^*, there is a unique element $\mu = \log b^* \in (-1)^{1/2} \mathfrak{b}^*$ such that $b^*(\exp H) = d(b^*) e^{\mu(H)}$ $(H \in \mathfrak{b})$; we fix a positive system P as before and put $\lambda(b^*) = \log b^* + \rho$. The affine action $s, b^* \mapsto s[b^*]$ of W on B^* is then defined by $s[b^*] = sb^*\, \xi_{s\rho-\rho}$ where $\xi_{s\rho-\rho}$ is the character of B defined by $s\rho - \rho$. $B^{*\prime}$ is then the set of

$b^* \in B^*$ with $\varpi(\lambda(b^*)) \neq 0$; it is stable under the above affine action of W, and we have a unique bijection

$$W\backslash B^{*\prime} \xrightarrow{\sim} \hat{G}_d$$

such that if $b^* \in B^{*\prime}$ and $\omega = \omega(b^*)$ is the corresponding class in $\hat{G}_d$,

$$\Theta_\omega(b) = (-1)^q \operatorname{sgn} \varpi(\lambda(b^*)) \frac{\sum_{s\in W} \varepsilon(s)s[b^*](b)}{\prod_{\alpha\in P}(1 - \xi_{-\alpha}(b))}$$

for all regular $b \in B$ (regular means as usual that $\xi_\alpha(b) \neq 1$ for all roots α). The distribution obtained by omitting the sign factors in the above expression is denoted by Θ_{b^*}. It is possible to characterize it as the unique tempered invariant eigendistribution for $\mathfrak{Z}$ whose values at the regular points $b \in B$ are given by

$$\Theta_{b^*}(b) = \frac{\sum_{s\in W} \varepsilon(s)s[b^*](b)}{\prod_{\alpha\in P}(1 - \xi_{-\alpha}(b))}$$

In all these statements we are treating the characters as point functions on G. This is of course permissible in view of the celebrated regularity theorem of Harish Chandra which assests that an invariant eigendistribution for $\mathfrak{Z}$ on G is a locally summable function which is analytic on the set of regular elements.

The matrix coefficients of the representations in discrete classes satisfy orthogonality relations that imitate those in the theory of compact groups. More precisely, let $\omega \in \hat{G}_d$; π, a representation from the class ω acting in a Hilbert space H; then there is a number $d(\omega) > 0$, called the formal degree of ω, such that for all $\varphi, \varphi', \psi, \psi' \in H$,

$$\int_G (\pi(x)\varphi,\psi)\overline{(\pi(x)\varphi',\psi')}dx = \frac{1}{d(\omega)}(\varphi,\varphi')\overline{(\psi,\psi')}$$

Note that the value of $d(\omega)$ depends on the normalization of dx. If ${}^\circ L^2(G)$

is the discrete part of the regular representation, it is the Hilbert space span of the matrix coefficients of the discrete classes; and if ${}^{\circ}E$ is the orthogonal projection $L^2(G) \to {}^{\circ}L^2(G)$, we have, for some constant $c > 0$ and all $f \in C_c^{\infty}(G)$,

$$\|{}^{\circ}Ef\|^2 = c \sum_{\omega \in \hat{G}_d} d(\omega)\Theta_{\omega}(\tilde{f} * f)$$

($\tilde{f}(x) = \overline{f(x^{-1})}$). This is obviously the "discrete part" of the Plancherel formula for G. In our case, there is a constant $c > 0$ such that

$$d(\omega(b^*)) = c|\varpi(\lambda(b^*))|d(b^*) \qquad (b^* \in B^{*\prime})$$

There is no need to treat this expansion in more detail since we shall subsume it under a more general Plancherel formula presently.

A given group G of class $\mathcal{H}$ does not always have a discrete series. To construct the series of representations of such groups that enter the Plancherel formula we proceed as follows. Let A be a CSG of G, stable under θ; we can then write $A = A_I A_R$, where $A_I = A \cap K$ is the maximal compact subgroup of A and A_R is a vector group with $\theta(a) = a^{-1}$ for all $a \in A_R$. There exist parabolic subgroups (psgrps) P whose Langlands decompositions are of the form $P = MA_R N$ (cf. Varadarajan [1], Part II, § 6). The group M is then of class $\mathcal{H}$, and A_I is a compact CSG of M; in particular, $rk(M) = rk(K_M)$ where $K_M = K \cap M = K \cap P$ is the maximal compact subgroup of M fixed by the Cartan involution $\theta_M = \theta|_M$. Thus $\hat{M}_d \neq \emptyset$. If $\omega \in \hat{M}_d$ and $\nu \in \mathfrak{a}_R^*$ where $\mathfrak{a}_R$ = Lie algebra of A_R, then one can start with the representation

$$ma_R n \mapsto \sigma(m)e^{i\nu(\log a_R)} \qquad (m \in M,\ a_R \in A_R,\ n \in N)$$

where σ is a representation of M in the class ω and $\log : A_R \to \mathfrak{a}_R$ is the inverse of $\exp : \mathfrak{a}_R \to A_R$, and obtain a representation of G by inducing from

P. Note that since G/P does not admit a G-invariant measure we must use the so-called unitary induction that takes unitaries of P to unitaries of G. Let us write $\pi_{P,\omega,\nu}$ for the unitary representation of G thus obtained. It can be shown that its class is independent of the choice of P. It can further be proved that if $\nu \in \mathfrak{a}_R^*$ is regular in the sense that $\langle \nu, \beta \rangle \neq 0$ for all roots β of $(\mathfrak{g}, \mathfrak{a}_R)$, $\pi_{P,\omega,\nu}$ is even irreducible. Let us write $\Theta_{\omega,\nu}$ for the character of $\pi_{P,\omega,\nu}$. If $W(G,A) = \mathfrak{X}/A$ where $\mathfrak{X}$ is the normalizer of A in G, it is easy to see that $W(G/A)$ operates in a natural manner on $\hat{M}_d$ as well as $\mathfrak{a}_R^*$; and one has the symmetry

$$\Theta_{s\omega,s\nu} = \Theta_{\omega,\nu} \qquad (\omega \in \hat{M}_d,\ \nu \in \mathfrak{a}_R^*,\ s \in W(G,A))$$

The procedure outlined above associates with each conjugacy class of CSG's of G a parametrized subset of $\hat{G}$, actually of $\hat{G}_t$; in the case when $\mathrm{rk}(G) = \mathrm{rk}(K)$, exactly one of these conjugacy classes consists of compact CSG's, and the associated series of representations is $\hat{G}_d$. We note that in all cases there are only finitely many conjugacy classes of CSG's. From the representation theoretic point of view Harish Chandra's Plancherel formula asserts that the regular representation of G can be decomposed as a direct integral of the representations described above and further that the measure involved in this decomposition is mutually absolutely continuous with respect to the Lebesgue measure $d\nu$ on $\mathfrak{a}_R^*$. More precisely, we have the following theorem (cf. Harish Chandra [8], Corollary to Theorem 19, p. 545, [5], Theorem 27.3.). Let $\{A_1, \ldots, A_r\}$ be a complete set of representatives in the various conjugacy classes of CSG's. We assume that each A_i is θ-stable and write $M_{1,i} = M_i A_{i,R}$ $(1 \leq i \leq r)$ for the Levi component of the psgrps associated with A_i. Since $A_{i,I}^*$ has a natural map onto a lattice in $(-1)^{1/2}\mathfrak{a}_{i,I}^*$ with fibers of cardinality at most $[A_{i,I} : A_{i,I}^o]$ it is clear that the notion of a function on $(\hat{M}_i)_d \times \mathfrak{a}_{i,R}^*$ of at most polynomial growth makes sense. We give $(\hat{M}_i)_d$ the discrete topology.

Theorem. There are unique continuous functions C_i on $(\hat{M}_i)_d \times \mathfrak{a}^*_{i,R}$ $(1 \leq i \leq r)$ with the following properties:

(i) each C_i is of at most polynomial growth

(ii) $C_i(s\omega : s\nu) = C_i(\omega : \nu)$ $(\omega \in (\hat{M}_i)_d,\ \nu \in \mathfrak{a}^*_{i,R},\ s \in W(G,A_i))$

(iii) for all functions f in the Schwartz space $\mathcal{C}(G)$

$$f(1) = \sum_{1 \leq i \leq \nu} \sum_{\omega \in (\hat{M}_i)_d} \int_{\mathfrak{a}^*_{i,R}} C_i(\omega : \nu)\Theta_{\omega,\nu}(f)d\nu$$

The functions $C_i(\omega : \nu)$ are ≥ 0 and the series on the right converges absolutely.

The series of representations $\pi_{\omega,\nu}$ $(\omega \in (\hat{M}_i)_d,\ \nu \in \mathfrak{a}^*_{i,R},\ 1 \leq i \leq \nu)$ are known as the principal series of representations of G.

We look at some examples.

1) $G = SL(2,\mathbb{R})$. There are two Cartan subgroups, B and L. B is the group of rotations $u_\theta = \begin{pmatrix} \cos\theta & \sin\theta \\ -\sin\theta & \cos\theta \end{pmatrix}$ $(\theta \in \mathbb{R})$ and L is the group of diagonal matrices: $L = A \cup \gamma A$ where $\gamma = \begin{pmatrix} -1 & 0 \\ 0 & -1 \end{pmatrix}$, and A is the group of all matrices $a_t = \begin{pmatrix} e^t & 0 \\ 0 & e^{-t} \end{pmatrix}$. The characters of the discrete classes are the Θ_n $(n \in \mathbb{Z},\ n \neq 0)$, which are the distributions defined by the invariant locally L^1 functions on G (with respect to standard Haar measure) given by

$$\Theta_n(u_\theta) = -\mathrm{sgn}(n)\frac{e^{in\theta}}{e^{i\theta} - e^{-i\theta}}\ (\theta \neq 0,\pi)$$

$$\Theta_n(za_t) = \varepsilon_n(z)\frac{e^{-|n||t|}}{|e^t - e^{-t}|}\ (t \neq 0),\ z \in \{1,\gamma\},\ \varepsilon_n(1) = 1,\ \varepsilon_n(\gamma) = (-1)^{n+1}$$

The formal degree of the class corresponding to Θ_n is $|n|$. Turning to L we find that $M = \{1,\gamma\}$ and so $\hat{M} = \hat{L}_I$ consists of the trivial character and the character ε that sends γ to -1. We combine these with the character

$a_t \mapsto e^{i\nu t}$ $(\nu \in \mathbb{R})$ and use the construction via induced representations to get the characters $T_{\sigma,\nu}(\sigma \in \hat{M})$. A simple calculation gives

$$T_{\sigma,\nu}(u_\theta) = 0 \qquad (\theta \neq 0,\pi)$$

$$T_{\sigma,\nu}(za_t) = \sigma(z)\frac{e^{i\nu t} + e^{-i\nu t}}{|e^t - e^{-t}|}$$

The Plancherel formula then becomes (cf. Lang [1], p. 174; Harish Chandra [10])

$$2\pi\delta = \sum_{\substack{n\in\mathbb{Z}\\ n\neq 0}} |n|\Theta_n + \frac{1}{2}\int_0^\infty \nu \tanh\frac{\pi\nu}{2} T_{1,\nu}d\nu + \frac{1}{2}\int_0^\infty \nu \coth\frac{\pi\nu}{2} T_{\varepsilon,\nu}d\nu$$

2) G is complex and simply connected. Let $\mathfrak{g}$ be the Lie algebra of G regarded as a Lie group over $\mathbb{R}$. The complex structure of G induces a complex Lie algebra structure on $\mathfrak{g}$. We write $\mathfrak{g}^C$ for this complex Lie algebra and J for multiplication by $(-1)^{1/2}$ in $\mathfrak{g}^C$. K is a compact form of G. If $\mathfrak{l}_I \subset \mathfrak{k}$ is a CSA of $\mathfrak{k}$ and $\mathfrak{l}_R = J\mathfrak{l}_I$, $\mathfrak{z}^C = \mathfrak{l}_I + \mathfrak{l}_R$ is a CSA of $\mathfrak{g}^C$. If L is the centralizer of $\mathfrak{z}$ in G, L is a CSG of G and is the only one upto G-conjugacy. Any element $\lambda \in \mathfrak{l}_R^*$ can be regarded as an element of the complex dual of $\mathfrak{z}^C$ by setting $\lambda(H + JH') = \lambda(H) + i\lambda(H')$ $(H, H' \in \mathfrak{l}_R)$. This allows us to introduce the lattice $\mathbb{L} \subset \mathfrak{l}^*$ of all $\lambda \in \mathfrak{l}_R^*$ for which $\exp H \mapsto e^{\lambda(H)}$ $(H \in \mathfrak{l}_I)$ is a character of L_I. But now $M = L_I$ and so the irreducible unitary representations of $MA = L = L_I L_R$ are of the form

$$\exp H \exp H' \mapsto e^{\xi(H)}e^{i\nu(H')} \qquad (H \in \mathfrak{l}_I,\ H' \in \mathfrak{l}_R,\ \xi \in \mathbb{L},\ \nu \in \mathfrak{l}_R^*)$$

The corresponding induced representations are all irreducible (cf. Wallach [1]). Let $T_{\xi,\nu}$ be their characters. Then the Plancherel formula takes the following form (cf. Harish Chandra [9]). There is a positive constant $c > 0$ such that for all f in the Schwartz space of G,

$$cf(1) = \sum_{\xi \in \mathbb{L}} \int_{I_R^*} p(\xi : \nu) T_{\xi,\nu}(f) d\nu$$

where $p(\xi : \nu)$ is the function on $\mathbb{L} \times I_R^*$ defined as follows: if $\langle \cdot, \cdot \rangle$ is the Cartan-Killing form of $\mathfrak{g}^C$ and P^C is a positive system of roots of $(\mathfrak{g}^C, \mathfrak{z}^C)$,

$$p(\xi : \nu) = \prod_{\alpha \in P^C} \{\langle \nu, \alpha \rangle^2 + \langle \xi, \alpha \rangle^2\}.$$

2. Eigenfunction expansions on G. The case of spherical functions.

1. We shall now introduce the central theme of these lectures, namely, the theory of eigenfunction expansions on $L^2(G)$ and its connection with the Plancherel formula. Let $U(\mathfrak{g}_c)$ be the universal enveloping algebra of $\mathfrak{g}_c$. We shall regard its elements as left invariant differential operators on G; if $f \in C^\infty(G)$, $a \in U(\mathfrak{g}_c)$, we write af for the action of a on f. Elements of $U(\mathfrak{g}_c)$ also act as right invariant differential operators, the corresponding action being written fa. We also write $f(x;a)$ for $(af)(x)$ and $f(a;x)$ for $fa(x)$; if $a = X_1 X_2 \cdots X_r$, $X_j \in \mathfrak{g}$,

$$f(x;a) = (\partial^r / \partial t_1 \cdots \partial t_r)_0 f(x \exp t_1 X_1 \cdots \exp t_r X_r)$$

$$f(a;x) = (\partial^r / \partial t_1 \cdots \partial t_r)_0 f(\exp t_1 X_1 \cdots \exp t_r X_r x)$$

where the suffix 0 indicates that the derivatives are evaluated for $t_1 = t_2 = \cdots = t_r = 0$. The adjoint representation Ad of G on $\mathfrak{g}$ extends to a representation (denoted by Ad again) of G on $U(\mathfrak{g}_c)$; we often write a^x instead of $\mathrm{Ad}(x)(a)$, for $a \in U(\mathfrak{g}_c)$, $x \in G$. The connection between the left and right differential actions of elements of $U(\mathfrak{g}_c)$ is then given by the formula

$$f(a;x) = f(x;\mathrm{Ad}(x^{-1})(a)) \qquad (x \in G,\ a \in U(\mathfrak{g}_c),\ f \in C^{\infty}(G)).$$

As in §1 we write $\mathfrak{Z}$ for the centralizer of G (via Ad) in $U(\mathfrak{g}_c)$; it is obvious that $\mathfrak{Z}$ is the center of $U(\mathfrak{g}_c)$. It is well known that $\mathfrak{Z} \approx \mathbb{C}[T_1,\ldots,T_\ell]$ where $\ell = \mathrm{rk}(G)$.

Let us now consider the right regular representation r of G in $L^2(G)$. Denote by $\mathfrak{D}$ the subspace of vectors which are differentiable for r. Elements of $\mathfrak{D}$ are all in $C^{\infty}(G)$. By differentiating we obtain a representation $a \mapsto r(a)$ of $U(\mathfrak{g}_c)$ on $\mathfrak{D}$; it is none other than the action $a,f \mapsto af$ of $U(\mathfrak{g}_c)$ on $C^{\infty}(G)$ introduced above. We now recall that $U(\mathfrak{g}_c)$ admits a canonical anti-linear antiautomorphism $\dagger$, the operation of <u>formal adjoints</u>, which is - id on $\mathfrak{g}$. It is well known that the operators $r(a)$ are all closeable and that $\mathrm{Cl}(r(a^{\dagger})) = \mathrm{Cl}(r(a))^{\dagger}$ where the $\dagger$ on the right refers to Hilbert space adjoints. In particular, if a is formally self adjoint (or normal), $r(a)$ is essentially self-adjoint (or normal) in the Hilbert space sense. The operators $\mathrm{Cl}(r(a))$ $(a \in \mathfrak{Z})$ are thus normal and mutually commuting and it makes sense to ask for their explicit spectral expansion. If we now observe that the matrix coefficients of irreducible unitary representations of G are eigenfunctions for $\mathfrak{Z}$, it is clear that the spectral decomposition of $r(\mathfrak{Z})$ is intimately related to the Plancherel expansion on G. However, in view of the infinite multiplicities that will arise on $L^2(G)$, it will be convenient to consider the spectral theory of $r(\mathfrak{Z})$ on subspaces of $L^2(G)$ that transform under right and left translations by elements of K according to fixed but arbitrary finite dimensional representations of K. Since the union of such subspaces is dense in $L^2(G)$, this is not a serious restriction.

2. The simplest of such K-finite subspaces is $L^2(G/\!/K)$ where we write $G/\!/K$ for the double co-set space $K\backslash G/K$. The elements of $L^2(G/\!/K)$ are

those that are biinvariant under K and are known as <u>spherical functions</u>. We may regard them as the K-invariant functions on G/K that are in $L^2(G/K)$ In this case there is an algebra that is slightly bigger than $\mathfrak{Z}$ which is the one to be considered for eigenfunction expansions. In fact, let $\mathfrak{Q}$ be the centralizer of K in $U(\mathfrak{g}_c)$. For each element $a \in \mathfrak{Q}$, $r(a)$ we have a G-invariant differential operator $r_{G/K}(a)$ on $C^\infty(G/K)$, and all the G-invariant differential operators on $C^\infty(G/K)$ may be obtained in this manner. Moreover $r_{G/K}(\mathfrak{Q})$ is abelian and is $\approx \mathbb{C}[T_1,\ldots,T_d]$ where d is the rank of the Riemannian symmetric space G/K. Furthermore $C^\infty(G/\!/K)$ is stable under $r_{G/K}(\mathfrak{Q})$ and one can therefore ask for the spectral theory of the commuting algebra of (unbounded) normal operators in $L^2(G/\!/K)$ determined by $r_{G/K}(\mathfrak{Q})$.

In this special case one can see very clearly the close relationship between the eigenfunction expansion problem for spherical functions and the representation theoretic problem of decomposing the natural representation of G in $L^2(G/K)$. We are concerned here with unitary representations of G of class 1, i.e., admitting nonzero K-invariant vectors. If π is an irreducible unitary representation of class 1 in a Hilbert space $H(\pi)$, the space $H(\pi)^K$ is of dimension 1 and defines a well determined spherical function φ_π on G by

$$\varphi_\pi(x) = (\pi(x)\psi,\psi) \qquad (x \in G,\ \psi \in H(\pi)^K,\ \|\psi\| = 1)$$

The function φ_π is in $C^\infty(G/\!/K)$, is normalized by $\varphi_\pi(1) = 1$, and is an eigenfunction for $r_{G/K}(\mathfrak{Q})$. Of course not all representations of class 1 are relevant for the decomposition of $L^2(G/K)$. (cf. remarks in §1 on the exceptional series of representations). Specializing the situation discussed in §1 we consider the psgrp $P = MAN$, where $G = KAN$ is now an Iwasawa decomposition of G and M is the centralizer of A in K (so that P is a minimal psgrp). If $\mathfrak{a}$ is the Lie algebra of A and $\log(A \to \mathfrak{a})$ is the map inverting $\exp(\mathfrak{a} \to A)$, then, for any $\nu \in \mathfrak{a}^*$, the representation $man \mapsto e^{i\nu(\log a)}$ of P is one dimensional and unitary; the corresponding representation of G obtained by unitary induction from P is of class 1 and is denoted by π_ν. It is irreducible for all $\nu \in \mathfrak{a}^*$ (cf. Kostant [1]; also Parthasarathy et al [1]). The basic result in this framework is due to Harish Chandra. It asserts that the π_ν for $\nu \in \mathfrak{a}^{*+}$ (the positive chamber of $\mathfrak{a}^*$) are inequivalent and π is a direct integral of the π_ν with respect to a measure class that is mutually absolutely continuous with respect to the Lebesgue measure on $\mathfrak{a}^*$ (cf. Harish Chandra [2]).

It is not difficult to set up an intertwining operator to go from $C_c^\infty(G/K)$ to the space of smooth sections of a Hilbert space bundle whose fibers are the spaces of the π_ν. For $f \in C_c^\infty(G/K)$ let

$$(Jf)(x : \nu) = \int_{A \times N} f(xan)e^{(\nu+\rho)(\log a)} da dn$$

Since for $a' \in A$, $n' \in N$,

$$(Jf)(xa'n' : \nu) = e^{-(\nu+\rho)(\log a')}(Jf)(x : \nu)$$

it is clear that for fixed ν, $(Jf)(\cdot : \nu)$ is an element of the space on which π_ν acts. The operator J commutes with left translations by elements of G and so gives us the intertwining operator we are seeking. If f is spherical, $(Jf)(\cdot : \nu)$ is completely determined by $(Jf)(1 : \nu)$, as $G = KAN$. For $(Jf)(1 : \nu)$ we have the expression

$$(Jf)(1 : \nu) = \int_A (Af)(a)e^{\nu(\log a)}$$

where Af is the <u>Abel transform</u> of f defined by

$$(Af)(a) = e^{\rho(\log a)} \int_N f(an)dn$$

The question of proving that J extends to an intertwining unitary isomorphism of $L^2(G/K)$ with a suitable direct integral of the π_ν then reduces to the question of proving that the map $f \mapsto (Jf)(1:\cdot)$ extends to a unitary isomorphism of $L^2(G/K)$ with $L^2(\mathfrak{a}^{*+}, \beta d\nu)$ for a suitable density function β on $\mathfrak{a}^{*+}$.

2. Let us write $\varphi(\nu:x)$ for the matrix coefficient φ_{π_ν}. A simple calculation shows that

$$(Jf)(1:\nu) = \int_G f(x)\varphi(\nu:x)dx$$

We think of $(Jf)(1:\cdot)$ therefore as a <u>transform</u> of f. Formally, for any spherical function f, the Harish Chandra transform $\mathcal{H}f$ of f is the function on $\mathfrak{a}^*$ given by

$$(\mathcal{H}f)(\nu) = \int_G f(x)\varphi(\nu:x)dx$$

Since the representations π_ν and $\pi_{s\nu}$ $(s \in W(G,A))$ are equivalent, we have the functional equation

$$\varphi(s\nu:x) = \varphi(\nu:x) \qquad (s \in W(G,A),\ x \in G)$$

So

$$(\mathcal{H}f)(s\nu) = (\mathcal{H}f)(\nu)$$

Finally, as $G = KAN$, the representation π_ν can be realized on $L^2(K/M)$ in such a manner that the constant function 1 is K-invariant. For $\varphi(\nu:x)$ we then obtain an integral representation over K given by

$$\varphi(\nu:x) = \int_K e^{(i\nu-\rho)(H(xk))}dk$$

Here dk is the Haar measure on K with $\int_K dk = 1$. The function H is from G to $\mathfrak{a}$ and is defined by the requirement that for any $x \in G$, $H(x) \in \mathfrak{a}$ is the unique element for which $x \in K \exp H(x) N$.

3. The fundamental theorem of Harish Chandra's theory asserts that the map $f \mapsto \mathcal{H}f$ $(f \in C_c^\infty(G/\!/K))$ extends uniquely to a unitary isomorphism of $L^2(G/\!/K)$ with $L^2(\mathfrak{a}^*, \beta d\nu)^{\mathfrak{w}}$ where β is a smooth nonnegative function on $\mathfrak{a}^*$ which is invariant under the Weyl group $W(G,A) = \mathfrak{w}$ and which has moderate growth, i.e., itself and each of its derivatives has atmost polynomial growth; the superfix $\mathfrak{w}$ indicates the subspace of functions invariant under $\mathfrak{w}$. Moreover (with $d\nu$ suitably normalized) we have an inversion formula

$$f(x) = \int_{\mathfrak{a}^*} (\mathcal{H}f)(\nu)\overline{\varphi(\nu : x)}\, \beta(\nu) d\nu$$

With these results established, it is natural to ask for an explicit determination of β. Harish Chandra's theory shows that β can be completely determined from the asymptotic behaviour of $\varphi(\nu : x)$. In fact it is possible to prove the existence of a meromorphic function on $\mathfrak{a}_c^*$, say $\underset{\sim}{c}$, analytic on $\mathfrak{a}^{*\prime}$, such that for any $\nu \in \mathfrak{a}^{*\prime}$, $H \in \mathfrak{a}^+$, we have the asymptotic relation (with exponentially decaying error terms)

$$e^{t\rho(H)}\varphi(\nu : \exp tH) \sim \sum_{s \in \mathfrak{w}} \underset{\sim}{c}(s\nu) e^{its\nu(H)}$$

Furthermore $\underset{\sim}{c}$ has no zeros on $\mathfrak{a}^*$,

$$|\underset{\sim}{c}(s\nu)| = |\underset{\sim}{c}(\nu)| \qquad (s \in \mathfrak{w},\ \nu \in \mathfrak{a}^*)$$

and

$$\beta(\nu) = |\underset{\sim}{c}(\nu)|^{-2} = \underset{\sim}{c}(\nu)^{-1}\, \underset{\sim}{c}(-\nu)^{-1} \qquad (\nu \in \mathfrak{a}^*)$$

Furthermore, the function $\underset{\sim}{c}$ has a remarkable integral representation

$$\underset{\sim}{c}(\nu) = \int_{\overline{N}} e^{-(i\nu+\rho)(H(\overline{n}))} d\overline{n} \qquad (\overline{N} = \theta(N))$$

This integral converges if $\nu \in \mathfrak{a}_c^*$ and

$$\operatorname{Im}\langle\nu,\alpha\rangle < 0 \quad \forall \text{ roots } \alpha > 0 \text{ (of } (\mathfrak{g},\mathfrak{a}))$$

It was evaluated explicitly by Gindikin and Karpelevič [1]. This explicit evaluation implies in particular the product formula

$$\underset{\sim}{c} = \gamma \prod_{\alpha\in\Delta^{++}} \underset{\sim}{c}_\alpha$$

where $\gamma > 0$ is a constant, Δ^{++} is the set of short positive roots, and $\underset{\sim}{c}_\alpha$ is the c-function associated with the rank one subgroup of G whose only short positive root is α. In particular

$$\beta = \gamma^2 \prod_{\alpha\in\Delta^{++}} \beta_\alpha$$

The $\underset{\sim}{c}_\alpha$ and hence the β_α can be determined by explicit calculation in the rank one case (more on this later).

These results in the spherical set up are the prototypes of the general case. Therefore it may be worthwhile to take a closer look at them as a preview for the general theory. I shall do this in the next lecture.

3. Asymptotic behaviour of spherical eigenfunctions

1. We begin the deeper study of the spherical theory by looking at the eigenfunctions $\varphi(\nu:\cdot)$ and the eigenvalues to which they belong. The $\varphi(\nu:\cdot)$ are given by the formula

$$\varphi(\nu:x) = \int_K e^{(i\nu-\rho)(H(xk))}dk \qquad (x \in G) \tag{1}$$

for any ν in the complex vector space $\mathfrak{a}_c^*$. The corresponding eigenvalue is a homomorphism of $\mathfrak{Z}$ into $\mathbb{C}$. This homomorphism can be explicitly determined by using the representation (1); in particular one finds that for a given element $u \in \mathfrak{D}$, its value at ν depends polynomially on ν. Indeed, we have a homomorphism

(2) $$\gamma = \gamma_{\mathfrak{g}/\mathfrak{a}} : \mathfrak{D} \to U(\mathfrak{a}_c)$$

such that

(3) $$\varphi(\nu : x; v) = \gamma(v)(i\nu)\varphi(\nu : x) \qquad (v \in \mathfrak{D},\ x \in G,\ \nu \in \mathfrak{a}_c^*)$$

Since $\varphi(\nu : \cdot)$ is invariant under the Weyl group $\mathfrak{w}$ as a function of ν, it follows that $\gamma(v)$ is $\mathfrak{w}$-invariant. It is a basic fact that every element of $U(\mathfrak{a}_c)^{\mathfrak{w}}$ can be thus obtained:

(4) $$\gamma(\mathfrak{D}) = I(\mathfrak{w}) = U(\mathfrak{a}_c)^{\mathfrak{w}}$$

It is also true that γ factors through to an isomorphism of $\gamma_{G/K}(\mathfrak{D})$ with $I(\mathfrak{w})$. For fixed $\nu \in \mathfrak{a}_c^*$, $\varphi(\nu : \cdot)$ is the unique eigenfunction for $\mathfrak{D}$ in $C^\infty(G/\!/K)$ which is 1 at the identity element corresponding to the eigenhomomorphism $v \mapsto \gamma(v)(i\nu)$; there are no other eigenfunctions of $\mathfrak{D}$ in $C^\infty(G/\!/K)$ (normalized to be 1 at 1). Finally, $\varphi(\nu : \cdot) = \varphi(\nu' : \cdot)$ if and only if ν and ν' lie in the same $\mathfrak{w}$-orbit; this is immediate from (4) since the homomorphisms of $I(\mathfrak{w})$ into $\mathbb{C}$ are naturally in bijection with $\mathfrak{w}\backslash\mathfrak{a}_c^*$. The $\varphi(\nu : \cdot)$ are called the <u>elementary</u> spherical functions of G.

It is not difficult to give an explicit description of γ. We note first that the Iwasawa decomposition $\mathfrak{g} = \mathfrak{k} \oplus \mathfrak{a} \oplus \mathfrak{n}$ implies that $U(\mathfrak{g}_c) = U(\mathfrak{a}_c) \oplus (\mathfrak{k}U(\mathfrak{g}_c) + U(\mathfrak{g}_c)\mathfrak{n})$, and hence it makes sense to speak of the projection $E_\mathfrak{a} : U(\mathfrak{g}_c) \to U(\mathfrak{a}_c) \bmod (\mathfrak{k}U(\mathfrak{g}_c) + U(\mathfrak{g}_c)\mathfrak{n})$; then one has

(5) $$\gamma(v)(\nu) = E_\mathfrak{a}(v)(\nu - \rho) \qquad (v \in \mathfrak{D},\ \nu \in \mathfrak{a}_c^*)$$

For instance, let ω be the Casimir element in $U(\mathfrak{g}_c)$. Denote by $\omega_\mathfrak{a}$ the Laplacian on $\mathfrak{a}$: $\omega_\mathfrak{a} = H_1^2 + \cdots + H_r^2$ where $(H_i)_{1 \le i \le r}$ is an orthonormal basis of $\mathfrak{a}$ relative to the Killing form (which is positive definite on $\mathfrak{a}$). Then

(6) $$\gamma(\omega) = \omega_{\mathfrak{a}} - \langle \rho, \rho \rangle .$$

2. The basic method of studying the differential equations (3) is to use polar coordinates. The map $(k_1, h, k_2) \mapsto k_1 h k_2$ from $K \times A^+ \times K$ into G is submersive everywhere and so we obtain a (unique) homomorphism δ^+ of $\mathfrak{D}$ into the algebra of differential operators on A^+ with analytic coefficients such that

(7) $$f(h;v) = f(h;\delta^+(v)) \quad (v \in \mathfrak{D},\ h \in A^+,\ f \in C^\infty(G//K))$$

It is natural to call $\delta^+(v)$ <u>the radial component of</u> v on A^+. By a simple calculation one can show that

(8) $$\delta^+(v) = E_{\mathfrak{a}}(v) + \sum_{1 \le i \le p} g_i' v_i' \qquad (v \in \mathfrak{D})$$

where the v_i' are in $U(\mathfrak{a}_c)$ and have degree $< \deg(v)$, while the g_i' are functions which are in the ring $\mathcal{R}(A^+)$ (without unit element) generated by the functions $(e^{2\alpha} - 1)^{-1}$ $(\alpha \in \Delta^+)$. Thus, for the Casimir element ω, an explicit calculation gives

(9) $$\delta^+(\omega) = \omega_{\mathfrak{a}} + 2H_\rho + \sum_{\alpha \in \Delta^+} n(\alpha)(e^{2\alpha} - 1)^{-1} H_\alpha$$

where, for ant $\lambda \in \mathfrak{a}^*$, H_λ is the element of $\mathfrak{a}$ defined by $\langle H_\lambda, H \rangle = \lambda(H)$ $(H \in \mathfrak{a})$, $\langle \cdot, \cdot \rangle$ being the Killing form; $n(\alpha)$ is the dimension of the root space corresponding to the root α.

The crucial consequences of the formula (8) are as follows:

(i) if we consider only the highest degree terms, $\delta^+(v)$ is a constant coefficient operator

(ii) If $h \in A^+$ goes to infinity in such a way that $\alpha(\log h) \to +\infty$ for all roots $\alpha \in \Delta^+$, the coefficients g_i' tend to zero (exponentially), and hence the lower order terms in $\delta^+(v)$ may be thought of as perturbative for the question of studying asymptotics when h goes to infinity in the sense described above

(iii) since the leading term of $\delta^+(v)$ is $E_a(v)$ which is a translate of $\gamma(v)$ by ρ (cf. (5)) and since $\gamma(v)$ is an <u>arbitrary</u> element of $I(\mathfrak{w})$, one would expect that

$$\varphi(\nu : h) = \sum_{s \in \mathfrak{w}} c_s e^{(is\nu - \rho)(\log h)} + \text{terms decaying exponentially.}$$

In a certain sense the entire analytical theory of the differential equations (3) comes out of an attempt to make the above remarks more precise.

Let $\mathcal{L}$ be the lattice generated by the simple roots $\alpha_1, \ldots, \alpha_r$ in Δ^+ and let $\mathcal{L}^+$ be the subset of elements of the form $m_1\alpha_1 + \cdots + m_r\alpha_r$ where the m_j are all integers ≥ 0 with $m_1 + \cdots + m_r > 0$. It is then clear that each element of the ring $\mathcal{R}(A^+)$ has an expansion of the form $\sum_{q \in \mathcal{L}^+} c_q e^{-q}$, $c_q \in \mathbb{C}$. A variation of the classical Frobenius method of finding power series solutions of ordinary differential equations applied to the system (7) of differential equations now shows that $\varphi(\nu : h)$ has an infinite series representation of the form

$$\varphi(\nu : h) e^{\rho(\log h)} = \sum_{s \in W} c_s(\nu) e^{is\nu(\log h)} \sum_{q \in \mathcal{L}^+ \cup \{0\}} \Gamma_q(\nu : h) e^{-q(\log h)}$$

where $\nu \in \mathfrak{a}^*$ and the $\Gamma_q(\nu : h)$ depend polynomially in $\log h$ of degree at most d, $d \geq 0$ being a fixed integer. Taking $\nu = 0$ and writing

$$\Xi(x) = \varphi(0 : x) \tag{10}$$

we find for the spherical function Ξ the estimate

$$0 < \Xi(h) \leq C e^{-\rho(\log h)} (1 + \|\log h\|)^d \qquad (h \in \mathrm{Cl}(A^+)) \tag{11}$$

From the formula (1) we get, for all $\nu \in \mathfrak{a}^*$,

$$(12) \qquad |\varphi(\nu : h)| \leq Ce^{-\rho(\log h)}(1 + \|\log h\|)^d \qquad (h \in Cl(A^+))$$

For the derivatives of $\varphi(\nu : \cdot)$ we have similar estimates; the constant C of these estimates in general depends on ν but is of at most polynomial growth in ν. The estimates (11) and (12), together with the corresponding estimates of the derivatives of $\varphi(\nu : \cdot)$, are the fundamental initial estimates of the theory.

Write $\mathcal{R}$ for $\mathcal{R}(A^+)$. For any compact $L \subset \mathfrak{a}^+$ let

$$(13) \qquad A^+[L] = \{\exp tH \mid t \geq 1, H \in L\}.$$

The functions of $\mathcal{R}$ go to zero exponentially fast on the conic sets of the form $A^+[L]$. Hence on $A^+[L]$, $\delta^+(v)$ is a nice perturbation of $E_{\mathfrak{a}}(v)$. But the $\gamma(v)$ exhaust only $I(\mathfrak{w})$ and not $U(\mathfrak{a}_c)$ so that the unperturbed system is not of the first order. As in the classical cases where we go from n^{th} order equations to first order systems of vectors with n components, we can get a first order system as follows. We observe that $U(\mathfrak{a}_c)$ is a finite module over $I(\mathfrak{w})$. It is in fact a free module of rank $w = |\mathfrak{w}|$, having a basis $u_1 = 1, u_2, \ldots, u_w$ of homogeneous elements. Using this basis we get a representation of $U(\mathfrak{a}_c)$ by $w \times w$ matrices of elements of $I(\mathfrak{w})$,

$$u \mapsto \Gamma(u) = (\gamma_{ij}(u))_{1\leq i,j\leq w}, \qquad \gamma_{ij}(u) \in I(\mathfrak{w})$$

such that

$$uu_j = \sum_{1\leq \ell \leq w} \gamma_{j\ell}(u)u_\ell$$

We now replace $\varphi(\nu : \cdot)$ by the $w \times 1$ vector function Φ on $\mathfrak{a}_c^* \times A$ whose j^{th} component is $\varphi(\nu : h;\, u_j \circ e^\rho)$. The equations (7) now become

$$(14) \qquad u\Phi(\nu : \cdot) = \Gamma(u : i\nu)\Phi(\nu : \cdot) + \Omega(u : \nu : \cdot)\Phi(\nu : \cdot)$$

where $\Omega(u:\cdot:\cdot)$ is a $w \times w$ matrix of elements from $\mathcal{R}[X]$. The system (14) is obviously a perturbation of a first order linear system since we can take $u = H_i$, $(H_i)_{1\le i\le r}$ being a basis of $\mathfrak{a}$.

The eigenvalues of the matrix $\Gamma(u:i\nu)$ are clearly important in the perturbation problem (14). One knows that these are the numbers

$$u(is^{-1}\nu) \qquad (s \in \mathfrak{w},\ \nu \in \mathfrak{a}_c^*).$$

If we take $u = H_i$ and $\nu \in \mathfrak{a}^*$, <u>these are all purely imaginary</u>. <u>The system (14) is thus stable for $\nu \in \mathfrak{a}^*$; this is the central analytical fact of the theory</u>.

The spectral theory of the commuting family of matrices $\Gamma(u:i\nu)$ ($u \in U(\mathfrak{a}_c)$, $\nu \in \mathfrak{a}_c^*$ fixed) is also important in the study of (14). If π is the polynomial function on $\mathfrak{a}_c^*$ which is the product of all short positive roots it can be proved that $\pi(\nu)E_s(\nu)$ ($s \in w$) depends polynomially on ν where $E_s(\nu)$ is the spectral projection on the one dimensional space where $\Gamma(u:i\nu)$ has eigenvalue $u(is^{-1}\nu)$ for all $u \in I(\mathfrak{w})$.

Simple arguments from the theory of differential equations now allow us to conclude that on every conic region of the form $A^+[L]$ we have

$$(15)\quad \varphi(\nu:h)e^{\rho(\log h)} = \sum_{s\in\mathfrak{w}} \underset{\sim}{c}(s\nu)e^{is\nu(\log h)} + O(e^{-\beta(\log h)}(1+\|\nu\|)^m)$$

for all $h \in A^+[L]$, $\nu \in \mathfrak{a}^{*\prime}$ (= subset of regular elements in $\mathfrak{a}^*$); the O is uniform in ν and h, $\beta(H) = \min_{\alpha\in\Delta^+}\alpha(H)$ $(H \in \mathfrak{a})$; $m \geq 0$ is a constant. Moreover, $\underset{\sim}{c}$ is the Harish Chandra c-function. It is meromorphic on $\mathfrak{a}_c^*$ and if $\underset{\sim}{b} = \pi\underset{\sim}{c}$, then $\underset{\sim}{b}$ is holomorphic and everywhere nonvanishing on a suitable tube domain in $\mathfrak{a}_c^*$ containing $\mathfrak{a}^*$, with both $\underset{\sim}{b}$ and $\underset{\sim}{b}^{-1}$ having at most polynomial growth therein

For estimates on <u>all</u> of $Cl(A^+)$ this is not enough; we must consider conical sets $A^+[L]$ where L is still compact but is allowed to be in

$Cl(\mathfrak{a}^+)$. The above considerations can be generalized suitably to handle this more general situation. The result obtained may be described as follows.

Fix $H_0 \in Cl(\mathfrak{a}^+)$, $H_0 \neq 0$. Let M_0 be the centralizer of H_0 in G. The elements $\nu \in \mathfrak{a}_c^*$ parametrize the elementary spherical functions of M_0 also; we shall denote these by $\theta(\nu : \ )$. Let ρ_0 be the analogue of ρ for the group M_0, so that $\rho_0 = 1/2\ \Sigma_{\alpha>0,\alpha(H_0)=0}\, n(\alpha)\alpha$. Write $\mathfrak{w}_0$ for the stablizer of H_0 in $\mathfrak{w}$ and let $s_1 = 1, s_2, \ldots, s_m$ be a complete set of representatives for $\mathfrak{w}/\mathfrak{w}_0$. Let $\underset{\sim}{c}_0$ be the counterpart of $\underset{\sim}{c}$ for the group M_0. Then we can find a compact neighborhood L of H_0 in $Cl(\mathfrak{a}^+)$ such that for all $\nu \in \mathfrak{a}^{*\prime}$ and all h in the conic set $A^+[L]$, we have the following approximation for $\varphi(\nu : h)$:

$$\begin{aligned} e^{\rho(\log h)}\varphi(\nu : h) = e^{\rho_0(\log h)} \sum_{1\leq j\leq m} (\underset{\sim}{c}/\underset{\sim}{c}_0)(s_j^{-1}\nu)\theta(s_j^{-1}\nu : h) \; + \\ + \; O(e^{-\varepsilon_0\rho(\log h)}(1 + \|\nu\|)^p) \end{aligned} \tag{16}$$

Here $\varepsilon_0 > 0$, $p \geq 0$ are constants and the O is uniform in h and ν.

4. Wave packets in spherical Schwartz space

1. The estimates (15) and (16) of §3 allow us to study the behaviour, as x goes to infinity on G, of the "wave packets"

$$\int_{\mathfrak{a}^*} a(\nu)\varphi(\nu : h)d\nu$$

The approximation (15) of §3 suggests that we take a to be in the usual Schwartz space $\mathcal{C}(\mathfrak{a}^*)$ of the real vector space $\mathfrak{a}^*$; however, as the function $\underset{\sim}{c}$ has poles and only $\underset{\sim}{b} = \pi\underset{\sim}{c}$ is holomorphic, we need to replace a with πa. The approximation (16) of §3 coupled with induction on $\dim(G)$ now shows that the wave packet

$$\varphi_a'(x) = \int_{\mathfrak{a}^*} \pi(\nu)a(\nu)\varphi(\nu : x)d\nu \qquad (a \in \mathcal{C}(\mathfrak{a}^*))$$

is a well-defined element of $C^\infty(G/\!/K)$ and is rapidly decreasing in the following sense: for any integer $m \geq 0$, elements $u_1, u_2 \in U(\mathfrak{g}_c)$, there is a constant $C = C(m : u_1 : u_2) > 0$ such that

$$|\varphi_a'(u_1;h;u_2)| \leq Ce^{-\rho(\log h)}(1 + \|\log h\|)^{-m}$$

for all $h \in C\ell(A^+)$.

Motivated by the above remark let us write $\mathcal{C}(G)$ for the space of all $f \in C^\infty(G)$ which are rapidly decreasing in the above sense. Equipped with the obvious family of seminorms it becomes a Frechet space. It is the <u>Harish Chandra Schwartz space</u> of G, and is contained in $L^2(G)$. It is stable under left and right translations by elements of G which actually define representations of G. $\mathcal{C}(G)$ is furthermore closed under convolution and in fact the map $f,g \mapsto f*g$ from $\mathcal{C}(G) \times \mathcal{C}(G) \to \mathcal{C}(G)$ is continuous, making $\mathcal{C}(G)$ a topological algebra. Let us put $\mathcal{C}(G/\!/K) = \mathcal{C}(G) \cap C^\infty(G/\!/K)$.

If we now observe that for $\nu \in \mathfrak{a}^*$

$$|\underset{\sim}{c}(\nu)|^{-2} = \underset{\sim}{c}(\nu)^{-1}\underset{\sim}{c}(-\nu)^{-1} = \pi(\nu)\pi(-\nu)\underset{\sim}{b}(\nu)^{-1}\underset{\sim}{b}(-\nu)^{-1}$$

we can prove the following result which I call <u>the first wave packet theorem</u> (for spherical functions).

<u>Theorem</u>. For any $a \in \mathcal{C}(\mathfrak{a}^*)^{\mathfrak{w}}$, the wave packet φ_a defined by

$$(1) \qquad \varphi_a(x) = \int_{\mathfrak{a}^*} a(\nu)\varphi(\nu : x)|\underset{\sim}{c}(\nu)|^{-2}d\nu$$

is an element of $\mathcal{C}(G/\!/K)$; and the map $a \mapsto \varphi_a$ is continuous.

The next step is obviously to form

$$\int_G \varphi_a(x)\, \overline{\varphi(\nu : x)}dx$$

and attempt to show that it is $\gamma \cdot a(\nu)$ where γ is a nonzero constant independent of a. It is not difficult to show that there is a unique $\mathfrak{w}$-invariant continuous function $\gamma(\cdot)$ on $\mathfrak{a}^*$ such that for all $\mathfrak{w}$-invariant $a \in \mathcal{C}(\mathfrak{a}^*)$,

$$\int \varphi_a(x)\varphi(-\nu : x)dx = \gamma(\nu)a(\nu) \qquad (\nu \in \mathfrak{a}^*) \tag{2}$$

(Here we must note that for $\nu \in \mathfrak{a}^*$, $\overline{\varphi(\nu : \cdot)} = \varphi(-\nu : \cdot)$). On the other hand, using $\bar{N} = \theta(N)$ instead of N for computing the Harish Chandra transform, one finds that for any $f \in \mathcal{C}(G//K)$

$$\int_G f(x)\varphi(-\nu : x)dx = \int_A e^{-i\nu(\log h)} \cdot \left(e^{\rho(\log h)} \int_{\bar{N}} f(\bar{n}h)d\bar{n}\right)dh$$

So, using Fourier inversion on A, with the measure dh being dual to $d\nu$, the formula (2) reduces to

$$e^{\rho(\log h)} \int_{\bar{N}} \varphi_a(\bar{n}h)d\bar{n} = \int_{\mathfrak{a}^*} \gamma(\nu)a(\nu)e^{i\nu(\log h)}d\nu \tag{3}$$

where we restrict a to be in $C_c^\infty(\mathfrak{a}^*)^W$ for obvious convergence reasons.

We shall now evaluate the left side of (3) directly, with a restricted to be in $C_c^\infty(\mathfrak{a}^{*\prime})$, and obtain for such a the same formula (3) with γ replaced by a constant. This will then prove that γ is a constant. Our calculation will explicitly evaluate the constant too for a very specific normalization of $d\bar{n}$.

I shall now describe in a <u>very heuristic way</u> Harish Chandra's beautiful method for evaluating the integral

$$e^{\rho(\log h)} \int_{\bar{N}} \varphi_a(\bar{n}h)d\bar{n}$$

We first remark that the study of the asymptotic behaviour of $\varphi(\nu : \cdot)$ leads to the following approximation. Write, for $x \in G$, $\nu \in \mathfrak{a}^{*\prime}$,

$$d(x) = e^{\rho(H(x))}, \quad \varphi_\infty(\nu : x) = d(x)^{-1} \sum_{s \in \mathfrak{w}} \underset{\sim}{c}(s\nu)e^{is\nu(H(x))} \tag{4}$$

Then, if $a \in A^+$ and goes to infinity on a conic region $A^+[L]$ where L is a compact in $\mathfrak{a}^+$ (rather than in $C\ell(\mathfrak{a}^+)$), we have

$$\varphi(\nu : xa) - \varphi_\infty(\nu : xa) \to 0 \tag{5}$$

in the $C^\infty(G)$ topology, uniformly in ν as long as ν varies over compact subsets of $\mathfrak{a}^{*\prime}$. As a first step we form the "truncated wave packet"

$$\varphi_{\infty, a}(x) = \int_{\mathfrak{a}^*} \varphi_\infty(\nu : x)a(\nu)|\underset{\sim}{c}(\nu)|^{-2}d\nu \tag{6}$$

where $a \in C_c^\infty(\mathfrak{a}^{*\prime})$, and evaluate

$$e^{\rho(\log h)} \int_{\overline{N}} \varphi_{\infty, a}(\overline{n}h)d\overline{n} \tag{7}$$

We find, proceeding formally,

$$d(h) \int_{\overline{N}} \varphi_{\infty, a}(\overline{n}h)d\overline{n} = d(h) \int_{\overline{N}}\int_{\mathfrak{a}^*} \underset{\sim}{c}(\nu)^{-1}\underset{\sim}{c}(-\nu)^{-1} \sum_s \underset{\sim}{c}(s\nu) \cdot e^{(is\nu-\rho)(H(\overline{n}h))}a(\nu)d\nu$$

$$= \int_{\mathfrak{a}^*} a(\nu)\left(\int_{\overline{N}} \underset{\sim}{c}(\nu)^{-1}\underset{\sim}{c}(-\nu)^{-1} \sum_s e^{is\nu(\log h)}\underset{\sim}{c}(s\nu) \cdot e^{(is\nu-\rho)(H(\overline{n}))}d\overline{n}\right)d\nu$$

To evaluate the inner integral we use the integral formula for $\underset{\sim}{c}$; the integral coming from the element $s \in \mathfrak{w}$ is then $\underset{\sim}{c}(-s\nu)$ (cf. §2). So, remembering that

$$\underset{\sim}{c}(\nu)\underset{\sim}{c}(-\nu) = \underset{\sim}{c}(s\nu)\underset{\sim}{c}(-s\nu)$$

we get

$$e^{\rho(\log h)} \int_{\overline{N}} \varphi_{\infty, a}(\overline{n}h)d\overline{n} = w\int_{\mathfrak{a}^*} a(\nu)e^{i\nu(\log h)}d\nu \tag{8}$$

The rigorous proof of (8) is a little more delicate since the integrals over

$\bar{N}$ do not converge when $\nu \in \mathfrak{a}^*$ and so one has to go into the complex domain where they converge.

The formula (8) is exactly of the same form as (3) with $\gamma(\nu) = w$ and $\varphi_{\infty,a}$ in place of φ_a. To complete the proof of (3) one is left with proving the remarkable fact that

$$(9) \qquad \int_{\bar{N}} (\varphi_a - \varphi_{\infty,a})(\bar{n}h)d\bar{n} = 0$$

for all h. The idea for proving this is to study $\varphi(\nu:\cdot) - \varphi_\infty(\nu:\cdot)$ on $\bar{N}A$ through the system of differential equations satisfied by $\varphi(\nu:\cdot)$. It turns out that the difference

$$\varphi(\nu:\bar{n}h) - \varphi_\infty(\nu:\bar{n}h)$$

can be expressed as a linear combination (with coefficients depending on h) of derivatives of $\varphi(\nu:\cdot)$, the derivatives being from $\bar{\mathfrak{n}}$. Since integrals over $\bar{N}$ of derivatives with respect to $\bar{\mathfrak{n}}$ are zero, we obtain (9). However, the actual carrying out of this argument is very delicate. Since an interchange of integrations is involved, one needs absolute convergence; but this is not available, and it becomes necessary to smooth things by convolutions with elements of $C_c^\infty(\bar{N})$ (causing no problems since these convolution operators commute with integration over $\bar{N}$). The actual result, which I call <u>the second wave packet theorem</u> (for spherical functions), is then as follows.

<u>Theorem</u>. Let $a \in C(\mathfrak{a}^*)^{\mathfrak{w}}$. Then

$$e^{\rho(\log h)} \int_{\bar{N}} \varphi_a(\bar{n}h)d\bar{n} = w \int_{\mathfrak{a}^*} a(\nu)e^{i\nu(\log h)}d\nu$$

Here the measure $d\bar{n}$ is the one that occurs in the integral representation of $\underset{\sim}{c}$, and may be specified by the condition $\int_{\bar{N}} e^{-2\rho(H(\bar{n}))}d\bar{n} = 1$.

From this theorem and (2) we then obtain

$$(10) \qquad \int_G \varphi_a(x)\overline{\varphi(\nu : x)}dx = w\, a(\nu)$$

For obtaining completeness and hence the Plancherel formula it only remains to prove that the map $a \mapsto \varphi_a$ is <u>onto</u> $\mathcal{C}(G/\!/K)$. A simple argument reduces this to showing that the Harish Chandra transform $f \mapsto \mathcal{H}f$ is <u>injective</u> on $\mathcal{C}(G/\!/K)$; in fact, let $g \in \mathcal{C}(G/\!/K)$ and let $a = \frac{1}{w}\mathcal{H}g$. Then the formula (10) shows that $\mathcal{H}(g - \varphi_a) = 0$, and the injectivity of $\mathcal{H}$ on $\mathcal{C}(G/\!/K)$ would give $g = \varphi_a$. The injectivity is proved by an argument using induction on $\dim(G)$ which reduces the assumption that $\mathcal{H}f = 0$ to the conclusion that for some character Θ of the discrete series, $\Theta(f) \neq 0$. This is a contradiction since the discrete series representations are never of class 1. This argument is a special case of a more general one which will occur again. The main theorem of the spherical theory may now be formulated as follows.

<u>Theorem</u>. The map

$$f \mapsto \mathcal{H}f, \ (\mathcal{H}f)(\nu) = \int_G f(x)\overline{\varphi(\nu : x)}dx$$

is an isomorphism of the convolution algebra $\mathcal{C}(G/\!/K)$ onto the multiplication algebra $\mathcal{C}(\mathfrak{a}^*)^{\mathfrak{w}}$, both being regarded as Frechet algebras. For the inverse map we have

$$f(x) = w^{-1}\int_{\mathfrak{a}^*}(\mathcal{H}f)(\nu)\varphi(\nu : x)|\underset{\sim}{c}(\nu)|^{-2}d\nu \qquad (x \in G)$$

Here $w = |\mathfrak{w}|$, $dx = e^{2\rho(\log a)}dk\,dh\,dn$ where $\int_K dk = 1$, and dn is the measure that goes over via θ to the measure $d\bar{n}$ on $\bar{N}$ for which $\int_{\bar{N}} e^{-2\rho(H(\bar{n}))}d\bar{n} = 1$; and dh and $d\nu$ are dual to each other. Furthermore, we have,

$$\int_G |f(x)|^2dx = w^{-1}\int_{\mathfrak{a}^*} |\mathcal{H}f(\nu)|^2|\underset{\sim}{c}(\nu)|^{-2}d\nu$$

5. Asymptotic behaviour of tempered eigenfunctions for $\mathfrak{Z}$. The constant term

1. In order to treat the problem of eigenfunction expansions on G in full generality it is clearly necessary to go beyond the spherical case discussed in §§3 and 4. To this end Harish Chandra introduced the very elegant notion of eigenfunctions for $\mathfrak{Z}$ on G, with values in a bimodule of K and transforming covariantly. More precisely, let U be a finite dimensional complex vector space. By a double representation of K in U is meant a representation of $K \times \check{K}$ in U where $\check{K}$ is the group opposite to K, i.e., a pair $(\tau_1,\tau_2) = \tau$ where τ_1 is a representation of K in U, and τ_2 a representation of $\check{K}$ in U. It will be convenient to let $\tau_1(k)$ act (on the vectors of U) from the left and to let $\tau_2(k)$ act from the right:

$$\tau(k_1,k_2)u = \tau_1(k_1)u\,\tau_2(k_2) \qquad (u \in U,\ k_1,k_2 \in K)$$

We say τ is unitary if τ_1 and τ_2 are. A function $f : G \to U$ is called τ-spherical if

$$f(k_1xk_2) = \tau_1(k_1)f(x)\tau_2(k_2) \qquad (k_j \in K,\ x \in G)$$

$C^\infty(G:\tau)$ will be the space of smooth τ-spherical f; $\mathcal{C}(G:\tau)$ and $C_c^\infty(G:\tau)$ its obviously defined subspaces.

To see the naturalness of this concept let π be an irreducible representation of G in a Hilbert space H, possessing an infinitesimal character $\chi_\pi(\mathfrak{Z} \to \mathbb{C})$. We assume that $\pi_K = \pi|_K$ is unitary. For any $\mathfrak{d} \in \hat{K}$ let $H_\mathfrak{d}$ be the isotypical subspace of H of vectors transforming like $\mathfrak{d}$; and for any finite set $F \subset \hat{K}$ let $H_F = \oplus_{\mathfrak{d}\in F} H_\mathfrak{d}$. We know that each H_F is finite dimensional. Let E_F be the orthogonal projection $H \to H_F$. We then define U to be the algebra of endomorphisms of the Hilbert space H_F, regarded as a Hilbert space itself under the scalar product $(u,v) = \mathrm{tr}(uv^\dagger)$. Write

$\pi_F(k) = E_F\pi(k)E_F \quad (k \in K), \quad \tau_1(k)u = \pi_F(k)u, \quad uT_2(k) = u\pi_F(k) \quad (k \in K,\ u \in U),$

so that τ is a unitary double representation of K in U. If we put

$$\pi_F(x) = E_F\pi(x)E_F \qquad (x \in G)$$

then $\pi_F \in C^\infty(G:\tau)$ and π_F is an eigenfunction of $\mathfrak{Z}$ for the eigenhomomorphism χ_π:

$$z\pi_F = \chi_\pi(z)\pi_F \qquad (z \in \mathfrak{Z})$$

The fundamental problem in doing harmonic analysis is to determine the behaviour of τ-spherical eigenfunctions at infinity on the group. Suppose $G = KA_0N_0$ is an Iwasawa decomposition of G. Then $G = K\,C\ell(A_0^+)K$ and so, if f is any τ-spherical eigenfunction on G, it is a question of studying the behaviour of $f(h)$ when $h \in C\ell(A_0^+)$ and goes to infinity. Now, when $h \to \infty$, it is not necessary that $\alpha(\log h) \to \infty$ for all positive roots α. So we fix a psgrp $P = MAN$ containing $P_0 = M_0A_0N_0$ ("standard") and study the behaviour of $f(h)$ when $h \to \infty$ relative to P, i.e., when $\alpha(\log h) \to \infty$ for all roots α of P_0.

The central observation of the theory is the following. Since $G = K(MA)K$, one can reduce the differential equations satisfied by f on G to a system of differential equations on MA, or at least on a suitable open subset of it; these equations, when $a \in A$ and $a \to \infty$ relative to P, may be regarded as perturbations of the eigenfunction equations on the group MA. Consequently it is reasonable to expect that f may be approximated asymptotically by a suitable eigenfunction or linear combination of eigenfunctions from the group MA, the approximation being valid on conic subsets of A_0^+ on which $\min_{\alpha \text{ a root of } P} \alpha(\log h)$ is of the same order of magnitude as $\|\log h\|$.

Basically this is the same situation that presents itself in the study of elementary spherical functions. There is however a crucial additional complication. On converting the perturbation problem to one involving only first order equations we find (unlike in the spherical case) that the unperturbed linear system admits eigenvalues <u>which do not all have real parts</u> ≤ 0, so that the <u>problem is not stable</u>. Fortunately it turns out that the assumption of <u>temperedness</u> acts as a substitute for the eigenvalue property mentioned above.

2. We shall now make these remarks precise. We begin by introducing the so-called Harish Chandra homomorphism. Given any CSA $\mathfrak{h} \subset \mathfrak{g}$, this is a homomorphism

$$\mu_{\mathfrak{g}/\mathfrak{h}} : \mathfrak{z} \to U(\mathfrak{h}_c) = S(\mathfrak{h}_c)$$

which is actually an isomorphism of $\mathfrak{z}$ with the algebra of elements of $U(\mathfrak{h}_c)$ invariant under the Weyl group $W(\mathfrak{g}_c, \mathfrak{h}_c)$:

$$\mu_{\mathfrak{g}/\mathfrak{h}} : \mathfrak{z} \xrightarrow{\sim} U(\mathfrak{h}_c)^{W(\mathfrak{g}_c, \mathfrak{h}_c)}$$

If $\mathfrak{h}_1 \subset \mathfrak{g}$ is another CSA of $\mathfrak{g}$ and y is an element of the complex adjoint group G_c such that $y(\mathfrak{h}_c) = \mathfrak{h}_{1c}$, we have the commutative diagram

$$\begin{array}{ccc} \mathfrak{z} & \xrightarrow{\mu_{\mathfrak{g}/\mathfrak{h}}} & U(\mathfrak{h}_c) \\ & \searrow^{\mu_{\mathfrak{g}/\mathfrak{h}_1}} & \downarrow \wr \; y \in G_c \\ & & U(\mathfrak{h}_{1c}) \end{array}$$

Suppose $\mathfrak{m}_1 \subset \mathfrak{g}$ is a reductive subalgebra with $\mathrm{rk}(\mathfrak{m}_1) = \mathrm{rk}(\mathfrak{g})$. Let $\mathfrak{z}(\mathfrak{m}_1)$ be the center of $U(\mathfrak{m}_{1c})$. If $\mathfrak{h} \subset \mathfrak{m}_1$ is a CSA of $\mathfrak{m}_1$, it is a CSA of $\mathfrak{g}$ also. So, as the inclusion $W(\mathfrak{g}_c, \mathfrak{h}_c) \supset W(\mathfrak{m}_{1c}, \mathfrak{h}_c)$ gives the inclusion

$$U(\mathfrak{h}_c)^{W(\mathfrak{g}_c, \mathfrak{h}_c)} \subset U(\mathfrak{h}_c)^{W(\mathfrak{m}_{1c}, \mathfrak{h}_c)}$$

we have a unique injection $\mu_{\mathfrak{g}/\mathfrak{m}_1}$ making the following diagram commutative:

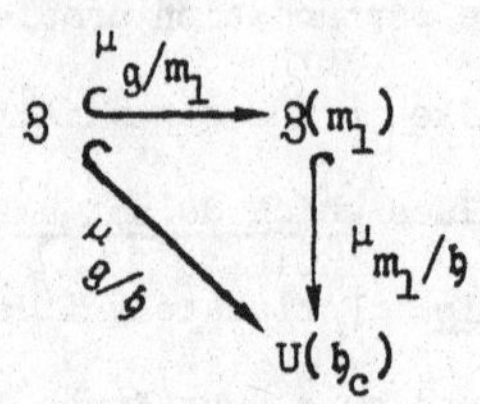

It is also clear from the previous diagram that $\mu_{\mathfrak{g}/\mathfrak{m}_1}$ is independent of the choice of $\mathfrak{h}$. Finally, using well known results from the theory of finite reflexion groups we have

<u>Proposition A</u>. $\mathfrak{z}(\mathfrak{m}_1)$ is a free module over $\mu_{\mathfrak{g}/\mathfrak{m}_1}(\mathfrak{z})$ of rank = $[W(\mathfrak{m}_{1c},\mathfrak{h}_c) : W(\mathfrak{g}_c,\mathfrak{h}_c)]$.

Next, we consider a psgrp $P = MAN$. We write $\mathcal{P}(A)$ for the (finite) set of psgrps $Q = MAN_Q$ with the same A (and hence M). We write d_Q for the homomorphism of MA into the positive reals given by

(1) $$d_Q(ma) = e^{\rho_Q(\log a)} \qquad (m \in M,\ a \in A)$$

where

(2) $$\rho_Q(H) = \frac{1}{2}\operatorname{tr}(\operatorname{ad}H\big|_{\mathfrak{n}_Q}) \qquad (H \in \mathfrak{a})$$

($\mathfrak{a}$, $\mathfrak{n}_Q$ are the respective Lie algebras of A, N_Q, and $\log : A \to \mathfrak{a}$ inverts $\exp : \mathfrak{a} \to A$). Given P, we define $\bar{P} \in \mathcal{P}(A)$ by

(3) $$\bar{P} = MA\bar{N}, \quad \bar{N} = \theta(N), \quad \bar{\mathfrak{n}} = \theta(\mathfrak{n}).$$

Let $\mathfrak{m}_1$ be the centralizer of A (or $\mathfrak{a}$) in $\mathfrak{g}$. Then $\mathfrak{m}_1$ is the Lie algebra of MA and is a reductive subalgebra of $\mathfrak{g}$ of the same rank as $\mathfrak{g}$. We often write μ_P instead of $\mu_{\mathfrak{g}/\mathfrak{m}_1}$ although it is independent of the choice of P within $\mathcal{P}(A)$. For our purposes the following computation is crucial:

<u>Proposition B</u>. Let $P = MAN$ and let notation be as above. Given $z \in \mathfrak{Z}$, $\exists$ unique $z_1 \in U(\mathfrak{m}_{1c})$ such that

$$z \equiv z_1 \mod(U(\mathfrak{g}_c)\mathfrak{n});$$

moreover, we have the following properties for the map $z \mapsto z_1$:

(i) $z_1 \in \mathfrak{Z}(\mathfrak{m}_1)$

(ii) $z \equiv z_1 \mod \theta(\mathfrak{n})U(\mathfrak{g}_c)\mathfrak{n}$

(iii) $z_1 = d_P^{-1} \circ \mu_P(z) \circ d_P$

(iv) If $u \mapsto u^t$ is the antiantomorphism of $U(\mathfrak{g}_c)$ which is $-id$ on $\mathfrak{g}_c$, then $\mathfrak{Z}^t = \mathfrak{Z}$ and $\mu_P(z^t) = \mu_P(z)^t$ for all $z \in \mathfrak{Z}$.

For any $z \in \mathfrak{Z}$ let us write

(4) $$r_P(z) = z - z_1 = z - d_P^{-1} \circ \mu_P(z) \circ d_P.$$

Suppose f is a smooth function on G. Then, for $m_1 \in MA$,

(5) $$f(m_1;z) = f(m_1;d_P^{-1} \circ \mu_P(z) \circ d_P) + f(m_1;r_P(z))$$

We shall now make an estimate of the term

$$f(m_1;r_P(z)).$$

Since $r_P(z) \in \theta(\mathfrak{n})U(\mathfrak{g}_c)$, it is enough to estimate $f(m_1;\theta(X)b)$ where $X \in \mathfrak{n}$, $b \in U(\mathfrak{g}_c)$. We have the following lemma.

<u>Lemma C</u>. Let $\varepsilon_P(m_1) = \|Ad(m_1)_{\theta(\mathfrak{n})}\|$ where the suffix denotes restriction, $\|\cdot\|$ being defined relative to the Hilbert space structure on $\mathfrak{g}$ given by $(X,X') = -(X,\theta X')(X,X' \in \mathfrak{g})$. Let $\theta(X_j)$ $(1 \leq j \leq p)$ be an orthonormal basis for $\theta(\mathfrak{n})$. Let $\eta_1,\eta_2 \in U(\mathfrak{m}_{1c})$, $b \in U(\mathfrak{g}_c)$ and $X \in \mathfrak{n}$. Then $\exists$ elements $\zeta_j \in U(\mathfrak{m}_{1c})$ depending on X and η_2 with the following property. For any smooth function g and any $m_1 \in MA$,

$$|g(\eta_1;m_1;\eta_2\theta(X)b)| \leq \varepsilon_P(m_1) \sum_{1\leq i,j\leq p} |g(\eta_1\theta(X_i);m_1;\zeta_j b)|.$$

For, since ad m_1 stabilizes $\theta(n)$, we can write

$$\eta_2\theta(X) = \sum_{1\leq j\leq p} \theta(X_j)\zeta_j$$

for suitable $\zeta_j \in U(m_{1c})$. Furthermore, as MA stabilizes $\theta(n)$

$$\theta(X_j)^{m_1} = \sum_{1\leq i\leq p} a_{ij}(m_1)\theta(X_i)$$

and

$$|a_{ij}(m_1)| \leq \varepsilon_P(m_1).$$

The lemma follows now from the following calculation:

$$\begin{aligned} g(\eta_1;m_1;\eta_2\theta(X)b) &= \sum_j g(\eta_1;m_1;\theta(X_j)\zeta_j b) \\ &= \sum_j g(\eta_1\theta(X_j)^{m_1};m_1;\zeta_j b) \\ &= \sum_i\sum_j a_{ij}(m_1)g(\eta_1\theta(X_i);m_1;\zeta_j b). \end{aligned}$$

<u>Remark</u>. Given a variable element $a \in A$, let us write

$$a \overset{P}{\to} \infty \quad \text{(a goes to infinity along } P)$$

if

$$\alpha(\log a) \to \infty \quad \text{for each root } \alpha \text{ of } (P,A).$$

If we put

$$\beta_P(\log a) = \min_{\alpha \in \text{set of roots of } (P,A)} \alpha(\log a)$$

this means $\beta_P(\log a) \to \infty$. For $(\mathrm{Ad}(a))_{\theta(n)}$ the eigenvalues are $e^{-\alpha(\log a)}$ and hence we have the estimate

$$\varepsilon_P(ma) \leq \varepsilon_P(m)e^{-\beta_P(\log a)} \qquad (m \in MA, a \in A).$$

This estimate, in conjunction with Lemma C, reveals that the second term on the right of (5) may be thought of as a perturbation term relative to the first.

Let

$$\chi : \mathfrak{Z} \to \mathbf{C}$$

be a homomorphism and let f be an eigenfunction for $\mathfrak{Z}$ with eigenhomorphism χ, i.e.,

$$zf = \chi(z)f \qquad (z \in \mathfrak{Z}).$$

By Proposition A we can find elements

$$v_1 = 1, v_2, \ldots, v_r \in \mathfrak{Z}(\mathfrak{m}_1)$$

such that the v_i form a basis for $\mathfrak{Z}(\mathfrak{m}_1)$ regarded as a module over $\mu_P(\mathfrak{Z})$. We now go over from f to the vector Φ with r components, the jth component Φ_j being given by

(6) $$\Phi(m_1)_j = f(m_1; v_j \circ d_P).$$

The regular representation of $\mathfrak{Z}(\mathfrak{m}_1)$ may now be identified with an $r \times r$ matrix representation with elements in $\mathfrak{Z}$; indeed, for $v \in \mathfrak{Z}(\mathfrak{m}_1)$, $\exists$ unique $z_{ij}(v) \in \mathfrak{Z}$ such that

$$vv_j = \sum_{1\leq i\leq r} v_i\mu_P(z_{ij}(v)).$$

Clearly

$$v \mapsto Z(v) = (z_{ij}(v))_{1 \le i,j \le r}$$

is the representation described above. We now have, for $m_1 \in MA$ and $v \in \mathfrak{Z}(m_1)$,

$$\begin{aligned}\Phi(m_1;v)_j &= f(m_1;vv_j \circ d_P)\\ &= \sum_{1\le i\le r} f(m_1;v_i\mu_P(z_{ij}(v)) \circ d_P)\\ &= \sum_{1\le i\le r} \{f(m_1;v_i \circ d_P \circ z_{ij}(v)) - f(m_1;v_i \circ d_P \circ r_P(z_{ij}(v)))\}\end{aligned}$$

using (5). Let

(7) $$\gamma_{ji}(v) = \chi(z_{ij}(v)), \quad \Gamma(v) = (\gamma_{ji}(v))_{1\le i,j\le r} .$$

Then

(8) $$\begin{cases}\Phi(m_1;v) = \Gamma(v)\Phi(m_1) + \Psi_v(m_1)\\ \Psi_v(m_1)_j = -d_P(m_1) \displaystyle\sum_{1\le i\le r} f(m_1;v_i' r_P(z_{ij}(v))\end{cases}$$

where $v_i' = d_P^{-1} \circ v_i \circ d_P \in \mathfrak{Z}(m_1)$ once again.

The equations (8), taken together with Lemma C and the remark following it, constitute the perturbation problem on MA that I spoke about earlier. Since A is abelian and $\mathfrak{a} \subset \mathfrak{Z}(m_1)$, we may take $v = H_i$ $(1 \le i \le \dim(\mathfrak{a}))$ in (8) where the H_i form a basis of $\mathfrak{a}$; the resulting equations

(9) $$\Phi(m_1;H_i) = \Gamma(H_i)\Phi(m_1) + \Psi_{H_i}(m_1)$$

are of the first order in the unperturbed part.

It is well known that if $\mathfrak{h}$ is a CSA of $\mathfrak{m}_1$ containing $\mathfrak{a}$, we can write χ in the form

$$\chi(z) = \mu_{\mathfrak{g}/\mathfrak{h}}(z)(\Lambda) \tag{10}$$

for some $\Lambda \in \mathfrak{h}_c^*$. The $W(\mathfrak{g}_c, \mathfrak{h}_c)$-orbit of Λ is uniquely determined by χ, and the matrix $\Gamma(v)$ has eigenvalues

$$\mu_{\mathfrak{m}_1/\mathfrak{h}}(v)(s\Lambda) \qquad s \in W(\mathfrak{g}_c, \mathfrak{h}_c).$$

In particular, $\Gamma(H_i)$ has the eigenvalues

$$s\Lambda(H_i) \qquad (s \in W(\mathfrak{g}_c, \mathfrak{h}_c)).$$

Since it is the complex Weyl group that governs here, it is in general not possible to guarantee stability - by eigenvalue considerations.

3. To overcome the difficulty suggested above Harish Chandra works only with tempered eigenfunctions. Let us recall the Iwasawa decomposition $G = KA_0N_0$. Analogous to the estimates for the spherical functions $\varphi(\nu : \cdot)$ we say that a function f on G satisfies the weak inequality if for some $m \geq 0$, $C > 0$,

$$|f(k_1hk_2)| \leq Ce^{-\rho_0(\log h)}(1 + \|\log h\|)^m \tag{11}$$

for all $k_1, k_2 \in K$, $h \in Cl(A_0^+)$. If f is a K-finite $\mathfrak{Z}$-finite function, satisfying the weak inequality is equivalent to saying that f defines a tempered distribution, or, briefly, that f is tempered; in this case, each derivative afb $(a,b \in U(\mathfrak{g}_c))$ satisfies the weak inequality. Let τ be a double representation of K in a finite dimensional space U. We then denote by

$$\mathbb{A}(G:\tau)$$

the subspace of $C^\infty(G:\tau)$ consisting of all $\mathfrak{Z}$-finite tempered elements. For functions in $\mathbb{A}(G:\tau)$ one can develop a powerful asymptotic theory using the perturbative method outlined in the preceding paragraph. I shall describe the basic results of this theory now.

Let me set up some notation. If τ is a double representation and $Q = MAN$ is a psgrp, we write $\tau_M = \tau|_{K_M}$ where $K_M = K \cap M = K \cap Q$ is as usual the maximal compact subgroup of M fixed by θ. We write Ξ for $\varphi(0:\cdot)$. Since (see Varadarajan [1], Part II, §8, Proposition 17)

$$\Xi(h) \geq e^{-\rho_0(\log h)} \qquad (h \in Cl(A_0^+))$$

we can replace $e^{-\rho_0}$ by Ξ in the definition of the weak inequality. Also, let $\sigma(x)$ be the distance of xK from K in G/K considered as a Riemannian space with a G-invariant metric. Then the weak inequality becomes

$$(12) \qquad |f(x)| \leq C\,\Xi(x)(1 + \sigma(x))^m \qquad (x \in G)$$

Since we often vary the group G, we write Ξ_G for Ξ to be more precise.

<u>Theorem A</u>. Let $f \in \mathcal{A}(G:\tau)$ and $Q = MAN$ a psgrp. Then $\exists$ a unique $f_Q \in \mathcal{A}(MA:\tau_M)$ such that, for each $m \in MA$

$$d_Q(ma)f(ma) - f_Q(ma) \to 0 \quad \text{as} \quad a \overset{Q}{\to} \infty$$

f_Q is called <u>the constant term of f along Q</u>.

Let me now describe the most important of the properties of the map $f \mapsto f_Q$. First we have

$$(13) \qquad (zf)_Q = \mu_Q(z)f_Q \qquad f_{Q^k}(m^k) = \tau_1(k)f_Q(m)\tau_2(k)^{-1}$$

for $m \in MA$, $k \in K$. Next we have the transitivity. To formulate it we need some notation. Fix a psgrp $\bar{Q} = \bar{M}\bar{A}\bar{N}$. Then there is a canonical bijection

$$Q \rightleftarrows {}^*Q$$

between the set of psgrps $Q \subset \bar{Q}$ and psgrps ${}^*Q$ of $\bar{M}$; the correspondence is defined by

(14) $${}^*Q = Q \cap \bar{M};$$

and the Langlands decompositions

(15) $$Q = MAN, \quad {}^*Q = {}^*M{}^*A{}^*N$$

are related by

(16) $${}^*M = M, \quad {}^*A = A \cap \bar{M}, \quad {}^*N = N \cap \bar{M}, \quad A = {}^*A\bar{A}, \quad N = {}^*N\bar{N}, \quad Q = {}^*Q\bar{A}\bar{N}.$$

Theorem B. Let $\bar{Q} = \bar{M}\bar{A}\bar{N}$ be a psgrp of G and let $Q \rightleftarrows {}^*Q$ be the above correspondence. Fix $f \in \mathcal{A}(G:\tau)$. For any $\bar{a} \in \bar{A}$, let $f_{\bar{Q},\bar{a}}(\bar{m}) = f_{\bar{Q}}(\bar{m}\bar{a})$ $(\bar{m} \in \bar{M})$. Then

$$f_{\bar{Q},\bar{a}} \in \mathcal{A}(\bar{M} : \tau_{\bar{M}})$$

and

$$(f_{\bar{Q},\bar{a}})_{{}^*Q}({}^*m) = f_Q({}^*m\bar{a}) \qquad ({}^*m \in {}^*M{}^*A).$$

Finally we have the following approximation result justifying our terminology of f_Q as the constant term.

Theorem C. Let $Q_0 = M_0A_0N_0$ be a minimal psgrp contained in the psgrp $Q = MAN$. Write $\beta_Q(H) = \inf_{\alpha \text{ a root of } (Q,A)} \alpha(H)$ $(H \in \mathfrak{a}_0)$. Fix $f \in \mathcal{A}(G:\tau)$. Then $\exists$ constants $c > 0$, $\varepsilon_0 > 0$, $m \geq 0$ such that for all $h \in Cl(A_0^+)$

$$|d_Q(h)f(h) - f_Q(h)| \leq c\, \Xi_M(h)(1 + \sigma(h))^m e^{-\varepsilon_0\beta_Q(\log h)}.$$

In particular, for any $t > 0$, let $A_0^+(Q:t)$ be the conic set of all

$h \in Cl(A_0^+)$ for which $\beta_Q(\log h) \geq t\, \rho_0(\log h)$. Then $\exists\, C > 0,\ m \geq 0\ \ \varepsilon_1 > 0$ such that

$$|f(h) - d_Q(h)^{-1} f_Q(h)| \leq C\, e^{-(1+\varepsilon_1 t)\rho_0(\log h)} (1 + \sigma(h))^m$$

for all $h \in A_0^+(Q:t)$.

To illustrate the power of this method of studying the asymptotics we mention the characterization due to Harish Chandra of eigenfunctions in $L^2(G)$. It is a consequence of the fact that the above error estimates are square integrable.

Theorem. Fix $f \neq 0$ in $\mathcal{A}(G:\tau)$. Then the following are equivalent:

(a) $f \in L^2(G:\tau)$

(b) G has compact center and $f_Q = 0$ for all psgrps $Q \neq G$

(c) $f \in \mathcal{C}(G:\tau)$.

It is also natural to ask whether one can not only define the constant term along Q but associate an entire perturbative expansion along Q. This question is not completely settled but it has been a fruitful line of investigation (cf. Harish Chandra [11], Trombi-Varadarajan [1], Trombi [1], [2], [3], Eguchi [1] etc.).

For a detailed treatment of the ideas of this lecture see Varadarajan [1], Harish Chandra [3].

6. Wave packets in Schwartz space

1. The next step in doing harmonic analysis is to investigate the decay properties at infinity on G of wave packets of matrix coefficients of tempered representations of G depending on a continuous parameter θ. The representations π_θ are generally assumed to act in a single Hilbert space H, and to possess infinitesimal characters; it is also convenient to assume

that the restrictions $\pi_\theta|_K$ are admissible and do not depend on θ. If F is a finite subset of $\hat{K}$ and H_F is defined as in §5.1, then we have the family of τ_F-spherical functions

$$\pi_F(x:\theta) = E_F\pi_\theta(x)E_F \qquad (x \in G)$$

which are eigenfunctions for $\mathfrak{Z}$; and one may begin the study of the wave packets

$$\int a(\theta)\pi_F(x:\theta)d\theta.$$

Let $P = MAN$ be a psgrp which is cuspidal, let $\omega \in \hat{M}_d$ and let $\pi_{P,\omega,\nu}$ ($\nu \in \mathfrak{a}^*$) be the family of representations introduced in §1. We choose a realization σ of ω in a Hilbert space $H(\sigma)$ and denote by H the Hilbert space of all (equivalence classes of) functions

$$f : K \to H(\sigma)$$

such that

(i) for each $k' \in K_M = K \cap M$,

$$f(k'k) = \sigma(k')f(k)$$

for almost all $k \in K$.

(ii) $\|f\|^2 = \int_K |f(k)|^2 dk < \infty$.

Right translations by elements of K define a unitary representation π_K of K in H. By Frobenius reciprocity it is clear that π_K is admissible; in fact, if $\mathfrak{d} \in \hat{K}$,

$$[\pi_K : \mathfrak{d}] = \sum_{\mathfrak{t}\in\hat{K}_M} [\mathfrak{d} : \mathfrak{t}][\sigma : \mathfrak{t}] < \infty$$

Let us now fix $\nu \in \mathfrak{a}_c^*$ and introduce the space $\Phi(\nu)$ of equivalence classes of functions

$$\varphi : G \to H(\sigma)$$

such that for each $p = man \in P$,

$$\varphi(px) = e^{(i\nu+\rho_P)(\log a)}\sigma(m)\varphi(x)$$

for almost all x. The space $\Phi(\nu)$ is stable under right translations by elements of G. Since $G = PK$, any $\varphi \in \Phi(\nu)$ is determined by its restriction to K and so, if

$$\|\varphi\|^2 = \int_K |\varphi(k)|^2 dk$$

then the subspace $\Phi_2(\nu)$ of $\Phi(\nu)$ of all φ with $\|\varphi\|^2 < \infty$ is a Hilbert space with $\|\cdot\|$ as its norm. The above action of G on $\Phi(\nu)$ leaves $\Phi_2(\nu)$ stable and defines a representation of G. If $\nu \in \mathfrak{a}^*$, this is a unitary representation and is in fact $\pi_{P,\omega,\nu}$. As the map

$$\Phi_2(\nu) \ni \varphi \mapsto \varphi|_K$$

is a unitary isomorphism of $\Phi_2(\nu)$ with H that takes $\pi_{P,\omega,\nu}|_K$ to π_K, we may transfer $\pi_{P,\omega,\nu}$ to H and thus guarantee that for these representations of G, π_K is their restriction to K.

Fix a finite set $F \subset \hat{K}$ and let U be the finite dimensional Hilbert space of endomorphisms of H_F, regarded as a bimodule for K as in §5.1; let τ_F denote the corresponding double representation of K. Define

$$\phi_{\omega,\nu,F}(x) = E_F \pi_{P,\omega,\nu}(x) E_F \qquad (x \in G).$$

The $\phi_{\omega,\nu,F}$ are τ_F-spherical eigenfunctions for $\mathfrak{Z}$, and for each $\nu \in \mathfrak{a}^*$, they are tempered.

Let us now describe the eigenhomomorphism to which $\phi_{\omega,\nu,F}$ belongs.

We select a θ-stable CSG L of G such that L_I is a compact CSG of M and $L = L_I A$ (recall $P = MAN$ is cuspidal). In the parametrization of §1 $\omega = \omega(b^*)$ for some irreducible character $b^* \in L_I^{*'}$. Let $\lambda = \log b^* + \rho_I$ where ρ_I is half the sum of positive roots of $(m, \mathfrak{l}_I)$ (with respect to some arbitrary but fixed positive system). The element λ is regular and varies in a lattice in $(-1)^{1/2}\mathfrak{l}_I^*$ when ω varies; for fixed ω it is unique upto the action of an element of $W(M,A_I)$. We also identify $\mathfrak{l}_{Ic}^*$ and $\mathfrak{l}_{Rc}^* = \mathfrak{a}_c^*$ with subspaces of $\mathfrak{l}^*$. Then $\Lambda = \lambda + i\nu$ is a well defined element of $\mathfrak{l}_c^*$ and we have the following.

__Lemma A.__ For $z \in \mathfrak{Z}$, $\nu \in \mathfrak{a}_c^*$,

$$z\phi_{\omega,\nu,F} = \mu_{\mathfrak{g}/\mathfrak{l}}(z)(\lambda + i\nu)\phi_{\omega,\nu,F}$$

We are especially interested in the case $\nu \in \mathfrak{a}^*$. For this we have the following result.

__Lemma B.__ (a) $\lambda \in (-1)^{1/2}\mathfrak{l}_I^*$ is regular in the sense that $\langle\alpha,\lambda\rangle \neq 0$ for each imaginary root α of $(\mathfrak{g}_c, \mathfrak{l}_c)$.

(b) $\lambda + i\nu \in (-1)^{1/2}\mathfrak{l}^*$ if $\nu \in \mathfrak{a}^*$

(c) Suppose $\nu \in \mathfrak{a}^*$ and $\lambda + i\nu$ is regular. Then $\mathfrak{l}$ is determined upto conjugacy. More precisely, let $\mathfrak{l}_j$ $(j = 1,2)$ be two θ-stable CSA's of $\mathfrak{g}$, $\Lambda_j \in (-1)^{1/2}\mathfrak{l}_j^*$, and suppose that Λ_1 and Λ_2 are regular; if

$$\mu_{\mathfrak{g}/\mathfrak{l}_1}(z)(\Lambda_1) = \mu_{\mathfrak{g}/\mathfrak{l}_2}(z)(\Lambda_2)$$

for all $z \in \mathfrak{Z}$, then one can find $k \in K$ such that $k\mathfrak{l}_1 = \mathfrak{l}_2$.

The point is that the condition on Λ_1 and Λ_2 expressed by (c) is governed __apriori__ only by the __complex__ adjoint group. This lemma shows that the regular parts of the spectra coming from the various series of representations are <u>disjoint</u>. It is the foundation on which one can build an orthogonal decomposition of $L^2(G)$ in terms of the wave packets associated with the various series.

I had remarked that for fixed $\nu \in \mathfrak{a}^*$, the $\phi_{\omega,\nu,F}$ are tempered, i.e., they satisfy the weak inequality. Actually they do much more; the constants involved in these estimates grow at most polynomially on ν. More precisely, we have, for all $\nu \in \mathfrak{a}^*$, $x \in G$

$$|\phi_{\omega,\nu,F}(x)| \leq C(1 + |\nu|)^r \; \Xi(x)(1 + \sigma(x))^r$$

for suitable $C = C_F > 0$, $r = r_F \geq 0$; and furthermore, such estimates are valid for the derivatives (with respect to x as well as ν) also. Finally, as we vary ν in the complex domain, the growth in ν is also well behaved; we have estimates of the form

$$|\phi_{\omega,\nu,F}(x)| \leq C(1 + |\nu|)^r \; \Xi(x)(1 + \sigma(x))^r \; e^{c|\nu_I|\sigma(x)}$$

for all $x \in G$, $\nu \in \mathfrak{a}_c^*$.

2. Motivated by the above considerations we shall introduce the theory of wave packets in a very general context. Actually we are not as general as we should be; we have assumed throughout that the double representation of K involved is finite dimensional. Ultimately one should vary it and an elegant way to do this is to consider possibly infinite dimensional double representations systematically from the very beginning, as is done by Harish Chandra. I decided to keep to the simpler framework since the main ideas may be understood well enough already in that context.

Let $\mathfrak{h}$ be a Θ-stable CSA of $\mathfrak{g}$; as usual we put $\mathfrak{h}_I = \mathfrak{h} \cap \mathfrak{k}$, $\mathfrak{h}_R = \mathfrak{h} \cap \mathfrak{s}$ where $\mathfrak{g} = \mathfrak{k} \oplus \mathfrak{s}$ is the Cartan decomposition of $\mathfrak{g}$ determined by Θ. We fix $\lambda \in (-1)^{1/2}\mathfrak{h}_I^*$ and assume it is regular, i.e.,

(1) $\langle\alpha,\lambda\rangle \neq 0$ for each imaginary root α of $(\mathfrak{g}_c, \mathfrak{h}_c)$.

We write

(2) $$\mathfrak{F} = \mathfrak{h}_R^*, \quad \mathfrak{F}_c = (\mathfrak{h}_R)_c^* = \text{complexification of } \mathfrak{F}.$$

We fix a finite dimensional Hilbert Space U and a unitary double representation τ of K on it. By an <u>eigenfunction of type $II(\lambda)$</u> we mean a function

$$\phi : \mathfrak{F} \times G \to U$$

with the following properties:

(3)
(i) ϕ is C^∞

(ii) For any $\nu \in \mathfrak{F}$, the function $\phi_\nu = \phi(\nu) = \phi(\nu : \cdot)$ from G to U is τ-spherical and

$$z\phi_\nu = \mu_{\mathfrak{g}/\mathfrak{h}}(z)(\lambda + i\nu)\phi_\nu \qquad (z \in \mathfrak{Z}).$$

(iii) For any $a_1, a_2 \in U(\mathfrak{g}_c)$ and any $\partial \in S(\mathfrak{F}_c)$, there are constants $C = C(a_1, a_2, \partial) > 0$ and $r = r(a_1, a_2, \partial) \geq 0$ such that

$$|\phi(\nu;\partial : a_1;x;a_2)| \leq C\, \Xi(x)(1 + \sigma(x))^r(1 + |\nu|)^r$$

for all $x \in G$, $\nu \in \mathfrak{F}$.

Here we use the usual interpretation of ∂ as a differential operator on $\mathfrak{F}$.

Let ϕ be a function of type $II(\lambda)$. For fixed $\nu \in \mathfrak{F}$, $\phi_\nu \in \mathbb{A}(G : \tau)$ and so, given any psgrp $P = MAN$ it makes sense (cf. §5) to speak of the constant term $(\phi_\nu)_P$. We put

(4) $$\phi_{P,\nu} = \phi_P(\nu) = (\phi_\nu)_P.$$

In studying the behaviour of ϕ_ν the idea is to use the perturbation theory of §5 but taking care that all estimates are uniform in ν. This is possible because the estimates in (3)(iii) above asserting that $\phi(\nu) \in \mathbb{A}(G:\tau)$ are actually uniform in ν.

Let us write

$$(5) \qquad \mathfrak{F}'(\lambda) = \{\nu \in \mathfrak{F} \mid \lambda + i\nu \text{ is regular}\}.$$

If we fix $\nu \in \mathfrak{F}$, the equations (13) of §5 show that $\phi_P(\nu)$ satisfies on MA the differential equations

$$(6) \qquad \mu_P(z)\phi_P(\nu) = \mu_{\mathfrak{g}/\mathfrak{h}}(z)(\lambda + i\nu)\phi_P(\nu) \qquad (z \in \mathfrak{z}).$$

If $\nu \in \mathfrak{F}'(\lambda)$, it is not difficult to deduce from this that $\phi_P(\nu)$ can be written as a sum of eigenfunctions for $\mathfrak{z}(\mathfrak{m}_1)$ on MA. Apriori one would expect this sum to be over the complex Weyl group; however, the assumption that $\phi(\nu)$ is tempered implies that only the real Weyl group comes in. To formulate this very basic result let us introduce some notation. Let $\mathfrak{a}$ be the Lie algebra of A. We write $\mathfrak{w}(\mathfrak{h}_R|\mathfrak{a})$ for the (possibly empty) set of linear injections s of $\mathfrak{a}$ into $\mathfrak{h}_R$ such that $s = \mathrm{Ad}(k)|_{\mathfrak{a}}$ for some $k \in K$. If $\mathfrak{a} = \mathfrak{h}_R$, we write $\mathfrak{w}(\mathfrak{a}) = \mathfrak{w}(\mathfrak{h}_R)$ for $\mathfrak{w}(\mathfrak{h}_R|\mathfrak{h}_R)$; it is a finite subgroup of $GL(\mathfrak{h}_R)$.

<u>Proposition C</u>. Let $\nu \in \mathfrak{F}'(\lambda)$. Then $\exists$ unique functions $\phi_{P,s}(\nu) \in \mathbb{A}(MA,\tau_M)$, $s \in \mathfrak{w}(\mathfrak{h}_R|\mathfrak{a}) = \mathfrak{w}$ with the following properties:

(i) $\phi_P(\nu)(m) = \Sigma_{s\in\mathfrak{w}}\, \phi_{P,s}(\nu)(m) \qquad (m \in MA)$

(ii) $\zeta\phi_{P,s}(\nu) = \mu_{\mathfrak{m}_1^s/\mathfrak{h}}(\zeta^s)(\lambda + i\nu)\phi_{P,s}(\nu) \qquad (\zeta \in \mathfrak{z}(\mathfrak{m}_1))$.

(We remark that $\mathfrak{m}_1^s$ and ζ^s are defined respectively as $\mathfrak{m}_1^k$ and ζ^k where $k \in K$ defines s; they are independent of the choice of k, and $\mathfrak{m}_1^s \supset \mathfrak{h}$).

Example. Let $\mathfrak{a} = \mathfrak{b}_R = \mathfrak{a}_0$ ($G = KA_0N_0$ is an Iwasawa decomposition) and $\phi(\nu : x) = \varphi(\nu : x)$, the elementary spherical function. Take $P = P_0 = M_0A_0N_0$, the minimal psgrp; then

(7) $$\phi_{P,s}(\nu)(h) = \underset{\sim}{c}(s^{-1}\nu)e^{is^{-1}\nu(\log h)} \qquad (\nu \in \mathfrak{a}_0^{*\prime})$$

where the $\underset{\sim}{c}(\cdot)$ is the c-function.

We also remark the following immediate consequence of (ii):

(8) $$\phi_{P,s}(ma) = \phi_{P,s}(m)e^{i(\nu\circ s)(\log a)} \qquad (m \in MA,\ a \in A,\ s \in w).$$

The example (7) suggests that when $\nu \in \mathfrak{F}'(\lambda)$ tends to a boundary point, $\phi_{P,s}(\nu)$ may blow up. To avoid this inconvenience we introduce the concept of regulated elements of type $II(\lambda)$. A ϕ of type $II(\lambda)$ is said to be regulated if the following is true. Given any psgrp $P = MAN$ and any $s \in \mathfrak{w}(\mathfrak{b}_R|\mathfrak{a})$, the function $(\phi_{P,s})^s$ which is well defined on $\mathfrak{F}'(\lambda) \times (MA)^s$ by

(9) $$(\phi_{P,s})^s(\nu : m^k) = \tau(k)\phi_{P,s}(\nu : m)\tau(k)^{-1} \qquad (m \in MA,\ Ad(k)|_{\mathfrak{a}} = s)$$

extends (uniquely) to a function of type $II(\lambda)$ on $\mathfrak{F} \times (MA)^s$. A regulated ϕ of type $II(\lambda)$ (on $\mathfrak{F} \times G$) is said to be of type $II_{reg}(\lambda)$. We have the following

Proposition D. (i) Let ϕ be of type $II_{reg}(\lambda)$ on $\mathfrak{F} \times G$. Then, for any psgrp $P = MAN$ and any $s \in \mathfrak{w}(\mathfrak{b}_R\ \mathfrak{a})$, $(\phi_{P,s})^s$ is of type $II_{reg}(\lambda)$ on $\mathfrak{F} \times (MA)^s$.

(ii) If ϕ is of type $II_{reg}(\lambda)$ on $\mathfrak{F} \times G$, and $P = MAN$ as above,

$$\phi_P(\nu) = \sum_{s\in\mathfrak{w}(\mathfrak{b}_R|\mathfrak{a})} \phi_{P,s}(\nu) \qquad (\nu \in \mathfrak{F}).$$

(iii) Let ϕ be of type $II(\lambda)$ on $\mathfrak{F} \times G$. Let

$$\psi(\nu : x) = \varpi(\lambda + i\nu)\phi(\nu : x) \qquad (\nu \in \mathfrak{F},\ x \in G)$$

where $\varpi = \prod_{\alpha>0} H_\alpha$ is the product of coroots of $(\mathfrak{g}_c, \mathfrak{h}_c)$ in a positive system. Then ψ is of type $II_{reg}(\lambda)$ on $\mathfrak{F} \times G$.

If we take a psgrp $P = MAN$ for which $\mathfrak{w}(\mathfrak{h}_R|\mathfrak{a})$ is empty, then $\phi_P(\nu) = 0$. By the transitivity of the constant terms this implies that if $\dim A = \dim \mathfrak{h}_R$, then all further constant terms of $\phi_P(\nu)|_M$ are zero so that $\phi_P(\nu)|_M$ is a cusp form. Note that in this case $\phi_P(\nu : \cdot)$, regarded as a function on M with values in U, is in $L^2(M : U)$ (recall that U is a Hilbert space.) As a special case of this we may take $P = MAN$ where $\mathfrak{a} = \mathfrak{h}_R$. The following proposition shows that the constant terms relative to such P already contain much of the information.

<u>Proposition E</u>. Let ϕ be of type $II(\lambda)$. Let $\mathcal{P}(\mathfrak{h}_R)$ denote the set of all psgrps $P = MAN$ with $\mathfrak{a} = \mathfrak{h}_R$. We then have the following.

(i) Fix $\nu \in \mathfrak{F}$. If $\phi_P(\nu) = 0$ for all $P \in \mathcal{P}(\mathfrak{h}_R)$, then $\phi(\nu) = 0$.

(ii) For ϕ to be of type $II_{reg}(\lambda)$ it is necessary and sufficient that the following be valid: for any $P \in \mathcal{P}(\mathfrak{h}_R)$ and any $s \in \mathfrak{w}(\mathfrak{h}_R|\mathfrak{h}_R)$, if $f_{P,s}(\nu)$ is the restriction to M of $\phi_{P,s}(\nu)$ $(\nu \in \mathfrak{F}'(\lambda))$, then $\|f_{P,s}(\nu)\|$ (norm in $L^2(M : U)$) should be locally bounded on $\mathfrak{F}$, i.e., should be bounded on every subset of the form $\mathfrak{L} \cap \mathfrak{F}'(\lambda)$ where $\mathfrak{L}$ is a compact subset of $\mathfrak{F}$.

3. Using these properties of eigenfunctions of type $II(\lambda)$ and $II_{reg}(\lambda)$ in conjunction with the perturbation theory of §5 (developed with uniformity in ν) one can prove the first and second wave packet theorems which are analogues of the corresponding theorems for spherical functions.

<u>Theorem 1</u>. Fix a function ϕ of type $II_{reg}(\lambda)$ on $\mathfrak{F} \times G$. For any $\alpha \in \mathcal{C}(\mathfrak{F})$ (= Schwartz space of $\mathfrak{F}$) let

$$\phi_\alpha(x) = \int_{\mathfrak{F}} \alpha(\nu)\phi(\nu : x)d\nu \qquad (x \in G)$$

Then ϕ_α is well defined and belongs to $\mathcal{C}(G:\tau)$. And

$$\alpha \mapsto \phi_\alpha$$

is a continuous map of $\mathcal{C}(\mathfrak{F})$ into $\mathcal{C}(G:\tau)$.

For the second wave packet theorem, we fix a psgrp $P = MAN$, ϕ being, as above, a function of type $II_{reg}(\lambda)$ on $\mathfrak{F} \times G$. For any $\alpha \in \mathcal{C}(\mathfrak{F})$ we form the "truncated wave packet"

(10)
$$\phi_{P,\alpha}(m) = \int_{\mathfrak{F}} \alpha(\nu)\phi_P(\nu)(m)d\nu \qquad (m \in MA).$$

It follows essentially from Proposition D that

(11)
$$\phi_{P,\alpha} \in \mathcal{C}(MA : \tau_M).$$

We extend $\phi_{P,\alpha}$ to a function on G by setting

(12)
$$\phi_{P,\alpha}(kmn) = \tau(k)\phi_{P,\alpha}(m) \qquad (k \in K,\ m \in MA,\ n \in N).$$

Take $\bar{P} = \theta(P)$, $\bar{N} = \theta(N)$.

<u>Theorem 2</u>. Let $d\bar{n}$ be a Haar measure on $\bar{N}$. Define

$$\phi_\alpha^{(\bar{P})}(m) = d_P(m)\int_{\bar{N}} \phi_\alpha(\bar{n}m)d\bar{n} \qquad (m \in MA).$$

Then, for all $m \in MA$,

$$\phi_\alpha^{(\bar{P})}(m) = \int_{\bar{N}} e^{-\rho_P(H_P(\bar{n}))}\phi_{P,\alpha}(\bar{n}m)d\bar{n} \qquad (\alpha \in \mathcal{C}(\mathfrak{F})).$$

Here ρ_P and H_P have their usual meanings. Thus $\rho_P(X) = \frac{1}{2}\mathrm{tr}(\mathrm{ad}\ X)_{\mathfrak{n}}$ $(X \in \mathfrak{a})$ and $H_P(kman) = \log a$ $(k \in K,\ m \in M,\ a \in A,\ n \in N)$.

<u>Corollary</u>. $\phi_\alpha^{(\bar{P})} = 0$ unless $\mathfrak{a}^k \subset \mathfrak{b}_R$ for some $k \in K$.

For the theory discussed in this lecture, see Harish Chandra [4].

7. The Eisenstein integral and the c-functions

1. We shall now apply the theory of constant terms and wave packets by choosing for ϕ the so-called Eisenstein Integral. To define it, let $\mathfrak{h}$ be a θ-stable CSA of $\mathfrak{g}$ and let notation be as in §6. We write L for the CSG corresponding to $\mathfrak{h}$ so that $L_I = L \cap K$, $L_R = \exp \mathfrak{h}_R$.

As before we denote by $\mathcal{P}(\mathfrak{h}_R)$ the finite set of psgrps $P = ML_RN$. Let τ be a unitary double representation of K in a finite dimensional Hilbert space U. We put $\tau_M = \tau|_{K_M}$ where, as usual, $K_M = K \cap M$. With ρ_P and H_P having their usual meanings we define, for any $g \in C^\infty(M : \tau_M)$, $\nu \in \mathfrak{F}_c$ $(=(\mathfrak{h}_R)^*_c)$, the Eisenstein integral

$$E(P : g : \nu : x)$$

by

$$\text{(1)} \qquad E(P : g : \nu : x) = \int_K e^{(i\nu-\rho_P)(H_P(xk))} g(xk)\tau(k^{-1})dk$$

for $x \in G$; here, g is extended to the whole of G by

$$\text{(2)} \qquad g(kman) = \tau(k)g(m) \qquad (k \in K,\ m \in M,\ a \in L_R,\ n \in N)$$

and we write $u\tau(k)$ for $u\tau_2(k)$ when $u \in U$, $k \in K$. When there is no doubt as to what P or g is, we abbreviate the notation $E(P : g : \nu : x)$ to $E(g : \nu : x)$ or $E(\nu : x)$.

The formula (1) is analogous to (1) of §3. In fact, if P is the minimal psgrp, $U = \mathbb{C}$, τ is the trivial double representation of K, and $g = 1$, $E(P : g : \nu : x)$ is just $\varphi(\nu : x)$. So there is a strong analogy of the Eisenstein integral with elementary spherical functions. On the other hand, suppose Γ is a discrete subgroup of G such that G/Γ has finite volume and that U is a bimodule for $K \times \Gamma$, with K acting on the left and Γ on the right; then, averaging over Γ instead of K will give the sum

$$(3) \qquad \sum_{\gamma \in P \cap \Gamma \backslash \Gamma} e^{(i\nu - \rho_P)(H_P(x\gamma))} g(x\gamma)\tau(\gamma^{-1})$$

which (for suitable g, Γ) is what is known in the theory of automorphic forms as an <u>Eisenstein series</u>. It is this analogy that prompted Harish Chandra to refer to (1) as the Eisenstein integral. Indeed, the theory of the Eisenstein integral is illuminated to a remarkable extent by the two analogies mentioned just now.

The Eisenstein integral is well defined on $\mathfrak{F}_c \times G$ for all $g \in C^\infty(M : \tau_M)$; it is holomorphic in $\nu \in \mathfrak{F}_c$ for fixed $x \in G$; for fixed $\nu \in \mathfrak{F}_c$ it is in $C^\infty(G : \tau)$. A simple calculation gives, for each $z \in \mathfrak{Z}$,

$$(4) \qquad E(P : g : \nu : x; z) = E(P : \mu_P(z)(i\nu)g : \nu : x)$$

where $\mu_P(z)(i\nu)$ is an element of $\mathfrak{Z}(\mathfrak{m})$ and is the value of $\mu_P(z)$ at $i\nu$ via the interpretation of elements of $\mathfrak{Z}(\mathfrak{m}_1)$ as polynomials on $\mathfrak{F}$ with values in $\mathfrak{Z}(\mathfrak{m})$ (recall $\mathfrak{Z}(\mathfrak{m}_1) \approx \mathfrak{Z}(\mathfrak{m}) \otimes U((\mathfrak{h}_R)_c)$). In particular, if g is an eigenfunction of $\mathfrak{Z}(\mathfrak{m})$, $E(g : \nu : \cdot)$ is an eigenfunction for $\mathfrak{Z}$. More precisely we have the following result that sets the stage for applying the theory of §§5 and 6.

<u>Proposition A.</u> Fix a regular $\lambda \in (-1)^{1/2}\mathfrak{h}_I^*$ and let $g \in \mathcal{C}(M : \tau_M)$ be an eigenfunction for $\mathfrak{Z}(\mathfrak{m})$ such that

$$\zeta g = \mu_{\mathfrak{m}_1/\mathfrak{h}}(\zeta)(\lambda)g \qquad (\zeta \in \mathfrak{Z}(\mathfrak{m})).$$

Then $E(P : g : \cdot : \cdot)$ is an eigenfunction of type $II(\lambda)$ on $\mathfrak{F} \times G$. Moreover, for each $a_1, a_2 \in U(\mathfrak{g}_c)$, $\partial \in S(\mathfrak{F}_c)$, we can find constants $C > 0$, $r \geq 0$, $c > 0$ such that for all $\nu \in \mathfrak{F}_c$, $x \in G$,

$$|E(P:g:\nu;\partial:a_1;x;a_2)| \le C\,\Xi(x)(1+\sigma(x))^r(1+|\nu|)^r e^{c|\nu_I|\sigma(x)}$$

Eigenfunctions in the Schwartz space of M are of course matrix coefficients of the discrete series of representations of M. One can also use other types of matrix coefficients of M in the Eisenstein integral; as long as they are tempered, the Eisenstein integral will satisfy the weak inequality. The special case considered above is however the important one and is decisive for our purposes.

2. It is well known that for a given class $\mathfrak{d} \in \hat{K}$ there are only finitely many discrete classes $\omega \in \hat{G}_d$ such that $[\omega : \mathfrak{d}] > 0$ (cf. Harish Chandra [7], Varadarajan [1]). Applying this result to M instead of G we see that the space spanned by the eigenfunctions for $\mathfrak{Z}(\mathfrak{m})$ in $\mathcal{C}(M:\tau_M)$ is finite dimensional. From Harish Chandra's theory of the discrete series one knows that these are also all the $\mathfrak{Z}(\mathfrak{m})$-finite functions in $\mathcal{C}(M:\tau_M)$, that the eigenhomomorphisms are defined by regular $\lambda \in (-1)^{1/2}\mathfrak{h}_I^*$, and that this is also the space ${}^o\mathcal{C}(M:\tau_M)$ of τ_M-spherical cusp forms (cf. Harish Chandra [3], Varadarajan [1], Part II, §§ 15, 16).

Put

$$V = {}^o\mathcal{C}(M:\tau_M). \tag{5}$$

Then $\dim(V) < \infty$. For any regular $\lambda \in (-1)^{1/2}\mathfrak{h}_I^*$ let $V[\lambda]$ be the subspace of V of functions defined by

$$V[\lambda] = \{g \in V \mid \zeta g = \mu_{\mathfrak{m}_1/\mathfrak{h}}(\zeta)(\lambda)g \;\forall\, \zeta \in \mathfrak{Z}(\mathfrak{m})\}. \tag{6}$$

For any discrete class $\omega \in \hat{M}_d$, let $L^2(M)_\omega$ be the Hilbert space spanned by the matrix coefficients of ω; and let

$$V[\omega] = V \cap (L^2(M)_\omega \otimes U) \tag{7}$$

where we are naturally identifying $C(M:\tau_M)$ as a subspace of $L^2(M)\otimes U$. This identification, since U is also a Hilbert space gives a natural Hilbert space structure on V. We have the orthogonal decompositions

$$V = \coprod_{\omega} V[\omega] = \coprod_{\lambda} V[\lambda]. \tag{8}$$

These are all finite since V is finite dimensional.

The theory of §6 now leads to the following theorem introducing the c-functions.

Theorem B. Let P_1, P_2 be psgrps in $\mathcal{P}(\mathfrak{h}_R)$, $\mathfrak{w} = \mathfrak{w}(\mathfrak{h}_R|\mathfrak{h}_R)$. Let $\mathfrak{F}'$ be the set of points ν of $\mathfrak{F}$ such that $\langle\nu,\alpha\rangle \neq 0$ for each root α of (P, L_R), $P \in \mathcal{P}(\mathfrak{h}_R)$. Then, for any $\nu \in \mathfrak{F}'$ and $s \in \mathfrak{w}$, we have uniquely defined endomorphisms

$$c_{P_2|P_1}(s:\nu)$$

of V such that

$$E_{P_2}(P_1:\psi:\nu:ma) = \sum_{s\in\mathfrak{w}} (c_{P_2|P_1}(s:\nu)\psi)(m)e^{is\nu(\log a)}$$

$\forall\, \psi \in V$, $m \in M$, $a \in L_R$.

The functions $c_{P_2|P_1}(s:\cdot)$ are certainly C^∞ on $\mathfrak{F}'$. Actually they are much nicer. The point is that the estimates furnished by Proposition A for the Eisenstein integral when ν varies in $\mathfrak{F}_c$ allows us to do the perturbation theory of §5 not only for $\nu \in \mathfrak{F}$ but for ν in a sufficiently small tubular domain $\mathfrak{F}_c(\delta)$ containing $\mathfrak{F}$ where

$$\mathfrak{F}_c(\delta) = \{\nu \in \mathfrak{F}_c \mid |\nu_I| < \delta\} \tag{9}$$

Let $P \in \mathcal{P}(\mathfrak{h}_R)$ and let

$$\pi(\nu) = \prod_{1\le i\le q} \langle\alpha_i,\nu\rangle \qquad (\nu \in \mathfrak{F}_c) \tag{10}$$

where $\alpha_1,\ldots,\alpha_q$ are all the distinct roots of (P,L_R), m_i being the multiplicity of α_i (changing P changes π only by ± 1). Then we get the following consequences of doing the perturbation theory in $\mathfrak{F}_c(\delta)$ for some $\delta > 0$.

Theorem C. There is some $\delta > 0$ such that

$$\pi(\ )c_{P_2|P_1}(s:\ )$$

extends to a holomorphic function on $\mathfrak{F}_c(\delta)$, for all $P_1,P_2 \in \mathcal{P}(\mathfrak{h}_R)$ and $s \in \mathfrak{w}$.

For any $P \in \mathcal{P}(\mathfrak{h}_R)$, $P = ML_RN$, we have $G = K\exp(\mathfrak{m} \cap \mathfrak{s})L_RN$. For $x \in G$, we write $\kappa(x)$, $\mu(x)$, $\exp H_P(x)$, and $n(x)$ for the components of x in K, $\exp(\mathfrak{m} \cap \mathfrak{s})$, L_R and N respectively.

Theorem D. Fix $P \in \mathcal{P}(\mathfrak{h}_R)$ and let $\mathfrak{F}_c(P)$ be the set of all $\nu \in \mathfrak{F}_c$ such that $\langle \nu_I,\alpha\rangle > 0\ \forall$ roots α of (P,L_R). Then $c_{\bar{P}|P}(1:\nu)$ and $c_{\bar{P}|P}(1:-\nu)$ extend to holomorphic functions on $\mathfrak{F}_c(P)$; there, they are given by the following (convergent) integrals:

$$(c_{\bar{P}|P}(1:\nu)\psi)(m) = \int_{\bar{N}} \tau(\kappa(\bar{n}))\psi(\mu(\bar{n})m)e^{(i\nu-\rho_P)(H_P(\bar{n}))}\,d\bar{n}$$

$$(c_{P|P}(1:-\nu)\psi)(m) = \int_{\bar{N}} \psi(m\mu(\bar{n})^{-1})\tau(\kappa(\bar{n}))^{-1}\, e^{(i\nu-\rho_P)(H_P(\bar{n}))}\,d\bar{n}.$$

Here $\psi \in V$, $\nu \in \mathfrak{F}_c(P)$, $m \in N$, and $d\bar{n}$ is normalized by $\int_{\bar{N}} e^{-2\rho_P(H_P(\bar{n}))}\,d\bar{n} = 1$.

The proof of Theorem D resembles closely the proof of the analogous result for the c-functions that occur in the spherical theory. We start with an eigenfunction ϕ of type $II(\lambda)$ on $\mathfrak{F} \times G$ but assume that it is actually given on some $\mathfrak{F}_c(\delta_0) \times G$, holomorphic in ν, and satisfying estimates of the kind introduced in Proposition A. Let $P \in \mathcal{P}(\mathfrak{h}_R)$ and let $H_0 \in \mathfrak{h}_R$ be such that $\alpha(H_0) > 0 \;\forall$ roots α of (P, L_R). Then we can find $\delta > 0$ and $\varepsilon = \varepsilon(\delta) > 0$ such that

$$d_P(m_t)\phi(\nu : m_t) - \sum_{s \in \mathfrak{w}} \phi_{P,s}(\nu : m) e^{it\nu(sH_0)} = O(e^{-\varepsilon t})$$

as $t \to +\infty$, $\nu \in \mathfrak{F}_c'(\delta, \lambda)$ (= set of $\nu \in \mathfrak{F}_c(\delta)$ with $\varpi(\lambda + i\nu) \neq 0$), $m_t = m \exp t H_0$ where $m \in ML_R$. In particular, if $s_0 \in \mathfrak{w}_0$ is such that $\nu_I(s_0 H_0) < \nu_I(sH_0) \;\forall\, s \in w,\; s \neq s_0$,

(11) $$\lim_{t \to +\infty} d_P(m_t) e^{-it\nu(s_0 H_0)} \phi(\nu : m_t) = \phi_{P,s_0}(\nu : m)$$

giving a method of explicit determination for a particular $\phi_{P,s_0}(\nu : m)$. This technique is applied to $\phi = E(P : \psi : \nu)$ where $P \in \mathcal{P}(\mathfrak{h}_R)$, $\psi \in V$. To compute the limit (11) we first transform the integral over K/K_M that represents ϕ into an integral over $\bar{N}$. This gives, with

(12) $$\nu_+ = i\nu + \rho_P, \quad \nu_- = i\nu - \rho_P,$$

the formula, valid for $m \in ML_R$:

(13) $$e^{\nu_+(H_P(m))} \phi(m^{-1}) = \int_{\bar{N}} \psi(\bar{n} m^{-1} \mu(\bar{n}^m)^{-1}) \tau(\kappa(\bar{n}^m))^{-1} e^{\nu_-(H_P(\bar{n})) - \nu_+(H_P(\bar{n}^m))} d\bar{n}.$$

Choose $m = m_0^{-1} a^{-1}$ where $m_0 \in M$, $a \in L_R$, $a \overset{\bar{P}}{\to} \infty$. We get

(14) $$\lim_{\substack{\bar{P} \\ a \to \infty}} e^{-(i\nu + \rho_P)(\log a)} \phi(m_0 a) = \int_{\bar{N}} \tau(\kappa(\bar{n})) \psi(\mu(\bar{n}) m_0) e^{\nu_-(H_P(\bar{n}))} d\bar{n}.$$

(This needs some delicate analysis on $\bar{N}$). By (11), the left side is $\phi_{\bar{P},1}(\nu : m_0) = (c_{\bar{P}|P}(1 : \nu)\psi)(m_0)$.

<u>Corollary 1</u>. Fix $\omega \in \hat{M}_d$. Then $V[\omega]$ is stable under $c_{\bar{P}|P}(1 : \nu)$ and $c_{P|P}(1 : \nu)$.

<u>Corollary 2</u>. $\det c_{P|P}(1 : \nu)$ is not identically zero.

As a much deeper consequence we mention the fact that the second wave packet theorem now yields an explicit formula.

<u>Theorem E</u>. For $\psi \in V$, $\alpha \in C_c^\infty(\mathfrak{F}')$, $P_1, P \in \mathcal{P}(\mathfrak{h}_R)$ and let

$$\phi_\alpha = \int_{\mathfrak{F}} \alpha(\nu)E(P_1 : \psi : \nu : \cdot)d\nu.$$

Then, with $d\bar{n}$ on $\bar{N}$ being normalized as above, we have for $m \in M$, $a \in L_R$,

$$\phi_\alpha^{(\bar{P})}(ma) = \int_{\mathfrak{F}} e^{i\nu(\log a)} \sum_{s \in \mathfrak{w}} \alpha(s^{-1}\nu)(c_{\bar{P}|P}(1 : \nu)c_{P|P_1}(s : s^{-1}\nu)\psi)(m)d\nu.$$

If we had more information on the c-functions - such as functional equations satisfied by them, the above formula would be a very substantial step in the explicit determination of the Plancherel measure. This needs a deeper study of the Eisenstein integral. In the next lecture I shall sketch the outlines of this study and indicate how one can obtain an explicit Plancherel formula from it.

To sketch at least formally the argument for deriving Theorem E we proceed as follows. We have, for $m \in ML_R$, as $\rho_P(H_P(\bar{n})) \geq 0$,

$$\phi_\alpha^{(\bar{P})}(m) = \lim_{\varepsilon \to 0+} \int_{\bar{N}} e^{-(1+\varepsilon)\rho_P(H_P(\bar{n}))} \phi_{P,\alpha}(\bar{n}m)d\bar{n}$$

while

$$\phi_{P,\alpha}(\bar{n}m) = \int_{\mathfrak{F}} \alpha(\nu)E_P(P_1 : \psi : \nu : \bar{n}m)d\nu$$

$$= \int_{\mathfrak{F}} \alpha(\nu)\tau(\kappa(\bar{n}))E_P(P_1 : \psi : \nu : \mu(\bar{n})m \exp H_P(\bar{n}))d\nu$$

For fixed $\varepsilon > 0$, we observe that for any $\mu \in \mathfrak{F}$, $\mu_\varepsilon = \mu + i\varepsilon\rho_P$ is in $\mathfrak{F}_c(P)$. Hence, for $m \in M$, $a \in L_R$,

$$\int_{\bar{N}} e^{-(1+\varepsilon)\rho_P(H_P(\bar{n}))} \phi_{P,\alpha}(\bar{n}m)d\bar{n} =$$

$$= \int_{\mathfrak{F}} \alpha(\nu)\left(\int_{\bar{N}} \tau(\kappa(\bar{n}))\sum_s (c_{P|P_1}(s:\nu)\psi)(\mu(\bar{n})m)e^{is\nu(H_P(\bar{n}))+is\nu(\log a)-(1+\varepsilon)\rho_P(H_P(\bar{n}))} d\bar{n}\right)d\nu$$

$$= \int_{\mathfrak{F}} \alpha(\nu)\left(\int_{\bar{N}} \tau(\kappa(\bar{n}))\sum_s e^{(i(s\nu)_\varepsilon-\rho_P)(H_P(\bar{n}))+is\nu(\log a)} (c_{P|P_1}(s:\nu)\psi)(\mu(\bar{n})m)d\bar{n}\right)d\nu$$

$$= \int_{\mathfrak{F}} \alpha(\nu) \sum_s (c_{\bar{P}|P}(1:(s\nu)_\varepsilon)c_{P|P_1}(s:\nu)\psi)(m)e^{is\nu(\log a)}d\nu .$$

Letting $\varepsilon \to 0+$, we get

$$\phi_\alpha^{(\bar{P})}(ma) = \sum_s \int_{\mathfrak{F}} \alpha(\nu)(c_{\bar{P}|P}(1:s\nu)c_{P|P_1}(s:\nu)\psi)(m)e^{is\nu(\log a)}$$

which is equivalent to Theorem E.

8. The Plancherel measure and the Plancherel formula

1. Our first objective is to state precisely the main theorems of Harish Chandra's article [5] on the Plancherel formula.

To begin with we work with a fixed θ-stable CSA $\mathfrak{h} \subset \mathfrak{g}$. We put $\mathfrak{a} = \mathfrak{h}_R$ so that $\mathfrak{F} = \mathfrak{a}^*$, and use the notation of §§6 and 7 without further comment; in particular we recall that $\mathfrak{F}'$ is the set of $\nu \in \mathfrak{F}$ for which $\langle\nu,\alpha\rangle \neq 0$ for all roots α of (P,A) for some (hence every) psgrp $P \in \mathcal{P}(A)$. As usual $\mathfrak{w} = \mathfrak{w}(\mathfrak{a}|\mathfrak{a})$.

Theorem A. Let $P = MAN \in \mathcal{P}(A)$, $\omega \in \hat{M}_d$, $\nu \in \mathfrak{F}$. Then the equivalence class of the unitary representation $\pi_{P,\omega,\nu}$ is independent of P and does not change if (ω,ν) is replaced by $(s\omega,s\nu)$ for $s \in w$. Moreover, if $\nu \in \mathfrak{F}'$, $\pi_{P,\omega,\nu}$ is irreducible.

Theorem B. Let $P_1,P_2 \in \mathcal{P}(A)$, $s \in \mathfrak{w}$, $\nu \in \mathfrak{F}'$, $\omega \in \hat{M}_d$. Let τ be a finite dimensional unitary double representation of K and $V = {}^\circ\mathcal{C}(M:\tau_M)$.

Then $c_{P_2|P_1}(s:\nu)$ defines a dijection of $V[\omega]$ onto $V[s\omega]$. Moreover, there is a (unique) function $\mu > 0$ on $\hat{M}_d \times \mathfrak{F}'$ such that for all τ, P_1, P_2, $\omega \in \hat{M}_d$, $\nu \in \mathfrak{F}'$

$$c_{P_2|P_1}(s:\nu)^{\dagger}\, c_{P_2|P_1}(s:\nu) = \mu(\omega:\nu)^{-1} \cdot \mathrm{id}_{\omega}$$

on $V[\omega]$, where id_{ω} denotes the identity operator on $V[\omega]$; and

$$\mu(s\omega : s\nu) = \mu(\omega:\nu) \qquad (s \in \mathfrak{w},\ \omega \in \hat{M}_d,\ \nu \in \mathfrak{F}').$$

<u>Theorem C</u>. Let τ be as in the previous theorem. We then have the following.

(i) For $P_1, P_2 \in \mathcal{P}(A)$, $s \in \mathfrak{w}$, $c_{P_2|P_1}(s:\nu)$ extends to a meromorphic function on $\mathfrak{F}_c$.

(ii) Put

$$^{\circ}c_{P_2|P_1}(s:\nu) = c_{P_2|P_2}(1:s\nu)^{-1}\, c_{P_2|P_1}(s:\nu).$$

Then $^{\circ}c_{P_2|P_1}(s:\nu)$ is holomorphic and unitary everywhere on $\mathfrak{F}$, and maps $V[\omega]$ onto $V[s\omega]$. Moreover, if $P, P_1, P_2 \in \mathcal{P}(A)$, $s, t \in w$,

$$^{\circ}c_{P_2|P_1}(st:\nu) = {}^{\circ}c_{P_2|P}(s:t\nu)\, {}^{\circ}c_{P|P_1}(t:\nu).$$

(iii) The Eisenstein integrals satisfy the functional equations

$$E(Q : {}^{\circ}c_{Q|P}(s:\nu)\psi : s\nu) = E(P:\psi:\nu)$$

for $P, Q \in \mathcal{P}(A)$, $s \in \mathfrak{w}$, $\psi \in V$.

The next theorem describes the analytic properties of the function μ.

<u>Theorem D</u>. For any $\omega \in \hat{M}_d$, $\mu(\omega:\cdot)$ is the restriction to $\mathfrak{F}$ of a meromorphic function on $\mathfrak{F}_c$ (denoted by $\mu(\omega:\cdot)$ also). Moreover, $\exists\, \delta > 0$ such that

(i) $\mu(\omega:\cdot)$ is holomorphic on $\mathfrak{F}_c(\delta)$

(ii) for constants $b > 0$, $r \geq 0$, we have an estimate

$$|\mu(\omega : \nu)| \leq b(1 + |\nu_R|)^r \qquad (\nu \in \mathfrak{F}_c(\delta))$$

Let $\Gamma_1, \ldots, \Gamma_r$ be a complete set of θ-stable CSG's of G, no two of which are conjugate. Put $A_i = (\Gamma_i)_R$ and let S be the set $(A_1, \ldots, A_r)$.

<u>Theorem E</u>. There are constants $c(A) > 0$ $(A \in S)$ with the following property. Let τ be any finite dimensional unitary double representation of K. Then, for any $f \in C(G : \tau)$,

$$f(1) = \sum_{A \in S} c(A) \sum_{\omega \in \hat{M}_d} d(\omega) \int_{\mathfrak{F}} \mu(\omega : \nu)(\Theta_{\omega,\nu}, f) d\nu.$$

Here, for given $A \in S$, MA is the centralizer of A in G; for $\omega \in \hat{M}_d$, $d(\omega)$ denotes the formal degree of the discrete class ω (using some normalization of Haar measure on M); and

$$(\Theta_{\omega,\nu}, f) = \int_G \Theta_{\omega,\nu}(x)^{\text{conj}} f(x) dx.$$

If one chooses specific Haar measures, say standard ones, then the constants $c(A)$ can be explicitly evaluated. From this, the usual L^2-version of the Plancherel formula is quite easy to derive. A more careful treatment would show that this formula is true for all $f \in C(G)$ (not only for K-finite f).

It is clearly important and useful to investigate whether there are explicit formulae for μ. Actually there is a <u>product formula</u> for $\mu(\omega : \cdot)$ which is a far reaching generalization of the product formula for the spherical Plancherel measure that Gindikin and Karpelevič established. To formulate this we need some notation.

A root α of $(\mathfrak{g}, \mathfrak{a})$ is called <u>reduced</u> if $r\alpha$ is not a root for $0 < r < 1$ $(r \in \mathbb{R})$. For any psgrp $P \in \mathcal{P}(A)$ we write $\Sigma(P)$ for the set of reduced roots of (P, A). If $P_1, P_2 \in \mathcal{P}(A)$, we write $\Sigma(P_2|P_1)$ for the set $\Sigma(\bar{P}_2) \cap \Sigma(P_1)$ $(\bar{P}_2 = \theta(P_2))$. If $d(P_1, P_2) = [\Sigma(P_2|P_1)]$ ([ ] is the cardinality), d is a metric on $\mathcal{P}(A)$.

Suppose $P \in \mathcal{P}(A)$ and $\alpha \in \Sigma(P)$. Let $\mathfrak{a}(\alpha)$ be the hyperplane in $\mathfrak{a}$ of zeros of α; Z_α, the centralizer of $\mathfrak{a}(\alpha)$ in G; $A_\alpha = \exp \mathbb{R} H_\alpha$; $N_\alpha = \exp \Sigma_{k\geq 1} \mathfrak{g}_{k\alpha}$. Then $Z_\alpha = M_\alpha A_\alpha$ and ${}^*P_\alpha = M A_\alpha N_\alpha$ is a psgrp of M_α (here $M_\alpha = {}^0Z_\alpha$ as usual; see Varadarajan [1], Part II, p. 20 for the definition of the function $L \mapsto {}^0L$; see also Harish Chandra [3], §2). For $\omega \in \hat{M}_d$, let $\mu_\omega^{(\alpha)}$ be the function defined as above on $\mathfrak{a}_\alpha^*$ and let

$$\mu_\alpha(\omega : \nu) = \mu_\omega^{(\alpha)}(\nu_\alpha) \qquad \nu_\alpha = \nu|_{\mathfrak{a}_\alpha}$$

The product formula for $\mu(\omega : \cdot)$ is then given by the following theorem.

<u>Theorem F</u>. $\exists$ a constant $c_1 > 0$ such that

$$\mu(\omega : \nu) = c_1 \prod_{\alpha\in\Sigma(P)} \mu_\alpha(\omega : \nu) \qquad (\nu \in \mathfrak{F})$$

for all $P \in \mathcal{P}(A)$, $\omega \in \hat{M}_d$.

In view of this theorem, the problem of explicit calculation of the Plancherel measure reduces to the case when $\dim(A) = 1$, i.e., P is maximal. We distinguish two cases according to whether $\mathrm{rk}(G) > \mathrm{rk}(K)$ or $\mathrm{rk}(G) = \mathrm{rk}(K)$; the first alternative is equivalent (when $\dim A = 1$) to the CSA $\mathfrak{h}$ being fundamental in $\mathfrak{g}$. We recall that for a fundamental CSA there are no real roots.

<u>Theorem G_1</u>. Suppose $\mathfrak{h}$ is fundamental in $\mathfrak{g}$ (but $\dim A$ arbitrary). Let R^+ be a positive system of roots of $(\mathfrak{g}_c, \mathfrak{h}_c)$, and let R_c^+ be the subset of R^+ consisting of complex (= nonimaginary) roots. We can always choose R^+ so that R_c^+ is stable under complex conjugation. With R^+ chosen thus, let $[R_c^+] = 2p$, and let $\varpi_+ = \prod_{\alpha\in R_c^+} H_\alpha$; then

$$(-1)^p \varpi_+(\Lambda) \text{ is real and } \geq 0$$

for all $\Lambda \in i\mathfrak{h}^*$. Moreover, there is a constant $c_2 > 0$ such that for all $\omega \in \hat{M}_d$, $\nu \in \mathfrak{F}$,

$$\mu(\omega : \nu) = (-1)^p c_2 \varpi_+(\lambda(\omega) + i\nu)$$

where $\lambda(\omega) \in (-1)^{1/2}\mathfrak{h}_I^*$ corresponds to ω in the usual parametrization of $\hat{M}_d$.

Remark. When $\mathfrak{h}$ is fundamental, the series of representations $\pi_{P,\omega,\nu}$ $(P \in \mathcal{P}(A))$ is called the fundamental series of G. These representations were proved by Harish Chandra to be irreducible for all $\nu \in \mathfrak{F}$ in [5]. For complex groups this had been established much earlier by Wallach [1] and Zhelebenko [1], the case of $SL(n,C)$ going back to Gel'fand and Neumark [1].

Recent work on representations from the infinitesimal point of view has led to an alternative approach to the fundamental series (cf. Varadarajan [3], Enright [1] and Enright-Wallach [1]). This approach makes it possible to treat the case of complex ν and to determine completely the k-multiplicities of these representations.

To complete the explicit determination of $\mu(\omega : \cdot)$ when $\dim(A) = 1$ it remains to consider the case $rk(G) = rk(K)$. We select a positive system R^+ of roots of $(\mathfrak{g}_c, \mathfrak{h}_c)$ such that the set R_c^+ of complex (= nonreal, non-imaginary) roots is stable under complex conjugation. We put, for $\omega \in \hat{M}_d$, $\nu \in \mathfrak{F}$,

$$\varpi_+(\omega : \nu) = \varpi_+(\lambda(\omega) + i\nu)$$

where $\varpi_+ = \Pi_{\alpha \in R_c^+} H_\alpha$ and $\lambda(\omega) \in (-1)^{1/2}\mathfrak{h}_I^*$ is the parameter corresponding to ω. If $[R_c^+] = 2p$, we have

$$(-1)^p \varpi_+(\omega : \nu) \geq 0.$$

For any irreducible character $a^* \in L_I^*$ (L is the CSG defined by $\mathfrak{h}$) let

σ_{a*} be a representation of L_I with character a^* and let

$$\mu_0(a^* : \nu) = d(a^*)^{-1} \operatorname{tr}\left\{\frac{\pi\nu_\alpha \sinh \pi\nu_\alpha}{\cosh \pi\nu_\alpha - \frac{(-1)^{\rho_\alpha}}{2}(\sigma_{a*}(\gamma) + \sigma_{a*}(\gamma^{-1}))}\right\}$$

where $\nu_\alpha = \frac{2\langle\nu,\alpha\rangle}{\langle\alpha,\alpha\rangle}$, α is the unique real root in R^+, and γ is a certain element of order atmost 2 in L_I; $d(a^*)$ is the dimension of σ_{a*}. Put

$$\mu_0(\omega : \nu) = [W(M,L_I)]^{-1} \sum_{s\in W(M,L_I)} \mu_0(sa^* : \nu).$$

Theorem G_2. Let $\dim(A) = 1$ but $rk(G) = rk(K)$. Then $\exists$ a constant $c_3 > 0$ such that for all $\omega \in \hat{M}_d$, $\nu \in \mathfrak{F}$,

$$\mu(\omega : \nu) = (-1)^p c_3\, \mu_0(\omega : \nu)\, \varpi_+(\omega : \nu)$$

with notation as above.

2. It is clearly not possible to discuss except in outline how these theorems are proved.

The starting point is to identify Eisenstein integrals with matrix coefficients of the $\pi_{P,\omega,\nu}$. To do this we introduce $C(K \times K) = C^\infty(K \times K)$ with the pre Hilbertian structure determined by

$$|v|^2 = \int |v(k_1 : k_2)|^2 dk_1 dk_2.$$

It carries the double representation τ defined by

$$(\tau(k')v\tau(k''))(k_1 : k_2) = v(k_1k' : k''k_2).$$

For any finite set $F \subset \hat{K}$

$$U = U_F = C_F(K \times K)$$

is the finite dimensional subspace of all v such that

$$v = \int \alpha_F(k)\tau(k)v\,dk = \int \alpha_F(k)v\tau(k)\,dk$$

where $\alpha_F(k) = \Sigma_{\mathfrak{d}\in F} \dim(\mathfrak{d})\overline{ch_{\mathfrak{d}}(k)}$, $ch_{\mathfrak{d}}$ being the character of $\mathfrak{d}$. U_F is stable under τ and we put $\tau_F = \tau|_{U_F}$. By varying F we obtain a family of double representations such that every finite dimensional double representation of K is contained as a direct summand of a direct sum of these.

Fix now $\omega \in \hat{M}_d$, choose a representative σ in ω acting in a Hilbert space $H(\sigma)$ and let H be the Hilbert space of the representation (of K) $\mathrm{Ind}_{K_M}^{K}(\sigma)$ as in §6.1. Fix a finite $F \subset \hat{K}$, take $U \doteq U_F$ as above, and define V as usual by $V = {}^{\circ}C(M : \tau_M)$. The endomorphisms T of H_F are then (uniquely) represented by C^{∞} kernels κ_T

$$\kappa_T : K \times K \to \mathrm{End}^{\circ}(H(\sigma))$$

($\mathrm{End}^{\circ}(H(\sigma))$ means $\mathrm{End}(H(\sigma))_R$ for a finite subset $R \subset \hat{K}_M$) so that for all $h \in H_F$,

$$(Th)(k_2) = \int \kappa_T(k_2 : k_1)h(k_1^{-1})dk_1;$$

apart from these smoothness and finiteness conditions, κ_T must satisfy the symmetry conditions

$$\kappa_T(m_2k_2 : m_1k_1) = \sigma(m_2)\kappa_T(k_2 : k_1)\sigma(m_1).$$

We can then form the function

$$\psi_T : M \to U$$

defined by

$$\psi_T(m)(k_1 : k_2) = \mathrm{tr}(\kappa_T(k_2 : k_1)\sigma(m)) \qquad (m \in M).$$

Then the basic result is that

$$\psi : T \mapsto \psi_T$$

is a linear bijection

$$\mathrm{End}(H_F) \xrightarrow{\sim} V[\omega]$$

and has the property

$$(1) \qquad E(P : \psi_T : \nu : x)(k_1 : k_2) = \mathrm{tr}(T\, \pi_{P,\omega,\nu}(k_1 x k_2))$$

for all $x \in G$, $k_1, k_2 \in K$.

All the representations $\pi_{P,\omega,\nu}$ act on H. This allows one to define intertwining operators between them as operators on H. If $P_1, P_2 \in \mathcal{P}(A)$, $\nu \in \mathfrak{F}_c$, the operator $J_{P_2|P_1}(\nu)$ (see §6.1 for the spaces $\Phi(\nu)$)

$$J_{P_2|P_1}(\nu) : \Phi_{P_1}(\nu) \to \Phi_{P_2}(\nu)$$

is defined <u>formally</u> by

$$(J_{P_2|P_1}(\nu)h)(x) = \gamma(P_2|P_1) \int_{N_2 \cap \bar{N}_1} h(\bar{n}x)\,d\bar{n}$$

The integrals converge only when ν is in suitable domains of $\mathfrak{F}_c$ and there is therefore the problem of analytic continuation; the $\gamma(P_2|P_1)$ are constants > 0 chosen so that the J's have the product property

$$J_{P_2|P_1}(\nu) = J_{P_2|P}(\nu)\, J_{P|P_1}(\nu)$$

if P is <u>between</u> P_1 and P_2 in the sense that $d(P_1, P_2) = d(P_1, P) + d(P, P_2)$. For any finite $F \subset \hat{K}$, we define

$$(2) \qquad {}_F j_{P_2|P_1}(\nu) = j_{P_2|P_1}(\nu) = J_{P_2|P_1}(\nu)\big|_{H_F}.$$

The main point is that the integral representation of the j-functions is essentially the same as the integral representations of the c-functions given in Theorem D of §7, for $U = U_F$ and $V = V_F$ as above. For instance, using the isomorphism ψ, we have, for all $T \in \mathrm{End}(H_F)$ and suitable ν,

$$c_{P|P}(1:\nu)\psi_T = c(A)\psi_{jT}$$

where $c(A)$ is a constant > 0 and

$$j = j_{\overline{P}|P}(\nu).$$

Moreover,

$$j^* = j_{P|\overline{P}}(\nu)$$

and

$$c_{\overline{P}|P}(1:\nu)\psi_T = c(A)\psi_{Tj^*}$$

(see §11, Harish Chandra [5]).

This link between intertwining operators and c-functions is fundamental. It allows on the one hand to analytically continue the intertwining operators since, as we mentioned in §7, the perturbation theory already gives analytic continuation of the c-functions. On the other hand, the j-functions have product properties which can now be transferred to the c-functions, and hence to the μ-functions. This circumstance is the source of the product representation of the Plancherel measure.

Before doing the explicit computations it is necessary to derive the functional equations for the Eisenstein integrals. Harish Chandra does this via what he calls the Maass-Selberg relations (originally obtained in the context of the theory of Eisenstein series). These relations are as follows. Let $f \in \mathcal{A}(G:\tau)$ and fix $\nu \in \mathfrak{F}'$; suppose that f has the following two properties (that are certainly possessed by the Eisenstein integrals corresponding to ν):

(a) if $P' = M'A'N'$ is a psgrp where A' is not conjugate to A, the constant term $f_{P'}$ is ~ 0 in the sense that $f_{P',a'}$ is orthogonal to all cusp forms of M' for each $a' \in A'$

(b) $f_P(ma) = \Sigma_{s\in \mathfrak{w}} f_{P,s}(m)e^{is\nu(\log a)}$ $(m \in M,\ a \in A)$ for suitable $f_{P,s} \in {}^{\circ}\mathcal{C}(M : \tau_M)$. Then

$$\|f_{P_1,s_1}\| = \|f_{P_2,s_2}\|$$

for arbitrary $P_1, P_2 \in \mathcal{P}(A)$, $s_1, s_2 \in \mathfrak{w}$. In particular, if we know that for two such f, say f_1, f_2, one knows that $(f_1)_{P,s} = (f_2)_{P,s}$ for <u>some</u> (P,s), this implies that $f_1 = f_2$ (actually, for $g = f_1 - f_2$, we have $g_Q = 0$ for all $Q \in \mathcal{P}(A)$; coupled with (a), one easily gets $g = 0$). The functional equations of the Eisenstein integral follow immediately. For, with the adjustment factors ${}^{\circ}c$, the function

$$f = E(P : \psi : \nu) - E(Q : {}^{\circ}c_{Q|P}(s : \nu)\psi : s\nu)$$

$(\psi \in V[\omega])$ has properties (a), (b) described above, and in addition $f_{Q,s} = 0$, so that f must be $= 0$.

The detailed information regarding the μ and c-functions and the Eisenstein integrals allows us to simplify the (second wave packet theorem) Theorem E of §7 (see Harish Chandra [5], Theorem 20.1). We put, with $P_1, P_2 \in \mathcal{P}(A)$

(3) $$\phi_\alpha = \int_{\mathfrak{F}} \mu(\omega : \nu)\alpha(\nu)E(P_1 : \psi : \nu)d\nu \qquad (\alpha \in C_c^\infty(\mathfrak{F}))$$

$(\psi \in V[\omega])$. Then $\phi_\alpha \in \mathcal{C}(G : \tau)$ and one has, for $m \in M$, $a \in A$,

(4) $$\phi_\alpha^{(P_2)}(ma) = c\int_{\mathfrak{F}} e^{i\nu(\log a)} \sum_{s\in\mathfrak{w}} \alpha(s^{-1}\nu)\,{}^{\circ}c_{P_2|P_1}(s : s^{-1}\nu)\psi\, d\nu$$

where $c > 0$ is a constant. If $\hat{\phi}_\alpha^{(P_2)}(\nu)$ denotes the Fourier transform,

(5) $$\hat{\phi}_\alpha^{(P_2)}(\nu) = c \sum_{s \in W} \alpha(s^{-1}\nu)^{\circ}c_{P_2|P_1}(s : s^{-1}\nu)\psi.$$

Formula (5) is very close to an inversion formula. Let us put

$$\hat{f}(\omega : \nu) = (\Theta_{\omega,\nu}, f) \qquad (f \in \mathcal{C}(G : \tau)).$$

Define the operator

$$\Pi_0 : U \to U$$

by

$$\Pi_0 u = \int_K \tau(k)u\tau(k^{-1})dk \qquad (u \in U).$$

Then from (5) one gets

(6) $$\hat{\phi}_\alpha(\omega' : \nu) = \hat{\phi}_\alpha(s\omega' : s\nu) = \begin{cases} 0 \quad \text{unless} \quad \omega' = s\omega \quad \text{for some} \quad s \in \mathfrak{w} \\ c\, d(\omega)^{-1}\, \Pi_0\psi(1) \sum_{s \in \mathfrak{w}(\omega)} \alpha(s\nu) \end{cases}$$

where $\mathfrak{w}(\omega)$ is the stabilizer of ω in $\mathfrak{w}$.

Formula (6) essentially completes the Fourier transform theory in one direction; it is necessary to extend it to $\alpha \in \mathcal{C}(\mathfrak{F})$ of course but this is easy once the growth properties of $\mu(\omega : \nu)$ described in Theorem D are in our possession. The fact that α is completely arbitrary in (6) leads to the uniqueness part of the Plancherel theorem stated in §1. To complete the transform theory in the inverse direction, we start with $f \in \mathcal{C}(G : \tau)$ and define $\hat{f}(\omega : \nu)$ (as before) as $(\Phi_{\omega,\nu}, f)$. We put

(7) $$\begin{cases} f_\omega = d(\omega)\int_{\mathfrak{F}} \mu(\omega : \nu)\hat{f}(\omega : \nu)d\nu \\ f_A = \sum_{\omega \in \hat{M}_d} f_\omega \end{cases}.$$

(This sum is only over a finite subset of $\hat{M}_d$). Then we find that for a constant $c(A) > 0$ independent of f,

$$(8) \qquad c(A)(E(P:\psi:\nu),f_A) = (E(P:\psi:\nu),f)$$

for all $\psi \in V$. So, if we consider

$$g = \sum_A c(A)f_A$$

where the summation is over a complete set of representatives A, then g and f have the same Fourier transform and a not too difficult argument yields $g = f$ (this is the argument that was first encountered in the spherical case). Evaluating the relation

$$f = \sum_A c(A)f_A$$

at 1 we get the Plancherel formula.

As remarked earlier, one still needs Theorem D giving the growth and holomorphy properties of μ. Using the product formula this comes down to the case when $\dim(A) = 1$. As described in §1 this is split into two cases, according as whether $rk(G) > rk(K)$ or $rk(G) = rk(K)$. The explicit Plancherel formula that one obtains here is by using methods entirely different from what we have been discussing so far. It is based on the Harish Chandra limit formula for orbital integrals on G (cf. Varadarajan [1], Part II, Theorem 13), and is analogous to the case when $rk(G/K) = 1$, treated earlier by Harish Chandra [12]. I cannot go into it here.

Acknowledgement. I wish to acknowledge the support of NSF Grant MCS 79-03184 during the preparation of this work. I am also grateful to Julie Honig for her typing and cooperation in the preparation of these notes.

REFERENCES

M. Eguchi

[1] Asymptotic exapnsions of Eisenstein integrals and Fourier transform on symmetric spaces, Jour. of Functional Analysis 34 (1979), 167-216.

L. Ehrenpreis and F. I. Mautner

[1] Some properties of the Fourier transform on semisimple Lie groups, I, Ann. Math. 61 (1955), 406-439; II, Trans. Amer. Math. Soc. 84 (1957), 1-55; III, Trans. Amer. Math. Soc. 90 (1959), 431-484.

T. J. Enright

[1] On the fundamental series of a real semisimple Lie algebra: their irreducibility, resolutions, and multiplicity formulae, Ann. Math. 110 (1979), 1-82.

T. J. Enright and N. R. Wallach

[1] The fundamental series of semisimple Lie algebras and semisimple Lie groups (preprint).

I. M. Gel'fand and M. A. Naǐmark

[1] Unitäre Darstellungen der Klassischen Gruppen, Akademic-Verlag, Berlin 1957.

S. G. Gindikin and F. I. Karpelevič

[1] Plancherel measure for symmetric Riemannian spaces of non positive curvature, Dok. Akad. Nauk. SSSR. 145 (1962), 252-255.

Harish Chandra

[1] Spherical functions on a semisimple Lie group, I. Amer. Jour. Math. 80 (1958), 241-310.

[2] Spherical functions on a semisimple Lie group II, Amer. Jour. Math. 80 (1958), 553-613.

[3] Harmonic Analysis on real reductive groups I. The theory of the constant term. Jour. of Functional Analysis 19 (1975), 104-204.

[4] Harmonic Analysis on real reductive groups II. Wave packets in the Schwartz space, Inv. Math. 36 (1976), 1-55.

[5] Harmonic Analysis on real reductive groups III. The Maass-Selberg relations and the Plancherel formula, Ann. Math. 104 (1976), 117-201.

[6] Discrete series for semisimple Lie groups, I, Acta Math. 113 (1965), 251-318.

[7] Discrete series for semisimple Lie groups, II, Acta Math. 116 (1966), 2-111.

[8] Harmonic Analysis on semisimple Lie groups, Bull. AMS 76 (1970), 529-551.

[9] The Plancherel formula for complex semisimple Lie groups, Trans. AMS 76 (1954), 485-528.

[10] Plancherel formula for the 2×2 real unimodular group, Proc. Nat. Acad. Sci. USA 38 (1952), 337-342.

[11] Some results on differential equations and their applications, Proc. Nat. Acad. Sci. USA 45 (1959), 1763-1764.

[12] Two theorems on semisimple Lie groups, Ann. Math. 83 (1966), 74-128.

A. W. Knapp

[1] Commutativity of Intertwining operators II, Bull. AMS 82 (1976), 271-273.

B. Kostant

[1] On the existence and irreducibility of certain series of representations, Lie groups and their representations, Edited by I. M. Gel'fand, Halsted Press, 1975.

S. Lang

[1] "$SL_2(R)$", Addison-Wesley, Reading, Mass., 1975.

K. R. Parthasarathy, R. Ranga Rao, and V. S. Varadarajan

[1] Representations of complex semisimple Lie groups and Lie algebras, Ann. Math. 85 (1967), 383-429.

I. E. Segal

[1] An extension of Plancherel's formula to separable unimodular groups, Ann. Math. 52 (1950), 272-292.

P. C. Trombi

[1] Asymptotic expansions of matrix coefficients: the real rank one case, Jour. of Functional Analysis 30 (1978), 83-105.

[2] Harmonic analysis of $C^p(G:F)$ $(1 \leq p < 2)$, (preprint).

[3] Invariant harmonic analysis on split rank one groups with applications, (preprint).

P. C. Trombi and V. S. Varadarajan

[1] Spherical transforms on semisimple Lie groups, Ann. Math 94 (1971), 246-303.

V. S. Varadarajan

[1] Harmonic Analysis on real reductive groups, Lecture Notes in Mathematics #576, Springer Verlag, 1977.

[2] Lie groups, Lie Algebras, and their representations, Prentice Hall, 1974.

[3] Infinitesimal theory of representations of semisimple Lie groups, Lectures given at the Nato Advanced Study Institute at Liege, Belgium on Representations of Lie groups and Harmonic Analysis, 1977.

N. R. Wallach

[1] Cyclic vectors and irreducibility for principal series of representations, Trans. AMS 158 (1971), 107-112.

G. Warner

[1] Harmonic Analysis on semisimple Lie groups, I, II. Springer Verlag, 1972.

D. P. Zhelebenko

[1] The analysis of irreducibility in the class of elementary representations of a complex semisimple Lie group, Math - USSR Izvestra 2 (1968), 105-128.

CENTRO INTERNAZIONALE MATEMATICO ESTIVO

(C.I.M.E.)

ERGODIC THEORY, GROUP REPRESENTATIONS, AND RIGIDITY*

ROBERT J. ZIMMER
University of Chicago

*Partially supported by a Sloan Foundation Fellowship and NSF Grant MCS 79-05036

These notes represent a mildly expanded version of lectures delivered at the C.I.M.E. summer session on harmonic analysis and group representations in Cortona, Italy, June-July 1980. The author would like to express his thanks and appreciation to the organizers of the conference, Michael Cowling, Sandro Figà-Talamanca, and Massimo Picardello, for inviting him to deliver these lectures and for their most warm and generous hospitality during his stay in Italy. We would also like to thank the other participants of the conference for their interest in these lectures. Finally, we would like to thank Terese S. Zimmer for (among innumerable other things that we need not go into here) helping with the translation of [29].

Contents

1. Basic Notions

In these lectures we discuss some topics concerning the relationship of ergodic theory, representation theory, and the structure of Lie groups and their discrete subgroups.

In studying the representation theory of groups, the assumption of compactness on the group essentially allows one to reduce to a finite dimensional situation, in which case one often can obtain complete information. For non-compact groups, of course, no such reduction is possible and the situation is much more complex. When studying general actions of groups, a somewhat similar situation arises. In the compact case every orbit will be closed, the space of orbits will have a reasonable structure, and one can often find nice (with respect to the action) neighborhoods of orbits. A large amount of information about actions of finite and compact groups has been obtained by topological methods. However, once again, if the compactness assumption on the group is dropped, one faces many additional problems. In particular, one can have orbits which are dense (for example, the irrational flow on the torus) and the orbit space may be so badly behaved as to have no continuous functions but constants. Furthermore, moving from a point to a nearby point may produce an orbit which doesn't follow closely to the original orbit. If one wishes to deal with actions in the non-compact case, this phenomenon of complicated orbit structure must be faced. For many actions, e.g., diffentiable actions on manifolds, there are natural measures that behave well with respect to the action. A significant part of ergodic theory is the study of group actions on measure spaces. In particular, ergodic theory aims to understand the phenomenon of bad orbit structure in the presence of a measure.

Throughout these lectures, G will be a locally compact, second countable group. Let (S,μ) be a standard measure space, and assume we have an action $S \times G \to S$ which is a Borel function. Then μ (which is always assumed to be

σ-finite) is invariant if $\mu(Ag) = \mu(A)$ for all $A \subset S$ and $g \in G$, and quasi-invariant if $\mu(Ag) = 0$ if and only if $\mu(A) = 0$.

Definition 1.1: The action is called ergodic if $A \subset S$ is G-invariant implies $\mu(A) = 0$ or $\mu(S - A) = 0$.

Clearly any transitive action is ergodic, or, more generally, any essentially transitive action (i.e., transitive on the complement of a null set). We can then write $S = G/G_0$ where $G_0 \subset G$ is as closed subgroup. An ergodic action that is not essentially transitive will be called properly ergodic.

Example 1.2. Let $S = \{z \in \mathbb{C} \mid |z| = 1\}$ and $T: S \to S$ be $T(z) = e^{i\alpha}z$ where $\alpha/2\pi$ is irrational. Then T generates a $\mathbb{Z}$-action. If $A \subset S$ is invariant, let $\chi_A(z) = \sum a_n z^n$ be the L^2-Fourier expansion of its characteristic function. Then by invariance $\chi_A(z) = \chi_A(e^{i\alpha}z) = \Sigma a_n e^{in\alpha} z^n$. Thus $a_n e^{in\alpha} = a_n$ and so $a_n = 0$ for $n \neq 0$. This implies χ_A is constant, so the action is properly ergodic.

Remark: If S is a (second countable) topological space and μ is positive on open sets, then proper ergodicity implies almost every orbit is a dense null set. This is one sense in which proper ergodicity is a reflection of complicated orbits. Another is the following.

Propositon 1.3 [12]. Let G act continuously on S where S is metrizable by a complete separable metric. Then the following are equivalent: (We say the action is "smooth" if they hold.)

i) Every G-orbit is locally closed

ii) S/G is T_0 in the quotient topology

iii) The quotient Borel structure on S/G is countably separated and generated. (I.e., there is a countable family $\{A_i\}$ separating points and generating the Borel structure.)

iv) Every quasi-invariant ergodic measure is supported on an orbit.

Proof. (i) $\Longrightarrow$ (i) $\Longrightarrow$ (iii) are elementary. To see (iii) $\Longrightarrow$ (iv), let $p: S \to S/G$ the projection, and μ an ergodic probability measure on S. Then $\nu = p_*(\mu)$ is a measure on S/G with the property that for any Borel set $B \subset S/G$, $\nu(B) = 0$ or 1. Since S/G is countably separated and generated, ν is supported on a point, so μ is supported on an orbit. The implication (iv) $\Longrightarrow$ (i) is difficult (and we will not be using it).

We will be making constant use of the implication (i) $\Longrightarrow$ (iv). For example:

Corollary 1.4. Every ergodic action of a compact group is essentially transitive.

If the action is on a metric space, this follows immediately. However, a theorem of Varadarajan [45] implies that any action can be so realized.

Corollary 1.5. Every ergodic algebraic action of a real (or p-adic) algebraic group (more precisely, the real or p-adic points) on an algebraic variety is essentially transitive.

This follows from the theorem of Borel and Borel-Serre that orbits are locally closed [3] [6].

While the decompositon of a general action into orbits may not be satisfactory there is always a good decompositon into ergodic components.

Proposition 1.6. Let (S,μ) be a G-space. Then there is a standard measure space (E,ν), a conull G-invariant set $S_0 \subset S$, and a G-invariant

Borel map $\varphi: S \to E$ with $\varphi_*(\mu) = \nu$ such that, writing $\mu = \int^{\oplus} \mu_y \, d\nu(y)$ where μ_y is supported on $\varphi^{-1}(y)$, we have μ_y is quasi-invariant and ergodic under G for almost all y.

(E,ν) is called the space of ergodic components of the action (and is essentially uniquely determined by the above conditions.)

We now discuss some notions of "isomorphism".

Definition 1.7 Let (S,μ), (S',μ') be ergodic G-spaces. Call them conjugate if modulo null sets there is $\varphi: S \to S'$ with

i) φ a bijective Borel isomorphism.

ii) $\varphi_*(\mu) \sim \mu'$ (i.e., same null sets).

iii) $\varphi(sg) = \varphi(s)g$.

If $A \in \mathrm{Aut}(G)$ and S is a G-space, we have a new G-action on S by defining $s \circ g = s \cdot A(g)$.

Definition 1.8. Call two actions automorphically conjugate if they become conjugate when modified by some automorphism.

An a priori much weaker notion is simply to ask for the orbit pictures to be the same. Here, we can compare actions of different groups.

Definition 1.9. Suppose (S,μ) is a G-space, (S',μ') a G'-space. Call the actions orbit equivalent if (modulo null sets) there exists $\varphi: S \to S'$ with

i) φ a bijective Borel isomorphism.

ii) $\varphi_*(\mu) \sim \mu'$.

iii) φ(G-orbit) = G'-orbit.

If $\varphi:(X,\mu) \to (Y,\nu)$ is a measure class preserving G-map of G-spaces we call X an extension of Y or Y a factor of X. Observe that we automatically have $\nu(Y - \varphi(X)) = 0$. If $H \subset G$ is a subgroup, and X is an ergodic G-space, we can restrict to obtain an action of H, which of course no longer need be ergodic. In the other direction, we can induce. Namely, suppose S is an ergodic H-space and $H \subset G$ is a closed subgroup. Then we obtain a naturally associated G-space as follows. Let H act on $S \times G$ by $(s,g)h = (sh,gh)$ and let $X = (S \times G)/H$. Then G acts on $S \times G$ by $(s,g)\tilde{g} = (s,\tilde{g}^{-1}g)$, and this action commutes with the H-action. Hence there is an induced action of G on X which will be ergodic with its natural measure class.

<u>Definition 1.10.</u> X is called the ergodic G-space induced from the G-action, and we denote it by $\text{ind}_H^G(S)$.

For example, if $H = Z$, $G = \mathbf{R}$, then X can be identified with $(S \times [0,1])/\sim$ where $\sim$ identifies $(s,1)$ with $(Ts,0)$. Under the induced $\mathbf{R}$-action a point simply flows up along the line it is in with unit speed.

Given an ergodic G-space X, it is useful to know when it is induced from an action of a subgroup. The following is helpful in this regard.

<u>Proposition 1.11</u> [52]. If X is an ergodic G-space and $H \subset G$ is a closed subgroup, then $X = \text{ind}_H^G(S)$ for some H-space S if and only if G/H is a factor of X, i.e., there is a measure class preserving G-map $X \to G/H$.

If X is a G-space, is there a unique (up to conjugacy) smallest closed subgroup from which it is induced? The answer in general is no, but we have the following

<u>Proposition 1.12</u> [54]. Suppose G is (the real points, or k-points, k a p-adic field) of an algebraic group and X an ergodic G-space. Then there is a

unique conjugacy class of algebraic subgroups such that $X = \mathrm{ind}_H^G(S)$ for H algebraic (and some S), if and only if H contains a member of this conjugacy class.

Definition 1.13 [54]. If H is in this class, call H the algebraic hull of the action. If this is all of G, call the action Zariski dense.

If $X = G/G_0$, then the algebraic hull is just the usual algebraic hull of the group G_0.

Proof of 1.12. There exist minimal such groups from the descending chain condition on algebraic subgroups. Suppose $H_1, H_2 \subset G$ are two such minimal algebraic groups. We have $\varphi_i : X \to G/H_i$. Let $\varphi = (\varphi_1, \varphi_2) : X \to G/H_1 \times G/H_2$. Then $\varphi_*(\mu)$ is an ergodic quasi-invariant measure on $G/H_1 \times G/H_2$. But the G-action on this product is algebraic, so $\varphi_*(\mu)$ is supported on an orbit. But as a G-space, an orbit is $G/(g_1 H_1 g_1^{-1} \cap g_2 H_2 g_2^{-1})$. By minimality assumptions, H_1 and H_2 are conjugate.

Theorem 1.14 (Borel Density Theorem [4]). If G is a connected semisimple real algebraic group with no compact factors, and S an ergodic G-space with finite invariant measure, then S is Zariski dense in G.

As an example of an ergodic action of such a group, we point out the following example. (One can show there are uncountably many inequivalent actions of such groups [with finite invariant measure].)

Example 1.15. Let $SL(n,\mathbb{Z})$ act on $\mathbf{R}^n/\mathbb{Z}^n$ by automorphisms. This is ergodic. The induced $SL(n,\mathbf{R})$ action will be properly ergodic, essentially free (i.e. almost all stabilizers trivial), and have finite invariant measure.

2. Ergodicity Theorems

A natural class of actions that arises in a variety of situations are actions on homogeneous spaces. Thus, if $H_1, H_2 \subset G$ are subgroups with H_2 closed, H_1 acts on G/H_2 and the question arises as to when this is ergodic. This is a special case of the following question. Suppose S is an ergodic G-space and $H \subset G$ is a subgroup. When will the restriction to H still be ergodic? In the special case in which S has a finite invariant measure, results about unitary representations can be directly applied. Namely, let $(U_g f)(s) = f(sg)$, where $f \in L^2(S)$. This defines a unitary representation of G on $L^2(S)$ and G is ergodic on S (assuming finite invariant measure) if and only if there are no non-zero invariant vectors in $L^2(S) \ominus C$. Thus, to settle the question about ergodicity of restrictions in this case, we have a representation U_g of G with no invariant vectors and we ask whether or not $U|H$ has invariant vectors. Let's consider some of the classical examples, when G is transitive on S.

Example 2.1. Suppose G is compact, $S = G$. Then $H \subset G$ is ergodic on S if and only if H is dense. This includes Example 1.2.

Now let N be a simply connected nilpotent Lie group, $\Gamma \subset N$ a lattice (i.e., Γ is discrete and N/Γ has finite invariant measure). For example,

$$N = \left\{ \begin{pmatrix} 1 & x & z \\ 0 & 1 & y \\ 0 & 0 & 1 \end{pmatrix} \middle| \; x,y,z \in \mathbf{R} \right\} \text{ and } \Gamma = N_Z, \text{ the subgroup with } x,y,z \in Z.$$

Then $[N,N] = \{A \in N \mid x = y = 0\}$, and $N/[N,N]\Gamma$ is a torus. The map $N/\Gamma \to N/[N,N]\Gamma$ exhibits the 3-manifold N/Γ as a circle bundle over the torus. In general $N/\Gamma \to N/[N,N]\Gamma$ will be a bundle over the torus with fiber $[N,N]/[N,N] \cap \Gamma$.

<u>Theorem 2.2</u> (L. Green [1]). $H \subset N$ is ergodic on N/Γ if and only if it is ergodic on $N/[N,N]\Gamma$.

As the latter is a torus, ergodicity can be determined as in Example 2.1. The proof of this depends on writing down the representations of N which appear in $L^2(N/\Gamma)$ and examining them with respect to restriction to subgroups. See [1] for details.

Results for 1-parameter subgroups acting on compact homogeneous spaces of solvable Lie groups have been obtained by Auslander [2] and Brezin and Moore [7].

If $G = SL(2,\mathbf{R})$, $\Gamma \subset G$ is a lattice in G and $H \subset G$ is the group of positive diagonal matrices, then G/Γ is in a natural way the unit tangent bundle of the finite volume negatively curved manifold D/Γ where $D = SO(2,\mathbf{R})\backslash G$ is the Poincare disk, and H is the geodesic flow. Thus, a classical result of Hedlund and Hopf says that H is ergodic on G/Γ [20] [21]. C.C. Moore generalized this to allow G to be a very general semisimple Lie group, and H to be an arbitrary subgroup.

<u>Theorem 2.3</u>. (C.C. Moore [32]). Let $G = \Pi G_i$ where G_i is a non-compact connected simple Lie group with finite center and let $\Gamma \subset G$ be an irreducible lattice. Then $H \subset G$ is ergodic on G/Γ if and only if $\overline{H}$ is not compact.

This theorem was proved by showing the following general result about arbitrary representations (not necessarily one appearing in $L^2(G/\Gamma)$). Let G be a non-compact connected simple Lie group with finite center and π a unitary representation of G with no non-zero invariant vectors. Then for any vector $x \neq 0$, $\{g \in G \mid \pi(g)x = x\}$ is compact. This result easily implies the theorem. A stronger result about such representations that we will need has subsequently come to light.

Theorem 2.4. If π is any unitary representation of a connected non-compact simple Lie group with finite center, then the matrix coefficients $f(g) = \langle\pi(g)x|y\rangle \to 0$ as $g \to \infty$, assuming there are no $\pi(G)$-invariant vectors.

A nice proof of this appears in a paper of Howe and Moore [22] although the basic idea is present in the work of Sherman [43]. (See also [49].) The idea of the proof is to let $G = KAK$ be a Cartan decomposition. Since K is compact, it suffices to see $f(a) \to 0$ as $a \to \infty$. Consider the example $G = SL(2,\mathbf{R})$, so that A is the positive diagonals. Let P be the upper triangular 2×2 matrices in G with positive diagonal entries. The representation theory of P is well known. There are 1-dimensional representations which factor through [P,P] and 2 infinite dimensional representations induced from [P,P]. For the latter, it is clear that the restriction of a representation to A is just the regular representation of A for which it is clear that matrix coeficients vanish at ∞. Thus it suffices to see that $\pi|P$ has a spectral decomposition which assigns measure 0 to the 1-dimensional representations. But if it assigned positive measure, [P,P] would have to leave a vector fixed, say v. Then $\varphi(g) = \langle\pi(g)v|v\rangle$ would be bi-invariant under $[P,P] = N$. G/N can be identified with $R^2 - \{0\}$, and the N orbits on G/N are the horizontal lines except for the x-axis, and single points on the x-axis. A continuous funtion on G/N constant on the orbits must clearly be constant on the x-axis as well. This translates into $\varphi(g) = 1$ for all $g \in P$, and since π is unitary, v is P-invariant. Thus φ is bi-invariant under P, and since P has a dense orbit on G/P, $\varphi(g) = 1$ for all $g \in G$, showing that v is G-invariant.

We thus have good information about some basic examples for the question of ergodicity of actions on homogeneous spaces of finite invariant measure. For the general homogeneous space we make use of the following observation.

Proposition 2.5. [49] If S is an ergodic G-space (general quasi-invariant measure) and $H \subset G$ is a closed subgroup, then H is ergodic on S if and only if G acts ergodically on the product $G/H \times S$.

To see this, suppose $A \subset G/H \times S$ is G-invariant. For each $x \in G/H$, let $A_x = \{s \in S \mid (x,s) \in A\}$. By quasi-invariance one easily sees that A and all A_x are simultaneously either null, of null complement, or neither. $A_{[e]}$ is an H-invariant set, and clearly any H-invariant set $B \subset S$ is of the form $B = A_{[e]}$ for some G-invariant A.

Corollary 2.6. [32] If $\Gamma, H \subset G$ are closed subgroups, then H is ergodic on G/Γ if and only if Γ is ergodic on G/H.

This enables us to use information about ergodicity of restrictions on spaces for which there is a finite invariant measure to obrtain results in the case no such measure exists.

Corollary 2.7. (Moore) $G = \Pi G_i$, Γ as in Theorem 2.3. If S is a transitive G-space, then Γ is ergodic on S if and only if the stabilizers in G of points in S are not compact.

Example 2.8. (Moore) SL(n,Z) is ergodic on $\mathbf{R}^n$, $n \geq 2$. This follows since $SL(n,\mathbf{R})$ is essentially transitive on $\mathbf{R}^n$ and the stabilizers in the orbit of full measure are not compact.

Example 2.9. Consider the action of $SL(2,\mathbf{R})$ on the Poincare disk $SL(2,\mathbf{R})/SO(2,\mathbf{R})$. This action extends to the boundary circle, and the boundary can be identified with $SL(2,\mathbf{R})/P$, where P is the upper triangular matrices in G. If $\Gamma \subset SL(2,\mathbf{R})$ is a torsion free lattice, then Γ acts in a properly discontinuous fashion on the disk, and the quotient space D/Γ is a Riemann surface of finite volume. On the other hand, since P is not compact, the action of Γ on the boundary will be properly ergodic. More generally, if G is any semisimple Lie group and $P \subset G$ is a minimal parabolic subgroup, then G/P is the unique compact G-orbit in the boundary of a natural compactification of the symmetric space $X = G/K$, $K \subset G$ maximal compact. Here again, Γ is ergodic on G/P. Thus these ergodic actions of Γ on homogeneous spaces of G arise very naturally in a geometric setting, and the study of these ergodic actions is extremely useful in understanding Γ.

Since this is such an important example, let us point out that for G/P compact (e.g. P a parabolic) that ergodicity of Γ on G/P can be demonstrated in a much less sophisticated fashion. Namely if there is a P-invariant vector in $L^2(G/\Gamma) \ominus C$ then there is a compact G-orbit in the Hilbert space. As is well known, this implies that there exist finite dimensional subrepresentations, which for G, it is also well known, must be the identity. This is impossible.

Corollary 2.7 deals with the restriction of transitive G-actions to Γ. We now deal with the properly ergodic case.

Theorem 2.10 [49]. If $G = \Pi G_i$, G_i connected non-compact simple Lie groups with finite center, $\Gamma \subset G$ an irreducible lattice and S is a properly ergodic G-space, then Γ is ergodic on S.

Proof. Suppose not. Let $A \subset S \times G/\Gamma$ be invariant. For each s, let $f_s \in L^2(G/\Gamma) \ominus \mathbf{C}$ be the image under orthogonal projection of the characteristic function of $A_s = \{x \in G/\Gamma \mid (s,x) \in A\}$. We can suppose $f_s \neq 0$ on a set of positive measure. Invariance of A is easily seen to imply that if we let $B \subset L^2(G/\Gamma) \ominus \mathbf{C}$ be the unit ball and let G act on the right in G via the unitary representation of G on $L^2(G/\Gamma)$, then $\Phi: S \to B$, $\Phi(s) = f_s$ is a G-map. Then $\Phi_*(\mu)$ is a quasi-invariant ergodic measure on B.

But by vanishing of the matrix coefficients (Theorem 2.4), for $w \in B$, $w \cdot g \to 0$ weakly as $g \to \infty$. This implies G-orbits in B are locally closed, i.e. the action is smooth. It follows that $\Phi_*(\mu)$ is supported on an orbit, so we can suppose $\Phi:S \to G/G_0$ where G_0 is the stabilizer of a point in this orbit. This implies $S = \mathrm{ind}_{G_0}^{G}(S_0)$ where S_0 is an ergodic G_0 space. But G_0 is compact, so G_0 is transitive on S_0. This implies G is transitive on S, which contradicts our hypotheses.

Similar results can be proven for other groups for which there is a vanishing theorem for matrix coefficients.

Theorem 2.11 [50]. Let G be an exponential solvable Lie group and S an ergodic G-space. Suppose [G,G] is ergodic on S. Then Γ is also ergodic on S for every cocompact $\Gamma \subset G$.

The proof uses the result of Howe and Moore [22] that for such a group, the matrix coefficients $\langle\pi(g)v|w\rangle \to 0$ as $g \to \infty$ in G/P_π where $P_\pi = \{g|\pi(g) \text{ is scalar}\}$. Here π is assumed irreducible.

3. Cocycles.

If X is a G-space and Y a Borel space, let F(X,Y) be the space of measurable functions $X \to Y$, two functions being identified if they agree off a null set. G acts on $X \times Y$ by $(x,y) \cdot g = (xg,y)$ and on F(X,Y) by $(g \cdot f)(x) = f(xg)$. If Y is also an H-space for some group H, we can define "twisted" actions. Namely, $(x,y) \cdot g = (xg, y \cdot \alpha(x,g))$ where $\alpha(x,g) \in H$, and for $f \in F(X,Y)$ (where for convenience we usually take H to be acting on the left), $(g{\cdot}f)(x) = \alpha(x,g)f(xg)$. For these to define actions, we need the following compatibility condition: $\alpha(x,gh) = \alpha(x,g)\alpha(xg,h)$. Such a Borel function $\alpha:X \times G \to H$ will be called a cocycle. (The question as to whether this holds everywhere or almost everywhere is an important technical point which we will not discuss. See [41].) When endowed with this action we shall denote $X \times Y$ by $X \times_\alpha Y$. If $Y = \mathscr{H}$ a Hilbert space, $H = U(\mathscr{H})$, the unitary group of the Hilbert space, and the measure on X is invariant, then the α-twisted action on $F(X,Y)$ restricts to $L^2(X;\mathscr{H})$ to yield a unitary representation U^α. If $\alpha,\beta: X \times G \to H$ are cocycles there is a certain relation which immediately implies equivalence of the actions or representations. Namely if we have a Borel map $\varphi: X \to H$ such that $\alpha(x,g) = \varphi(x)\, \beta(x,g)\, \varphi(xg)^{-1}$, this will be the case. We then call α and β equivalent, or cohomologous, and write $\alpha \sim \beta$.

To get some further feeling for this notion, consider the case $X = G/G_0$. If $\alpha: G/G_0 \times G \to H$ is a cocycle, then $\alpha|[e] \times G_0$ defines a homomorphism $G_0 \to H$. Equivalent cocycles yield conjugate homomorphisms. Furthermore, every homomorphism $G_0 \to H$ arises from a cocycle α in this way. Namely, let $\gamma: G/G_0 \to G$ be a Borel section. Then for $(x,g) \in G/G_0 \times G$, $\gamma(xg)$ and $\gamma(x) \cdot g$ are equal when projected to G/G_0. Thus $\gamma(x)g\gamma(xg)^{-1} \in G_0$. We can suppose $\gamma([e]) = e$, and then $(x,g) \to \gamma(x)g\,\gamma(xg)^{-1}$ is a cocycle $G/G_0 \times G \to G_0$ which when restricted to $[e] \times G_0$ yields the identity $G_0 \to G_0$. Thus if $\pi: G_0 \to H$ is a homomorphism, $\alpha(x,g) = \pi(\gamma(x)g\gamma(xg)^{-1})$ is the required cocycle. Thus we have

<u>Theorem 3.1</u>. $\alpha \to \alpha|\,[e] \times G_0$ defines a bijection between equivalence classes of cocycles $G/G_0 \times G \to H$ and conjugacy classes of homomorphisms $G_0 \to H$.

We remark that if $H = U(\mathscr{H})$, and $\pi: G_0 \to H$ is a unitary representation, we have an associated cocycle $\alpha: G/G_0 \times G \to H$, and then an associated representation U^α of G on $L^2(G/G_0;\mathscr{H})$. Of course $U^\alpha = \mathrm{ind}_{G_0}^{G}(\pi)$. See [45] for this approach to induced representations.

We now consider some other examples.

Example 3.2 a) If $h:G \to H$ is a homomorphism, X a G-space, then $\alpha(x,g) = h(g)$ is a cocycle. If $X = G/G_0$, this corresponds to a homomorphism $G_0 \to H$, which is simply $h|G_0$. Thus in general we shall sometimes call α the restriction of h to $X \times G$ and write $\alpha = h|X \times G$.

b) Suppose X is an ergodic G-space with quasi-invariant measure μ. Let $r_\mu(x,g) = d\mu(xg)/d\mu(x)$, the Radon-Nikodym derivative. The chain rule implies $r_\mu : X \times G \to R^+$ is a cocycle, called the Radon-Nikodym cocycle. If $\mu \sim \nu$, so $d\mu = f d\nu$, $f > 0$, then $d\mu(xg)/d\mu(x) = f(x)^{-1}(d\nu(xg)/d\nu(x))f(xg)^{-1}$, i.e. $r_\mu \sim r_\nu$. Therefore the cohomology class we obtain does not depend upon the measure, only the measure class. In particular, there is an equivalent σ-finite invariant measure if an only if the cocycle is trivial (i.e. equivalent to the identity $\alpha(x,g) = 1$).

c) Suppose X is a G-space, X' a free G'-space, and that the actions are orbit equivalent, with $\theta: X \to X'$ the orbit equivalence. Then for $(x,g) \in X \times G$, $\theta(x)$ and $\theta(xg)$ are in the same G'-orbit, say $\theta(x)\ \alpha(x,g) = \theta(xg)$ for $\alpha(x,g) \in G'$. Then $\alpha: X \times G \to G'$ is a cocycle. If $G = G'$, we have the following.

Proposition 3.3 [55] If α is equivalent to the restriction of an automorphism $A \in \mathrm{Aut}(G)$ to $X \times G$, then X and X' are automorphically conjugate. If this automorphism is inner, then X and X' are conjugate.

Proof. If $\alpha(x,g) = \lambda(x)A(g)\lambda(xg)^{-1}$, then $\theta_1(x) = \theta(x)\ \lambda(x)$ satisfies $\theta_1(xg) = \theta_1(x)A(g)$, so we have automorphic conjugacy. If $A(g) = hgh^{-1}$, let $\theta_2(x) = \theta_1(x)h$. This is then a G-map.

There are many other naturally arising situations in which cocycles appear, but we shall not have time to discuss them here. Instead, we turn to an important invariant attached to a cocycle, namely the Mackey range. Let $\alpha:S \times G \to H$ where H is also locally compact. Form the twisted G-action $S \times_\alpha H$ where we view H as acting on itself by right translations. H also acts on $S \times_\alpha H$ by $(s,h) \cdot h_0 = (s,h_0^{-1}h)$, and this H action commutes with the G-action. Note that if $i:G \to H$ is an embedding of G into a larger group and $\alpha = i|S \times G$, this is exactly the situation in the inducing procedure. As in the latter, we obtain an action of H on the space of G-orbits. But this space may not be a decent measure space, so instead, we let X be the space of G-ergodic components of the action of G of $S \times_\alpha H$. Then H will act on X as well, and this will be an ergodic H-action.

Definition 3.4. If $\alpha: S \times G \to H$ is a cocycle, the associated H space X will be called the Mackey range of α. This is a cohomology invariant of α.

Example 3.5 a) If $i:G \to H$ is an embedding of G as a closed subgroup of H, and $\alpha(s,g) = i(g)$, i.e. $\alpha = i|S \times G$, then the Mackey range of α is $\mathrm{ind}_G^H(G_0)$.

b) If $\theta:X \to X'$ is an orbit equivalence, $\alpha:X \times G \to G'$ the associated cocycle, then the Mackey range is the G'-space X'.

c) If $S = G/G_0$ and α corresponds to a homomorphism $\pi:G_0 \to H$, then the Mackey range of $\alpha:G/G_0 \times G \to H$ is the H-space $H/\overline{\pi(G_0)}$.

Finally, the following relates the Mackey range to the cohomology class of α.

Proposition 3.6. If $\alpha:S \times G \to H$, the following are equivalent.

i) $\alpha \sim \beta$ where $\beta(S \times G) \subset H_o$, $H_o \subset H$ a closed subgroup.

ii) H/H_0 is a factor of the Mackey range.

iii) $X = \mathrm{ind}_{H_0}^H(S_0)$ for some S_0, where X is the Mackey range.

For a proof, see [47], [52].

4. Generalized Discrete Spectrum

Suppose (S,μ) is an ergodic space with μ finite and invariant. In this lecture we try to see what the algebraic structure of the representation π of G on $L^2(S)$ says about the geometric structure of the action.

Definition 4.1. We say that the action has discrete spectrum if π is the direct sum of finite dimensional irreducible subrepresentations.

Example 4.2. Let K be a compact group, H a closed subgroup, and $\varphi: G \to K$ a homomorphism with $\varphi(G)$ dense in K. Let G act on K/H by $[k]\cdot g = [k\varphi(g)]$. Then this action has discrete spectrum.

Theorem 4.3. (von Neumann-Halmos-Mackey). These are all the examples. That is, if S is a G-space with discrete spectrum, then there exists a compact group K, a closed subgroup H, and a homomorphism $\varphi: G \to K$ with dense range such that S and K/H are conjugate G-spaces.

This was originally proved by von Neumann and Halmos for $G = \mathbb{Z}$ or $\mathbb{R}$, and by Mackey [26] for general G. We sketch Mackey's proof.

Let $L^2(S) = \Sigma^{\oplus} W_i$ where W_i are $\pi(G)$ - invariant and finite dimensional. Let $B = \Pi U(W_i)$, the product of the associated unitary groups, which is a compact subgroup of $U(L^2(S))$. Further, $\pi: G \to B$. Let $K = \overline{\pi(G)}$, so that K is also compact. Let M be the abelian von Neumann algebra on $L^2(S)$ consisting of multiplication by elements of $L^\infty(S)$. Then clearly $\pi(g)M\pi(g)^{-1} = M$, and by passing to the strong limit, we obtain $TMT^{-1} = M$ for all $T \in K$. From this one can deduce that each operator T in K is induced by a point transformation of S, and thus the G-action on S extends to an action of K. (There is some delicate measure theory we are ignoring here if G is not discrete.) Since the G-action is already ergodic, so is the K action. Since K is compact, K must act transitively, so we can identify $S \cong K/H$.

Theorem 4.3 can be generalized to extensions. Namely, suppose $X \to Y$ is an extension of ergodic G-spaces with finite invariant measure. The Hilbert space $L^2(X)$ not only has a natural representation of G on it, but $L^2(X)$ is also an $L^\infty(Y)$-module in a natural way. (Namely, lift a function on Y to a function on X and multiply.) Alternatively, we can express this by saying that there is a natural system of imprimitivity for π on $L^2(X)$ based on Y.

Definition 4.4. [47] We say that X has relatively discrete spectrum over Y is $L^2(X) = \Sigma^{\oplus} \overline{W_i}$ where W_i are G-invariant subspaces that are finitely generated as $L^\infty(Y)$ - modules.

Example 4.5. Suppose Y is an ergodic G-space with finite invariant measure, $\alpha: Y \times G \to K$ is a cocycle where K is compact, and $H \subset K$ is a closed subgroup. Then $X = Y \times_\alpha K/H$ is an extension of Y with relatively discrete spectrum. To see this, observe that $L^2(X) = L^2((Y); L^2(K/H))$. Write $L^2(K/H) = \Sigma^{\oplus} Z_i$ where Z_i are finite-dimensional and K-invariant. We then have $L^2(X) = \Sigma^{\oplus} L^2(Y; Z_i)$ and $L^2(Y; Z_i)$ will be G-invariant since G acts from fiber to fiber in X by an element of K, and Z_i is K-invariant. Clearly $L^2(Y; Z_i) = \overline{L^\infty(Y, Z_i)}$ and the latter is finitely generated over $L^\infty(Y)$.

Theorem 4.6. [47]. These are all the examples. That is, if $X \to Y$ is an ergodic extension with relatively discrete spectrum, then there exists a compact group K, a closed subgroup $H \subset K$, and a cocycle $\alpha: Y \times G \to K$, such that as extensions of Y, $X \cong Y \times_\alpha K/H$.

Thus Theorem 4.6 tells us how to recognize extensions of the form $Y \times_\alpha K/H$ from information about the unitary representation of the extension. There is now a larger class of actions whose "structure" we know.

Definition 4.7 [48]. We say that X has generalized discrete spectrum if X can be built from a point via the operations of taking extensions with relatively discrete spectrum and inverse limits. More precisely, there is a countable ordinal α and for each $\sigma \leq \alpha$ a factor X_σ of X such that

i) X_0 = point

ii) $X_{\sigma+1} \to X_\sigma$ is an extension with relatively discrete spectrum, for $\sigma < \alpha$.

iii) if σ is a limit ordinal, $X_\sigma = \varprojlim\{X_\beta, \beta < \sigma\}$

iv) $X_\alpha = X$.

In light of Theroem 4.6, we have an exact picture of the structure of such actions. We would now like to see which actions arise in this fashion.

Definition 4.8. If G acts continuously on a compact metric space X, G is called distal on X if $x,y \in X$, $x \neq y$, implies $\inf_{g \in G} d(xg,yg) > 0$.

Clearly any isometric action is distal. However, not every distal action admits an invariant metric. For example, if N is a nilpotent Lie group and $\Gamma \subset N$ is a lattice, then the action of N on N/Γ is distal. This was first shown in [1].

Definition 4.9 (Parry [38]) If (S,μ) is an ergodic G-space, call the action measure-distal if there is a decreasing sequence of sets of positive measure $\{A_i\}$ with $\mu(A_i) \to 0$, such that if $x,y \in S$, $xg_i, yg_i \in A_i$ for some sequence $g_i \in G$, then $x = y$. (We have ignored some measure theoretic issues in this definition, which arise if G is not discrete. See [48] for a more careful formulation.)

Any distal action with an invariant measure that is positive on open sets is clearly measure distal.

Theorem 4.10 [48] A finite measure preserving ergodic action (on a non-atomic measure space) is measure distal if and only if it has generalized discrete spectrum.

This is an analogue for measure theoretic actions of the Furstenberg structure theorem for mimimal (i.e. every orbit dense) distal actions on compact metric spaces [16].

Another situation in which actions with generalized discrete spectrum arise is the following.

Theorem 4.11. Let N be a nilpotent group. Suppose S is an ergodic N-space for which $L^2(S)$ is a direct sum of irreducible representations (not necessarily finite dimensional). Then S has generalized discrete spectrum (and the ordinal in Definition 4.7 can be taken to be finite.)

This theorem is false for solvable groups. Let us give an example of such a properly ergodic N-space. Let N be the Heisenberg group, $N_Z = \Gamma$ the integer points, so that Γ is a lattice. There is an injective homomorphism $\Gamma \to K$ where K is compact. For example, let $K = \Pi N_{Z/pZ}$, where the product is taken over all primes. Then K is a Γ - space with discrete spectrum. Let $X = \mathrm{ind}_\Gamma^N (K)$. Since a finite dimensional representation of Γ induced to N decomposes into a direct sum of irreducibles, and π on $L^2(X)$ can be expressed as $\pi = \mathrm{ind}_\Gamma^N(\sigma)$ where σ is the representation of Γ on $L^2(K)$, it follows that $L^2(X)$ is a direct sum of irreducibles.

It is natural to ask which groups have actions of the sort we have been discussing in a non-trivial way, say an effective or essentially free action. A group will have an effective or free action with discrete spectrum if and only if there are enough finite dimensional unitary representations to separate points. In the connected case, such groups are identified by a classical theorem of Freudenthal.

Theorem 4.12 (Freudenthal [14]). A connected group has a free (or effective) action with discrete spectrum if an only if it is isomorphic to $R^n \times K$ where K is compact.

To describe the analagous result for generalized discrete spectrum, we recall that a connected group is said to be of polynomial growth if for any compact neighborhood of the identity, W, the Haar measure $m(W^n)$ grows no faster than a polynomial in n. (If this is true for one compact neighborhood, it is true for all such neighborhoods.) For Lie groups, this condition is equivalent to the group being of type (R) [19] [23]. We recall that this means that every eigenvalue of Ad(g) lies on the unit circle for all $g \in G$. For example, nilpotent groups and euclidean motion groups are type (R), while semisimple groups and the ax + b group are not. The following is joint work with C.C. Moore.

Theorem 4.13. [34] A connected group has a free (or effective) ergodic action with generalized discrete spectrum if and only if it is of polynomial growth.

Proof. We indicate the proof for the ax + b group. The general proof is based on this argument and some structure theory for Lie groups, particularly that of solvable Lie groups.

Let G = AB be a semidirect product where $B = \mathbf{R}$ is normal and $A = \mathbf{R}^+$ acts on B by multiplication. If X is a G-space with generalized discrete spectrum, and X_1 is the factor of X with discrete spectrum, then B must act trivially on X_1 since all finite dimensional unitary representations of G are one dimensional and thus factor through B = [G,G]. It therefore suffices to show the following: suppose $\varphi: X \to Y$ is an extension of G-spaces with relatively discrete spectrum and suppose B acts trivially on Y; then B acts trivially on X. To prove this assertion, let $L^2(X) = \Sigma^{\oplus} \overline{W_i}$, W_i a finitely generated $L^\infty(Y)$-module which is G-invariant. Let μ, ν be the given measures on X and Y respectively, and decompose μ with respect to ν over the fibers of φ. Thus, we write $\mu = \int \mu_y \, d\nu(y)$ where μ_y is supported on $\varphi^{-1}(y)$. This gives us a direct integral decomposition $L^2(X) = \int_Y^{\oplus} L^2(\varphi^{-1}(y), \mu_y) d\nu$. For each

y,g we have $\alpha(y,g) : L^2(\varphi^{-1}(yg),\mu_{yg}) \to L^2(\varphi^{-1}(y),\mu_y)$ given by $[\alpha(y,g)f](z) = f(zg)$ for $z \in \varphi^{-1}(y)$. Fix i. Saying that W_i is finitely generated over $L^\infty(Y)$ and G-invariant means that there is $V_y \subset L^2(\varphi^{-1}(y))$, a finite dimensional subspace, such that $W_i = \int^{\oplus} V_y d\nu(y)$ and $\alpha(y,g)V_{yg} = V_y$. For $g \in B$, $yg = y$, so $\alpha|\{y\} \times B$ is a unitary representation of B on V_y. Say dim $V_y = n$. Then for each y, we have n elements in $\hat{B}$ = character group of B. Furthermore, G acts on $\hat{B}$ and one can check that the cocycle identity for α implies

(*) $(\alpha|\{y\} \times B) \cdot g \cong \alpha|\{yg\} \times B$, where $\cong$ means unitary equivalence.

Let $\hat{B}^n/S_n$ (S_n is the symmetric group on n letters) be the set of unordered n-tuples of elements of $\hat{B}$. We have a map $\Phi: Y \to \hat{B}^n/S_n$, and (*) implies that Φ is a G-map. The action of G on $\hat{B} \cong \mathbf{R}$ has three orbits, namely the origin and the 2 half lines. From this it is easy to see that every G - orbit in $\hat{B}^n$ (and hence in $\hat{B}^n/S_n$) is locally closed, and that the only compact G - orbit is the identity (i.e. the origin). But $\Phi_*(\mu)$ is a finite invariant ergodic measure on $\hat{B}^n/S_n$. By smoothness, this must be suported on an orbit and by finiteness and invariance, this must clearly be the zero orbit. Thus, $\alpha|\{y\} \times B$ is the identity for all y, so B acts trivially on each W_i. Therefore, B is trivial on $L^2(X)$ and hence on X as well.

Finally, we remark that the notion of generalized discrete spectrum yields a type of structure theorem for general actions with finite invariant measure that is sometimes useful. If S is an ergodic G - space, it is not always true that $S \times S$ is also ergodic, where G acts by $(s,t)g = (sg,tg)$. If this additional ergodicity property holds, the action is called weakly mixing. More generally, if $X \to Y$ is an ergodic extension of Y, the fibered product $X \times_Y X$ has a natural G - invariant measure on it [47], but this action no longer need be ergodic. Once again, if this extra ergodicity holds, the extension X is called relatively weakly mixing over Y. Given any ergodic G-space X, there is a unique maximal factor Z of X such that Z has generalized discrete spectrum and X is relatively weakly mixing over Z. Thus we break X up into a factor whose structure we know explicitly, and an extension with extra ergodicity properties. Of course, simply by knowing that an action or extension is weak mixing does not say very much about its detailed strucure, so for most questions, this is not a satisfactory structure theorem aside from the factor Z. Nevertheless, weak mixing does clearly have some information, and thus one can hope to find this decomposition useful in some circumstances. An example of this appears in recent work of Furstenberg. Szemeredi recently succeeded in proving a conjecture of Erdos which aserts that every set of positive integers of positive upper density contains arithmetic progressions of arbitrary (finite) length. In [17], Furstenberg gave another proof of Szemeredi's theroem, first by converting this to a statement about measure preserving integer actions, and then proving the latter by proving it first for actions with generalized discrete spectrum, and then showing the property is preserved by passing to relatively weakly mixing extensions.

5. Amenability

The notion of an amenable group can be described in a variety of ways. Here, we shall focus on the fixed point property.

Let E be a separable Banach space, E^* the dual, E_1^* the unit ball in E^*, and Iso(E) the group of isometric isomorphisms of E. Suppose $\pi: G \to \mathrm{Iso}(E)$ is a representation of G on E, and that $A \subset E_1^*$ is a compact convex G - invariant set. (Here G acts on E^* via the adjoint representation, $\pi^*(g) = (\pi(g^{-1}))^*$.

Definition 5.1. G is amenable if for all π and A as above, there is a fixed point for G in A.

For example, if G is amenable and G acts continuously on a compact metric space X, then there is a G-invariant probability measure on X. We simply apply the definition to $E = C(X)$ where $A \subset C(X)^*$ is the set of probability measures. In fact a standard convexity argument shows that G is amenable if and only if there is a G-invariant measure on every compact metric G-space.

Abelian groups are amenable by the Markov-Kakutani fixed point theorem, and compact groups are easily seen to be amenable. If $0 \to A \to B \to C \to 0$ is an exact sequence, then B is amenable if and only if A and C are amenable. Thus, groups with a cocompact solvable normal subgroup are amenable. Every connected amenable group is of this form, but this is no longer true among all discrete groups [18].

We now wish to define the notion of an amenable ergodic action of a group, originally introduced in [51]. This will include all actions of amenable groups, as well as some actions of non-amenable groups. We begin by describing certain classes of G-invariant compact convex sets that arise from an ergodic G-space S.

So suppose S is an ergodic G-space and $\alpha: S \times G \to \mathrm{Iso}(E)$ is a cocycle. We then have the adjoint cocycle $\alpha^*(s,g) = (\alpha(s,g)^{-1})^*$, and the α^*-twisted action on $L^\infty(S,E^*)$, given by $(g \cdot f)(s) = \alpha^*(s,g)f(sg)$, for $f \in L^\infty(S,E^*)$. We observe that $L^\infty(S,E^*) = (L^1(S,E))^*$, so that $L^\infty(S,E^*)$ is a dual space. We want to describe certain G - invariant compact convex sets in the unit ball of this dual space. One natural possibility is to take $A \subset E_1^*$ compact, convex, and satisfying the condition $\alpha^*(s,g)A = A$. Then $F(S,A)$ (= measurable functions $S \to A$) will be a compact convex G - invariant set in $L^\infty(S,A)$. However, it is also possible to vary the set A as we move from point to point in S. Thus, suppose $\{A_s\}$ is a collection of compact convex subsets $A_s \subset E_1^*$, which vary measurably in s, and satisfying the condition $\alpha^*(s,g)A_{sg} = A_s$. Then $F(S,\{A_s\}) = \{f \in L^\infty(S,E^*) \mid f(s) \in A_s\}$ is a compact convex G - invariant set.

Definition 5.2 [51] Call a set of the form $F(S,\{A_s\})$ a compact convex set over S. An ergodic action of G on S is called amenable if every compact convex G - invariant set over S has a fixed point.

Thus while amenability of G demands a fixed point in every compact convex set, amenability of the action demands a fixed point only in compact convex sets over the action. We also remark that the condition that one has a fixed point simply means $\alpha^*(s,g)\, f(sg) = f(s)$ for $f:S \to E^*$, $f(s) \in A_s$. As an example of how one can use this condition, suppose S is an amenable G-space and that X is a compact metric G-space. Let M(X) be the space of probability measures on X. We have a representation $\pi: G \to \mathrm{Iso}(C(X))$ and hence a cocycle $\alpha:S \times G \to \mathrm{Iso}(C(X))$ by restriction, i.e. $\alpha(s,g) = \pi(g)$. M(X) will be a G-invariant compact convex set, and thus we can take $A_s = M(X)$ for all s. (So for this example, we didn't have to vary the compact convex set in going from point to point.) Amenability of the action then implies that there is a function $f:S \to M(X)$ such that $\alpha^*(s,g)f(sg) = f(s)$, i.e., $\pi^*(g)f(sg) = f(s)$. Switching to a right action on M(X), we obtain that $f(sg) = f(s) \cdot g$. Thus, we conclude that if S is an amenable ergodic G - space, X a compact metric G-space, then there is a measurable G - map $f:S \to M(X)$.

We now list some basic properties. Proofs can be found in [51], [52].

Proposition 5.3.

a) If G is amenable, every ergodic G-space is amenable.

b) If S is an amenable ergodic G-space with finite invariant measure, then G is amenable.

c) If $S = G/H$, then G/H is an amenable G-space if and only if H is amenable.

d) If S is an amenable ergodic G-space, and $\Gamma \subset G$ is a closed subgroup, then the restriction of the action on S to Γ is amenable.

Example 5.4. Let $\Gamma \subset SL(2,\mathbf{R})$ be a lattice and consider the ergodic action of Γ on the boundary circle of the Poincare disk. This is just the action of Γ on $Sl(2,\mathbf{R})/P$ where P is the upper triangular subgroup. Since P is amenable, the Γ action on the boundary circle is amenable by (c) and (d) of the above proposition. More generally, let G be a semisimple Lie group, $\Gamma \subset G$ a lattice, and $P \subset G$ a minimal parabolic subgroup. Then P is amenable and so Γ acting on G/P is amenable.

This example indicates how natural and important examples of actions of non-amenable groups are amenable. Assertion (c) of the above proposition shows that any group has amenable transitive actions. More generally, we have the following.

Proposition 5.5 [52] If $H \subset G$ is a closed subgroup and S is an ergodic H-space, then S is an amenable H-space if and only if $\mathrm{ind}_H^G(S)$ is an amenable G-space.

This proposition raises the following question. Although one can have amenable actions of non-amenable groups, does every such action come in a simple way from an action of an amenable subgroup, namely just by inducing? In fact, the answer in general is no. One can show that if $\Gamma \subset SL(2,\mathbf{C})$ is a lattice, then Γ is amenable on $\mathbf{CP}'$ (this is just example 5.4) but that this action is not induced from an action of an amenable subgroup. This latter fact is not trivial, and a proof is given in [52]. On the other hand, we do have the following.

Theorem 5.6 [52], [55] Let G be a connected group. Then every amenable ergodic action of G is induced from an action of an amenable subgroup.

Proof. We give the proof for G semisimple. Let S be an amenable G-space and P G be a minimal parabolic subrougp. Since G/P is compact, by the remarks following Definition 5.2, there is a measurable G - map $f:S \to M(G/P)$, where the latter is the space of probability measures on G/P. The proof will now follow from two basic results about the action of G on M(G/P). The first is due to the author, the second to C.C. Moore [33].

Theorem 5.7 [52], [55] Let G be a connected semisimple Lie group and $P \subset G$ any parabolic subgroup. Then every element in M(G/P) has a locally closed orbit under the G - action.

Theorem 5.8 (C.C. Moore [33].) Let G be a connected semisimple Lie group with trivial center and $P \subset G$ a minimal parabolic subgroup. Let $\mu \in M(G/P)$ and G_μ the stabilizer of μ in G. Then G_μ is an amenable algebraic group. (By algebraic, we mean the intersection of G with an algebraic subroup of $\tilde{G}$, where the latter is an algebraic subgroup containing G as a subgroup of finite index.)

To conclude the proof of Theorem 5.6 given these results, we simply observe that Theorem 5.7 asserts the smoothness of the G-action on M(G/P), so if m denotes the measure on S, then $f_*(m)$ is quasi-invariant and ergodic on M(G/P) and hence supported on an orbit. Thus we can consider f as $f:S \to G/H$, where H is the stabilizer of an orbit. But by theorem 5.8, H is amenable so the result follows from Proposition 1.11.

Without giving a proof of Theorems 5.7 and 5.8, let us at least try to give some indication of why they are true. To this end, we state the following lemma of Furstenberg [15].

Lemma 5.9 (Furstenberg). Suppose $g_n \in SL(n,\mathbf{R})$, $g_n \to \infty$. Let $\mu \in M(\mathbf{P}^{n-1})$ and suppose $\mu \cdot g_n \to \nu$ for some $\nu \in M(\mathbf{P}^{n-1})$. Then ν is supported on a union of two proper projective subspaces.

Remarks i) Since the set of measures supported on a union of two proper projective subspaces is closed, this shows that the orbit of any μ not so supported is locally closed. This is the beginning of an inductive procedure for proving theorem 5.7.

ii) If $H \subset SL(n,\mathbf{R})$ leaves a measure fixed, then H is either compact or leaves the union of two subspaces invariant. Both of these conditions are "algebraic". Proceeding inductively, one should obtain either compactness or further splitting. This should result in an algebraic object which is a compact extension of a solvable group. In this way, Lemma 5.9 suggests Theorem 5.8 as well.

Proof of Lemma 5.9. Let $h_n = g_n/\|g_n\|$. Then $\|h_n\| = 1$, $\det(h_n) \to 0$, so we can assume $h_n \to h$, $h \neq 0$, $\det(h) = 0$. Let $V = \text{range}(h)$, $N = \ker(h)$, $[V]$, $[N]$ the corresponding projective subspaces in $\mathbf{P}^{n-1}$. Write $\mu = \mu_1 + \mu_2$ where $\text{support}(\mu_1) \subset [N]$, $\text{support}(\mu_2) \subset \mathbf{P}^{n-1} - [N]$. If $x \in \mathbf{P}^{n-1} - [N]$, then as $n \to \infty$, $h_n(x) \to h(x) \in [V]$. Passing to a subsequence, we can write

$$\nu = \lim \mu \cdot g_n = \lim \mu_1 \cdot g_n + \lim \mu_2 \cdot g_n,$$

and the previous sentence implies $\lim \mu_2 \cdot g_n$ is supported on $[V]$. We also have $\lim \mu_2 \cdot g_n$ supported on $[W]$ where $\lim [N] \cdot g_n = [W]$.

We now turn to the theory of orbit equivalence for amenable actions. We begin with an observation.

Proposition 5.10. For free ergodic actions amenability is an invariant of orbit equivalence.

The circle of ideas concerning orbit equivalence began in the late 1950's with the work of H. Dye [10] [11] and for amenable actions has been brought to complete form very recently. We begin with the fundamental theorem of Dye.

Theorem 5.11 (Dye [10]) All finite measure preserving (properly) ergodic Z-actions are orbit equivalent.

This was later extended to the σ-finite case by Krieger [25].

Theorem 5.12 (Krieger) All σ-finite (but not finite) measure preserving Z-actions are orbit equivalent.

Krieger also extended the theorem to the case of quasi-invariant measure without equivalent invariant measure. He showed that a measurement of the extent to which the action fails to be measure preserving is a complete invariant of orbit equivalence. Namely, let X be an ergodic G-space, and $r:X \times G \to \mathbf{R}^+$ be the Radon-Nikodym cocycle. Let $\Delta:G \to \mathbf{R}^+$ be the modular function of G, and let $m:X \times G \to \mathbf{R}^+$ be $m(x,g) = r(x,g)\Delta(g)^{-1}$. We call m the modular cocycle. The Mackey range of this cocycle will be an ergodic $\mathbf{R}^+$-action, which we call the modular flow or the modular range. For unimodular groups, the modular flow will be translation of $\mathbf{R}^+$ on $\mathbf{R}^+$ itself if and only if there is an invariant measure for the action (for then the Radon-Nikodym cocycle is trivial (Proposition 3.6)).

Theorem 5.13. (Krieger) For Z-actions with quasi-invariant measure, and not possessing (an equivalent) finite invariant measure, the modular flow is a complete invariant or orbit equivalence.

A good account of Krieger's work is [44].

Example 5.14. Let us see how to compute the modular flow in some examples. Let $\Gamma \subset SL(2,\mathbf{R})$ a lattice, P the upper triangular subgroup. The Radon-

Nikodym cocycle for the action of $G = SL(2,\mathbf{R})$ on G/P is the cocycle $\alpha: G/P \times G \to \mathbf{R}^+$ corresponding to the homomorphism $P \to \mathbf{R}^+$ given by Δ_p, the modular function of P. Clearly $r: G/P \times \Gamma \to \mathbf{R}^+$ is just $\alpha | G/P \times \Gamma$. Now the Γ action on $G/P \times \Gamma \to \mathbf{R}^+$ that appears in the construction of the Mackey range is just the resriction to Γ of the G action $G/P \times_\alpha \mathbf{R}^+$. Since $\Delta: P \to \mathbf{R}^+$ is surjective, as is well known, this is the G - action $G/P \times_\alpha P/\ker \Delta_P$ on which G acts transitively with stabilizer $\ker\Delta_P$. Thus as a Γ-space, $G/P \times_r \mathbf{R}^+$ is just the action of Γ on $G/\ker \Delta_P$. Since $\ker \Delta_P$ is not compact, by Moore's ergodicity theorem (section 2) Γ is ergodic on this space. Thus there is only one ergodic component, and so the modular flow is the action of $\mathbf{R}^+$ on a point. This computation can clearly be carried out on any semisimple non-compact Lie group. If Γ is a lattice in such a group G, and $P \subset G$ is a minimal parabolic, then the modular flow of the action of Γ on G/P will be the action of $\mathbf{R}^+$ on a point.

The Dye-Krieger theorems were extended over the years by a number of persons to include within its framework actions of larger classes of groups. (In fact Dye did not restrict himself to the integers.) This work has recently culminated with the following theorems.

Theorem 5.15 (Connes-Feldman-Ornstein-Weiss) [8], [37]

i) A free properly ergodic action of a discrete group is amenable if and only if it is orbit equivalent to a Z-action.

ii) The Dye-Krieger theorems (5.11-5.13) hold for the class of amenable properly ergodic actions of discrete groups.

Theorem 5.16 (Connnes-Feldman-Ornstein-Weiss) [8], [37]

i) A free properly ergodic action of a continuous group is amenable if and only if it is orbit equivalent to an $\mathbf{R}$-action.

ii) For such actions, the modular flow is a complete invariant of orbit equivalence. In particular, any two free properly ergodic actions of continuous amenable unimodular groups with invariant measure are orbit equivalent.

Example 5.17. If G is a semisimple non compact Lie group, $\Gamma \subset G$ a lattice, $P \subset G$ a minimal parabolic, we saw in Example 5.14 that the modular flow of Γ on G/P is independent of G and Γ. Since these actions are amenable, theorem 5.15 says that they are all orbit equivalent.

6. Rigidity: The Mostow-Margulis Theorem and a Generalization to Ergodic Actions.

In this lecture we shall describe the proof of the Mostow-Margulis rigidity theorem for lattices in semisimple Lie groups and indicate how this result can be extended to yield results about orbit equivalence for ergodic actions of semisimple groups and their lattices.

Theorem 6.1. (Mostow-Margulis Rigidity) Let G, G' be connected semisimple Lie groups with finite center, no compact factors, and Γ, Γ' irreducible lattices in G, G' respectively. Suppose $\mathbf{R}$-rank$(G) \geq 2$. Then

i) If Γ and Γ are isomorphic, then G and G' are locally isomorphic.

ii) In the center free case, any isomorphism $\Gamma \to \Gamma'$ extends to a rational isomorphism $G \to G'$.

This was first proved for cocompact lattices by Mostow [36] and for non-cocompact lattices by Margulis [27]. In an extraordinary and highly original and innovative paper, Margulis then gave an alternate proof in [28] which subsumed both cases, gave stronger results on the extension of homomorphisms from Γ to G, and which was powerful enough to prove the arithmeticity of lattices. In [55] we showed how Margulis' techniques could be incorporated into a proof of rigidity for ergodic actions. Theorem 6.1 is also true without the $\mathbf{R}$-rank assumption as long as $G \neq PSL(2,\mathbf{R})$. This is due to Mostow [36] and Prasad [39]. We now describe rigidity for ergodic actions.

Definition 6.2 [55]. Suppose G is a semisimple connected Lie group with finite center and no compact factors. An ergodic G space S is called irreducible if every non-central normal subgroup of G is also ergodic on S.

(Thus if Γ is a lattice, G/Γ is irreducible if and only if Γ is irreducible.)

Theorem 6.3 [55] (Rigidity for ergodic actions). Let G, G' be connected semisimple Lie groups with finite center and no compact factors, S, S' free irreducible ergodic G, G'-spaces, respectively, with finite invariant measure. Let $\mathbf{R}$-rank(G) ≥ 2. Suppose the actions are orbit equivalent. Then

i) G and G' are locally isomorphic,

ii) In the centerfree case, we can take G = G', and then the actions on S and S' are automorphically conjugate.

Thus, this theorem asserts that one has behavior that is diametrically opposed to the behavior of actions of amenable groups. Although theorems 6.1 and 6.3 look rather different, let us show that they are both direct consequences of the following theorem.

Theorem 6.4 [55]. Let G, G' be connected semisimple Lie groups with trivial center and no compact factors, and let S be an irreducible ergodic G space with finite invariant measure. Assume $\mathbf{R}$-rank(G) ≥ 2. Let $\alpha: S \times G \to G'$ be a cocycle whose Mackey range is Zariski dense in G'. Then α is equivalent to a cocycle β that is the restriction of a rational epimorphism $\pi: G \to G'$.

To deduce theorem 6.1 from theorem 6.4, one observes that when applied to the case $S = G/\Gamma$ and $\alpha: G/\Gamma \times G \to G'$ a cocycle corresponding to a homomorphism $\Gamma \to G'$, theorem 6.4 yields the following theorem of Margulis.

Theorem 6.5 (Margulis). G, G' as in Theorem 6.4, ($\mathbf{R}$-rank(G) ≥ 2), $\Gamma \subset G$ an irreducible lattice. Suppose $\pi: \Gamma \to G'$ is a homomorphism with $\pi(\Gamma)$ Zariski dense in G'. Then π extends to a rational epimorphism $G \to G'$.

Theorem 6.1 then follows. To deduce Theorem 6.3 from Theorem 6.4, one simply applies Theorem 6.4 to the cocycle $\alpha: S \times G \to G'$ coming from an orbit equivalence. The hypothesis of Theorem 6.4 are satisfied since the Mackey range of α is the G'-space S' and the Borel density theorem implies Zariski density. The conclusion of 6.4 implies that of 6.1 by Proposition 3.3.

Let us give an example of how to apply Theorem 6.3 to some natural examples.

Corollary 6.6 [55]. Let G, G' be connected simple non-compact Lie groups with finite center, $\Gamma, \Gamma' \subset G, G'$ lattices and suppose S, S' are free ergodic Γ, Γ' spaces with finite invariant measure. Suppose $\mathbf{R}$-rank$(G) \geq 2$, and that the Γ action on S and Γ'-action on S' are orbit equivalent. Then G and G' are locally isomorphic.

Proof. Let $X = \mathrm{ind}_\Gamma^G(S)$, $X' = \mathrm{ind}_{\Gamma'}^{G'}(S')$. Then one easily checks that the hypothesis of Theorem 6.3 is satisfied.

Example 6.7 [55]. As we vary n, $n \geq 2$, the natural actions of $SL(n,Z)$ on $\mathbf{R}^n/Z^n$ by automorphisms are mutually non-orbit equivalent.

We now turn to some proofs. We will not prove Theorem 6.4 here, but rather only Theorem 6.5, (which therefore gives us a proof of Theorem 6.1). The first part of the proof we present is different from Margulis' original argument. Instead, we present an argument which generalizes nicely when one attempts to prove Theorem 6.4. It is perhaps also more transparent then the original argument. The remainder of the proof will be that of Margulis, although we shall try to give some motivation. For the proof of Theorem 6.4, see [55].

Proof of Theorem 6.5. Let $P \subset G$, $P' \subset G'$ be minimal parabolic subgroups. We have a homomorphism $\pi: \Gamma \to G'$ and hence G'/P' becomes a compact metrizable Γ-space. On the other hand, as we observed in Example 5.4, the action of Γ on G/P is ergodic and amenable. By the remarks following Definition 5.2, there is a measurable Γ-map $\varphi: G/P \to M(G'/P')$, the latter space being the space of probability measures on G'/P'. By Theorem 5.7, the action of G' on $M(G'/P')$ is smooth, so $\hat{M}(G'/P') = [M(G'/P')]/G'$ is countably separated and generated. Since φ is a Γ-map, $\varphi(x\gamma) = \varphi(x)\pi(\gamma)$, so $\varphi(x\gamma) \equiv \varphi(x)$ in $\hat{M}(G'/P')$. By ergodicity of Γ on G/P, the projection of φ into $\hat{M}(G'/P')$ is essentially constant, i.e. $\varphi(G/P)$ can be assumed to lie in one G'-orbit in $M(G'/P')$. Thus, we can view φ as a Γ-map $\varphi: G/P \to G'/H'$, and by Theorem 5.8, H' is an amenable algebraic subgroup. What we have done is to obtain a Γ-map φ where the image is no longer an infinite dimensional space $M(G'/P')$ but an algebraic variety G'/H'. The existence of such a measurable map φ is the first main step in the proof. The second step is the following fundamental lemma of Margulis.

Lemma A. $\varphi: G/P \to G'/H'$ is a rational mapping of algebraic varieties.

Let us show why this lemma suffices to prove the theorem. Suppose $\varphi: G/P \to G'/H'$ is a rational mapping such that $\varphi(x\gamma) = \varphi(x)\,\pi(\gamma)$. Let $R(G/P, G'/H')$ be the space of rational mappings. Then $G \times G'$ acts on $R(G/P, G'/H')$ by

$$[(g,g')\cdot f](x) = f(xg)\cdot(g')^{-1}$$

The fact that φ is a Γ-map means φ is fixed under Γ, where Γ is identified with the subgroup of $G \times G'$ given by $\{(\gamma,\pi(\gamma))\}$. Let $\overline{\Gamma}$ be the algebraic hull of Γ in $G \times G'$. Then φ is also fixed under $\overline{\Gamma}$. We claim that $\overline{\Gamma}$ is the graph of homomorphism $G \to G'$. Since Γ is Zariski dense in G, $\overline{\Gamma}$ must project onto all of G. So suppose $(g,h_1), (g,h_2) \in \overline{\Gamma}$.

Then $\varphi(xg) = \varphi(x)h_1$ and $\varphi(xg) = \varphi(x)h_2$. Therefore $h_1h_2^{-1}$ leaves $\varphi(G/P)$ pointwise fixed. But $\pi(\Gamma)$ leaves $\varphi(G/P)$ invariant, and since $\pi(\Gamma)$ is Zariski dense in G', $\varphi(G/P)$ must be Zariski dense in G'/H'. Therefore $h_1h_2^{-1}$ leaves all G'/H' pointwise fixed, and since $\bigcap_{g \in G'} gH'g^{-1} = \{e\}$ (since it is an amenable normal subgroup), $h_1h_2^{-1} = e$. Therefore $\overline{\Gamma}$ is the graph of a function $G \to G'$, which is a homomorphism since $\Gamma \subset G$ is Zariski dense and the map is a homomorphism on Γ.

We now return to the proof of the lemma. We must show that a certain measurable mapping between varieties is actually rational. There is one well known situation in which a measurable map is known to have much stronger properties, namely if the map is a homomorphism. For example, any measurable homomorphism between Lie groups is C^∞, and similarly, any measurable homomorphism between real algebraic groups $R \to R'$, with R reductive, will be rational on all unipotent subgroups of R. Of course our map φ is defined on G/P which is not a group. However, up to a set of measure 0, it is a group. For example, consider $G = SL(n,\mathbf{R})$, P = upper triangular subgroup. Let U be the lower triangular unipotent matrices. Then the natural map $G \to G/P$ carries U onto an open subset of measure 1. Furthermore, this establishes an isomorphism of U with its image as algebraic varieties. Thus, we can view φ as a map $U \to G'/H'$. Now, although we have φ defined on a group, it is not a homomorphism (as the image is not even a group.) We do however, have some sort of algebraic relation, namely the fact that φ is a Γ-map. Thus we might hope to be able to force this algebraic relation to show that φ only depends upon a homomorphism of U. This, however, is not possible. As we shall see, when we try to force the algebra, we shall need some commutativity with U from elements of A, where A is the positive diagonals. But the centralizer of U in A is trivial. In the $\mathbf{R}$-rank 1 case, we can proceed no further, but in higher rank all is not lost. Let us fix an element $t \in A$, and consider the centralizer C_t. For example, in $SL(3,\mathbf{R})$, let

$$t = \begin{pmatrix} \lambda & 0 & 0 \\ 0 & \lambda & 0 \\ 0 & 0 & \lambda^{-2} \end{pmatrix}, \quad C_t = \left\{ \begin{pmatrix} M & \Big| & 0 \\ \hline 0\ 0 & \Big| & \alpha \end{pmatrix} \text{ where } \alpha = (\det M)^{-1} \right\}$$

Let $C_t^u = C_t \cap U = \left\{ \begin{pmatrix} 1 & 0 & 0 \\ a & 1 & 0 \\ 0 & 0 & 1 \end{pmatrix} \right\}$. Now $U \cong \mathbf{R}^3$, and $C_t^u \cong \mathbf{R}$. Thus C_t^u will give us one direction in U, and C_t is a reductive group that has a centralizer in A. As we shall see, this will be enough to show that $\varphi : U \to G'/H'$ depends rationally on C_t^u. But now if we vary $t \in A$, we can pick up the other directions in the same way. The following lemma of Margulis is now clearly relevent.

<u>Lemma B</u>. Let φ be a measurable function defined on $\mathbf{R}^n \times \mathbf{R}^k$. If φ is rational in x for almost all $y \in \mathbf{R}^k$ and rational in y for almost all $x \in \mathbf{R}^n$, then φ is rational.

The above remarks about SL(n, R) extend to general G. Thus, if we let U be the unipotent radical of the parabolic oposite to P, then $U \to G/P$ is an isomorphism of algebraic varieties with its image, the latter being open and of full measure. Let $A \subset P$ be a maximal abelian **R**-diagonalizable subgroup and $t \in A$, $t \neq 0$. Let C_t be the centralizer of t in G. Then C_t is reductive. Letting $C_t^u = U \cap C_t$, U can be built from the various C_t^u by varying t. Thus, using Lemma B, it suffices now to prove the following. View $\varphi : G/P \to G'/H'$ as a map $\varphi : G \to G'/H'$, with $\varphi(pg) = \varphi(g)$, for $p \in P$.

<u>Lemma C</u>. For almost all $g \in G$, $\varphi(cg)$ depends rationally on c for $c \in C_t^u$ (for any $t \in A$).

<u>Proof</u>. Let $C_t = C$. We want to study dependence of φ on C, so for each $g \in G$, define $\omega_g : C \to G'/H'$ by $\omega_g(c) = \varphi(cg)$.

Thus we have a map $\omega : G \to F(C, G'/H')$, the latter being the space of measurable maps $C \to G'/H'$. Let $T = \{t^n\}$. Then $\omega_{tg}(c) = \omega(ctg) = \omega(tcg)$

$\omega(cg)$ (since $t \in P$), and so we have $\omega_{tg}(c) = \omega_g(c)$. Thus we can view ω as a map $\omega: G/T \to F(C, G'/H')$. We now use Γ-invariance of φ:

$$\omega_{g\gamma}(c) = \varphi(cg\gamma) = \varphi(cg)\pi(\gamma) = \omega_g(c) \cdot \pi(\gamma).$$

Thus, $\omega_{g\gamma}$ and ω_g are in the same G'-orbit in $F(C,G'/H')$, where G' acts on the latter pointwise. We now need another smoothness result.

Lemma D. Every G'-orbit in $F(X, G'/H')$ is locally closed, where X is a measure space.

We observe that if X is finite, this is immediate from the fact that $F(X, G'/H')$ would then be a variety. Margulis observed that the lemma is true for any measure space X. For a simple proof, see the appendix of [55].

Returning now to the proof of Lemma C, we have that $\omega_{g\gamma} \equiv \omega_g$ when projected to $[F(C,G'/H')]/G'$. By Lemma D, this latter space is countably generated and separated, and by Moore's theorem Γ is ergodic on G/T. Therefore, all ω_g are equal when projected to $[F(C,G'/H')]/G'$, or equivalently, all ω_g lie in the same G'-orbit. So for $a, g \in G$, we have $\omega_{ag} = \omega_g \cdot h(a,g)$ where $h(a,g) \in G'$ and h is measurable. For any $f \in F(C, G'/H')$, let G'_f be the stabilizer, and N_f the normalizer of G'_f in G'. Clearly for $a \in C$, $G'_{\omega_{ag}} = G'_{\omega_g}$, so for $a \in C$, $h(a,g) \in N_{\omega_g}$. Suppose now that $a_1, a_2 \in C$. Then

$$\begin{aligned}
\omega_g(c) \cdot h(a_1a_2,g) &= \omega_{a_1a_2g}(c) \\
&= \varphi(ca_1a_2g) \\
&= \omega_{a_2g}(ca_1) \\
&= \omega_g(ca_1)\cdot h(a_2,g) \\
&= \varphi(ca_1g)\cdot h(a_2,g) \\
&= \omega_{a_1g}(c)\cdot h(a_2,g) \\
&= \omega_g(c)\cdot h(a_1,g)h(a_2,g).
\end{aligned}$$

Thus, for almost all g, $a \to h(a,g)$ is a measurable homomorphism

$$C \to N_{\omega_g}/G'_{\omega_g}.$$

We have $\varphi(cag) = \varphi(cg) \cdot h(a,g)$. Thus for a in any unipotent subgroup of C, $a \to \varphi(cag)$ depends rationally on a. Choosing this subgroup to be $c^{-1}C_t^u c$, we obtain $\varphi(bg)$ depends rationally on $b \in C_t^u$. This completes the proof.

7. Complements to the Rigidity Theorem for Ergodic Actions: Foliations by Symmetric Spaces, and Kazhdan's Property (T).

The rigidity theorem for ergodic actions stated in section 6 allowed us to distinguish ergodic actions of lattices on the basis of orbit equivalence if the actions had finite invariant measure (e.g. corollary 6.6 and example 6.7). However, some of the most interesting actions of lattices, e.g., the action of SL(n,Z) on $\mathbf{P}^{n-1}$ or other flag and Grassman varieties, do not have finite invariant measure. We now indicate how to extend the rigidity theorem to enable us to deal with this situation. The main step is to first extend the rigidity theorem to actions of general connected groups.

Let H be a connected group. Every locally compact group has a unique maximal normal amenable subgroup N. If H is connected H/N will be a product of non-compact connected simple Lie groups with trivial center. We shall say that an ergodic action of H is irreducible if the inverse image in H of each of these simple factors of H/N is still ergodic.

Theorem 7.1. [56] Let H, H' be connected locally compact groups, N, N' the maximal normal amenable subgroups. Suppose $\mathbf{R}$-rank(H/N) ≥ 2. Let S, S' be free ergodic irreducible H, H'-spaces with finite invariant measure, and suppose the actions are orbit equivalent. Then H/N and H'/N' are isomorphic, and N is compact if and only if N' is also compact.

Thus, for connected groups, orbit equivalence implies isomorphism of the semisimple parts of the groups. The proof of this result is an extension of the proof of the rigidity theorem for ergodic actions of semisimple groups. To see how to apply this to obtain results about actions of lattices without invariant measure, observe that the orbit space of Γ acting on G/H can be identified with the orbit space of H acting on G/Γ. Theorem 7.1 deals with the latter situation since now G/Γ has a finite H-invariant measure, so we can try to apply this to the action of Γ on G/H. One can then prove the following precise result.

Theorem 7.2 [56]. Let G, G' connected semisimple Lie groups with finite center, Γ, Γ'- irreducible lattices. Let $H \subset G$, $H' \subset G'$ be almost connected non-compact subgroups. Assume the actions of Γ on G/H and Γ' on G'/H' are essentially free and orbit equivalent. Let N, $N' \subset H$, H' be the maximal normal amenable subgroups, and suppose $\mathbf{R}$-rank$(H/N) \geq 2$. Then H/N and H'/N' are locally isomorphic.

Example 7.3 [56]. As we vary n, $n \geq 2$, the actions of $SL(n,Z)$ on $\mathbf{P}^{n-1}$ are mutually non-orbit equivalent. This follows by simply observing that the semisimple parts of the corresponding maximal parabolics in $SL(n,\mathbf{R})$ are not isomorphic. (Actually, Theorem 7.2 will not apply to compare the cases $n=2$ and $n=3$. However, the action of $SL(2, Z)$ on $\mathbf{P}^1$ is amenable, while the action of $SL(3,Z)$ on $\mathbf{P}^2$ is not.) In a similar fashion, one can read off a large number of results about actions of lattices on the flag and Grassman varieties.

A natural question that arises in light of Theorem 7.1 is how sensitive orbit equivalence is to the way in which H is built from N and H/N. For example, what is the relation of actions of $SL(n,\mathbf{R}) \times R^n$ to that of $SL(n,\mathbf{R}) \circledS R^n$, where the latter semidirect product just results from the natural action of $SL(n,\mathbf{R})$ on R^n? To answer this question, we recall Kazhdan's notion of property (T) for groups, and then indicate how to define this for actions.

Let G be a locally compact group, and I the one dimensional trivial representation. If π_n, π are unitary representations of G, then recall $\pi_n \to \pi$ means that for any unit vectors $v_1,\dots,v_k \in H_\pi$ there exist unit vectors $v_1^n,\dots,v_k^n \in H_{\pi_n}$ such that $\langle \pi_n(g)v_i^n | v_j^n \rangle \to \langle \pi(g)v_i | v_j \rangle$ uniformly on compact sets in G for each i,j. Kazhdan [24] defined a group to have property (T) if $\pi_n \to I$ implies $I \leq \pi_n$ for n sufficiently large.

Theorem 7.4 (Kazhdan) [24], [9]. i) Semisimple Lie groups with all simple factors having **R**-rank at least 2 have property (T). (Actually, Kazhdan proved this assuming **R**-rank ≥ 3. That one only need assume **R**-rank ≥ 2 was observed by a number of persons, e.g. [9].)

ii) Any lattice subgroup of a group with property (T) also has property (T).

We shall also need the following result of Wang.

Theorem 7.5 (Wang [46]). $SL(n,\mathbf{R}) ⓢ \mathbf{R}^n$ has property (T), and hence so does $SL(n,Z) ⓢ Z^n$ $(n \geq 3)$.

We now define property (T) for ergodic actions. For simiplicity, we restrict attention to actions of discrete groups. This notion for actions originally appeared in [57].

Let G be a discrete group, S an ergodic G-space. Let $\alpha: S \times G \to U(H)$ be a unitary group valued cocycle. Let $v,w: S \to H$ be Borel functions with $\|v\|_\infty = \|w\|_\infty = 1$. Let $f_{\alpha,v,w}: S \times G \to C$ be given by $f_{\alpha,v,w}(s,g) = \langle \alpha(s,g)v(sg) | w(s) \rangle$. We consider $f_{\alpha,v,w}$ as a function $G \to F(S,\mathbf{C})$, and we endow $F(S,\mathbf{C})$ with the topology of convergence in measure. For cocycles α_n, α, we say $\alpha_n \to \alpha$ if given $v_1,\dots,v_k: S \to H_\alpha$, $\|v_i\|_\infty = 1$, there exist $v_1^n,\ldots,v_k^n : S \to H_{\alpha_n}$ such that $f_{\alpha_n, v_i^n, v_j^n} \to f_{\alpha, v_i, v_j}$ pointwise on G (i.e. in measure on S for each $g \in G$) for all i,j.

Definition 7.6 [57]. The action of G on S has property (T) if $\alpha_n \to I$ implies $\alpha_n \geq I$ for n sufficiently large. Here I is the one dimensional trivial cocycle and $\alpha \geq I$ means $\alpha \sim \beta$ where $\beta(s,g)v = v$ for some non-zero vector v.

We then have the following results.

Theorem 7.7 [57]. a) If G has property (T), and S has a finite invariant measure then S has property (T).

b) If S has property (T), finite invariant measure and is weak mixing (i.e. there are no finite dimensional invariant subspaces in $L^2(S) \ominus C$), then G has property (T).

c) For free actions of discrete groups, property (T) is an invariant of orbit equivalence.

Combining 7.5 and 7.7, we have the following, showing that, in fact, orbit equivalence is quite sensitive to the way H is constructed from N and H/N.

Corollary 7.8 [57]. Let $n \geq 3$, and $\Gamma_1 = SL(n,Z) \times Z^n$, $\Gamma_2 = SL(n,Z) \circledS Z^n$. Then Γ_1 and Γ_2 do not have free orbit equivalent weakly mixing actions with finite invariant measure.

We shall now describe a geometric interpretation of the rigidity theorem for ergodic actions. We begin by recalling the geometric formulation of the Mostow-Margulis theorem. Let G be a connected semisimple Lie group with finite center and no compact factors, $K \subset G$ a maximal compact subgroup, and $\Gamma \subset G$ a torsion free lattice. Then G/K is a Riemannian symmetric space (diffeomorphic to Euclidean space), and Γ operates properly discontinuously on $X = G/K$. Thus $\Gamma\backslash X$ is locally symmetric space of finite volume, and $\pi_1(\Gamma\backslash X) \cong \Gamma$.

Theorem 7.9 (Mostow-Margulis rigidity, geometric form). Let M_1, M_2 be locally symmetric Riemannian manifolds of finite volume whose universal covers, X_1, X_2 are symmetric spaces of purely non-compact type, and whose fundamental groups $\pi_i(M_i)$ act as irreducible groups of isometries of X_i. Suppose further that the rank of M_1 is at least 2. Then any isomorphism $\pi_1(M_1) \to \pi_1(M_2)$ is induced by a diffeomorphism $M_1 \to M_2$ that is an isometry modulo normalizing scalar multiples.

Roughly speaking, this asserts that for a particular class of Riemannian manifolds, i.e. suitable locally symmetric spaces, that a purely topological invariant, namely the fundamental group, determines the Riemannian structure. We now describe an analogous geometric interpretation of the rigidity theorem for actions which will make an assertion about foliations by symmetric spaces.

Let G be a connected semisimple non-compact Lie group with finite center and no compact factors, $K \subset G$ a maximal compact subgroup, and (S,μ) a free ergodic G-space with finite invariant measure. Let $Y = S/K$. Then because K is compact, Y is a standard Borel space, and the orbits in S yield an equivalence relation $\mathcal{J}$ on Y in which each equivalence class can be identified with G/K. Thus, $(Y,\mathcal{J})$ is a "Riemannian measurable foliation", i.e., a measure space with an equivalence relation $\mathcal{J}$ in which each equivalence class (or "leaf") has the structure of a C^∞- Riemannian manifold, so that these structures vary measurably in a suitable sense [53] over Y. Given Y, Y', two spaces supporting Riemannian measurable foliations, we call them isometric if there is a measure space isomorphism between Y and Y' that carries leaves onto leaves isometrically (possibly after discarding null sets of leaves). By a transversal for $(Y, \mathcal{J})$, we mean a Borel set intersecting almost every leaf in a countable set. Such a set T will have a natural equivalence relation on it with countable equivalence classes (namely $\mathcal{F}|T$), and a natural measure class ν satisfying the condition that for $B \subset T$, $\nu(B) = 0$ if and only if the union of the leaves intersecting B has μ-measure 0 [13], [41]. We call two Riemannian measurable foliations transversally equivalent if they have isomorphic transversals. By an isomorphism of transversals, we mean isomorphism as measure spaces with equivalence relations, i.e., a measure space isomorphism carrying one equivalence relation onto the other. This is a purely measure-theoretic invariant of the foliation. The following is the geometric version of Theorem 6.3.

Theorem 7.10 [55] (Rigidity for foliations by symmetric spaces). Let G,G', S,S' be as in Theorem 6.3. Let $Y = S/K$, $Y' = S'/K'$ where $K,K' \subset G,G'$ are the maximal compact subgroups. Let $(Y, \mathcal{F})$, $(Y', \mathcal{F}')$ the associated Riemannian measurable foliations by symmetric spaces. If the foliations are transversally equivalent, then they are isometric, modulo normalizing scalar multiples (independent of the leaves).

Thus, roughly speaking, for suitable foliations in which the leaves are symmetric spaces, a purely measure theoretic invariant, namely the measure theory of the transversal, determines the Riemannian structure on almost every leaf.

As we have already remarked, the rigidity theorem for lattices holds in the **R**-rank 1 case as well as long as $G \neq PSL(2,\mathbf{R})$, although the proof we have given in section 6 does not apply, and one must use other techniques, for example those of Mostow [36] and Prasad [39]. It is natural to enquire as to what extent the rigidity theorem for ergodic actions holds in the **R**-rank 1 case as well. In [58] we proved the following result in this direction, applying basic results of Mostow [35] on quasi-conformal mappings.

Theorem 7.11. [58] Let S, S' be free ergodic $SO(1,n)/\{\pm I\}$-spaces with finite invariant measure, and assume $n \geq 3$. Let $(Y, \mathcal{F})$, $(Y', \mathcal{F}')$ be the associated measurable foliations by hyperbolic space (as in the discussion preceding Theorem 7.10). If $(Y, \mathcal{F})$ and $(Y', \mathcal{F}')$ are quasi-conformally equivalent, then they are isometric (modulo a normalizing scalar independent of the leaf), and the actions of $SO(1,n)/\{\pm I\}$ on S and S' are automorphically conjugate.

Here, of course, quasi-conformal equivalence asserts the existence of a measure space isomorphism taking (almost all) leaves to leaves quasi-conformally. We remark that the analogous statement for R^n actions can be shown to be false by many counterexamples.

8. Margulis' Finiteness Theorem.

In section 6, we saw how the analysis of the ergodic action of Γ on G/P led to Margulis' proof of the rigidity theorem. Margulis has also demonstrated some other deep properties of this ergodic action and used this to obtain very strong results about the structure of Γ. More precisely, he has shown the following.

Theorem 8.1 (Margulis [30],[31]). Let G be a connected semisimple Lie group with finite center and no compact factors, and assume $\mathbf{R}$-rank$(G) \geq 2$. Let $\Gamma \subset G$ be an irreducible lattice, and $H = \Gamma/N$ a non-amenable quotient group. Then $N \subset Z(G)$, the center of G, and in particular, is finite.

If we further assume that the $\mathbf{R}$-rank of every simple factor of G is at least 2, then Γ has property (T) of Kazhdan [24], and hence if $H = \Gamma/N$ is an amenable quotient, H must also have property (T) and hence is finite. Thus, we conclude the following.

Corollary 8.2. Let G be a connected semisimple Lie group with finite center and assume $\mathbb{R}$-rank of each simple factor of G is at least 2. Let $\Gamma \subset G$ be an irreducible lattice. Then every normal subgroup of G is either finite or of finite index.

Margulis' results are in fact significantly more general, both in terms of taking lattices in products of algebraic groups defined over various local fields and in terms of rank restrictions. The basic difficult step in the proof of Theorem 8.1 is the following result concerning the action of Γ on G/P. Let P' be another parabolic subgroup containing P. Then there is a Γ-map $G/P \to G/P'$, i.e. G/P' is a Γ-space factor of G/P.

Theorem 8.3 (Margulis [30]). Let G, Γ as in theorem 8.1, $P \subset G$ a minimal parabolic. Then any measurable factor of the Γ-space G/P is of the form $G/P \to G/P'$ for some parabolic $P' \supset P$.

In other words, every measurable Γ-factor of G/P is actually also a G-factor. This theorem is difficult and we will not prove it here. Instead, we show how to deduce theorem 8.1 from it.

Let $H = \Gamma/N$ be a non-amenable quotient. Then there is a compact metric H-space X so that there is no H-invariant measure on X. We can also view X as a compact metric Γ-space. Since the action of Γ on G/P is amenable, by the discussion following definition 5.2, there is a measurable Γ-map $\varphi: G/P \to M(X)$, where the latter is the space of probability meaures on X. If we let μ be a measure on G/P in the natural measure class, then $(M(X), \varphi_*(\mu))$ is a Γ-space factor of G/P. Thus, there is some parabolic P' so that as Γ-spaces, $(M(X), \varphi_*(\mu))$ is conjugate to G/P'. Since there are no fixed points in M(X) under Γ, $P' \neq G$. But N acts trivially on M(X) by definition, so N is trivial on G/P' which implies $N \subset \bigcap_{g \in G} gP'g^{-1}$, a proper normal subgroup of G. Dividing G by its center, it clearly suffices to observe that if $\Gamma \subset \prod_{i \in I} G_i$ is an irreducible lattice in a product of simple Lie groups with trivial center, that $N = \Gamma \cap \prod_{i \in J} G_i$ is trivial for $J \subset I$ a proper subset. But since N is normalized by Γ and $\prod_{I-J} G_i$, it is normalized by the product of these groups which is dense in G by irreducibility. The result follows.

9. Margulis' Arithmeticity Theorem.

(This section will require a bit more knowledge about algebraic groups than previous sections. We also caution the reader that in this section, by algebraic group, Zariski closure, etc., we shall mean with respect to the algebraically closed field, unless we explicitly declare otherwise in a given instance.)

In this section we describe the proof of Margulis' arithmeticity theorem for lattices in semisimple Lie groups. The proof of the rigidity theorem in section 6 was based on a result asserting that under suitable hypotheses, a homomorphism of Γ into a real algebraic group extended to a homomorphism of G. This result is also basic to the proof of the arithmeticity theorem. However, we shall also need results concerning homomorphisms of Γ into complex groups and algebraic groups over local fields. With some additional comments, the proof of theorem 6.5 can be applied to give us these needed results, so that the bulk of the work of the proof of arithmeticity has in fact already been done. But before passing to these arguments, let us recall the statement of the problem.

The first example of a lattice in a Lie group is the integer lattice $Z^n \subset R^n$. This is of course not the only lattice in R^n. However if L is any lattice there is an automorphism $A: R^n \to R^n$ such that $A(Z^n) = L$. Thus, L is "arithmetically" defined.

To get other examples of lattices, suppose $G \subset GL(n, C)$ is an algebraic group defined over **Q**, i.e. there is an ideal $I \subset Q[a_{ij}, \det(a_{ij})^{-1}]$ such that $G = \{a \in GL(n,C) \mid p(a) = 0 \text{ for all } p \in I\}$. As usual, if $B \subset C$ is any subring, we let

$$G_B = \{a \in G \mid a_{ij} \in B, \text{ for all } i,j \text{ and } \det(a_{ij})^{-1} \in B\}.$$

Theorem 9.1 (Borel-Harish-Chandra) [5]. If G is semisimple, then G_Z is a lattice in G_R.

For example, for $G = SL(n, C)$, we have $SL(n, Z)$ is a lattice in $SL(n, R)$. The question the arithmeticity theorem answers is to what extent this is a general construction, i.e. to what extent are lattices arithmetically defined? We now exhibit two ways of modifying a given lattice to obtain a new lattice.

Definition 9.2. Γ, Γ' discrete groups, then Γ and Γ' are called commensurable if $[\Gamma : \Gamma \cap \Gamma'] < \infty$ and $[\Gamma' : \Gamma \cap \Gamma'] < \infty$.

Proposition 9.3. If $\Gamma, \Gamma' \subset G$, Γ is a lattice and Γ, Γ' are commensurable, then Γ' is a lattice.

For example, given that $SL(n,z)$ is a lattice, $\{a \in SL(n,Z) \mid a = I \bmod p$ for a given prime $p\}$ is a commensurable lattice.

Here is another way to get new lattices.

Proposition 9.4. If $\Gamma \subset H$ is a lattice, $\varphi: H \to G$ a surjective homomorphism with compact kernel then $\varphi(\Gamma)$ is a lattice in G.

Margulis' theorem says that aside from these two types of rather trivial modifications, every lattice in a semisimple Lie group of higher **R**-rank arises as in Theorem 9.1. More precisely, let us make the following definition. (If H is a group, H^0 denotes the topologically connected component of the identity.)

Definition 9.5. Let G be a connected semisimple Lie group with trivial center and no compact factors. Let $\Gamma \subset G$ be a lattice. Then Γ is called arithmetic if there exists an algebraic group H defined over **Q**, and a surjective homomorphism $\varphi: H^0_{\mathbf{R}} \to G$ such that

i) kernel(φ) is compact;

ii) $\varphi(H_Z \cap H^0_{\mathbf{R}})$ is a lattice in G commensurable with Γ.

Theorem 9.6 (Margulis [28]). Let G be as in definition 9.5, and assume **R**-rank $(G) \geq 2$. Then any irreducible lattice in G is arithmetic.

As we indicated above, the proof is based on two further results about homomorhisms of Γ.

Theorem 9.7 (Margulis [28]). Let $\Gamma \subset G$ an irreducible lattice, G as above, R-rank(G) ≥ 2.

i) If H is a (complex) simple algebraic group, connected and with trivial center, then any homomorphism $\pi:\Gamma \to H$ with $\pi(\Gamma)$ Zariski dense in H either satisfies $\overline{\pi(\Gamma)}$ compact or extends to a rational endomorphism $\tilde{G} \to H$, where $\tilde{G}$ is the Zariski closure of G (embedding G in the linear transformations in the complexified Lie algebra for example).

ii) Any homomorphism $\pi:\Gamma \to H_K$ where H is a semisimple algebraic group over K, and K is a local totally disconnected field of characteristic 0, with $\pi(\Gamma)$ Zariski dense, satisfies $\overline{\pi(\Gamma)}$ is compact.

The proof we present is in the spirit of the proof we gave of Theorem 6.5 so as to be generalizable to cocycles defined on general ergodic G-spaces. We expect these generalized results to be of use in describing "arithmetic" features of an ergodic action, but we do not discuss this here.

Proof. i) The proof we gave of Theorem 6.5 can be applied if we can find a measurable Γ-map $\varphi:G/P \to H/H_0$ where H_0 is an algebraic subgroup of H such that $\bigcap_{h \in H} hH_0h^{-1} = \{e\}$. As in Theorem 6.5, we can let $P' \subset H$ be a minimal parabolic subgroup, use amenability to find a Γ - map $\varphi:G/P \to M(H/P')$ and prove that each orbit in M(H/P') under H is locally closed. Again, as in 6.5, we can then assume $\varphi:G/P \to H/H_1$ where H_1 is the stabilizer of a measure in M(H/P'). Unlike the real case however, this stabilizer need not be algebraic. For example, the group may be compact which in the real case implies that it is the real points of an algebraic group, while in the complex case, of course, a compact group will not be algebraic. However, we can suppose H is rationally represented on a finite dimensional complex space in such a way that P' is the stabilizer of a point in projective space. Let μ be the measure on H/P' stabilized by H_1. If H_1 is not compact, then using an argument as in Furstenberg's lemma, (lemma 5.9) we see that μ must be supported on the intersection of H/P' with the union of two proper projective subspaces. Choose a proper subspace V so that $\mu(H/P' \cap [V]) > 0$, and V has minimal dimension among all subspaces with this property. By the minimality property of [V] and H_1 -invariance of μ, the H_1 -orbit of [V] must clearly be a finite union of projective subspaces. Hence, if we let H_0 be the Zariski closure of H_1, then $H_0 \subset H$ is a proper algebraic subgroup. Since H is simple and with trivial center $\bigcap hH_0h^{-1} = \{e\}$, and as we remarked at the beginning of the proof, this suffices.

We must now consider the case in which H_1 is compact. We then have a Γ-map $\varphi:G/P \to H/H_1$, so that if we let $\nu = \varphi_*(\mu)$, ν is a quasi-invariant ergodic measure for the action of Γ on H/H_1. (Unlike the previous paragraph, μ is now the natural measure class on G/P.) Consider the Γ-map $\varphi \times \varphi: G/P \times G/P \to H/H_1 \times H/H_1$. It is well known that on G/P, the P-action is essentially transitive, the conull orbit having $P \cap \overline{P}$ as stabilizer, where $\overline{P}$ is the opposite parabolic to P. Thus as a G-space, $G/P \times G/P$ will be essentially transitive with stabilizer $P \cap \overline{P}$, which is non-compact. By Moore's ergodicity theorem (section 2), Γ is therefore ergodic on $G/P \times G/P$. It follows that Γ must also be ergodic on $(H/H_1, \nu) \times (H/H_1, \nu)$. Since H_1 is compact, the H-orbits on $H/H_1 \times H/H_1$ are closed. Since $(\nu \times \nu)$ is ergodic under Γ, and Γ-orbits are of course contained in H-orbits, smoothness of the H-action on $H/H_1 \times H/H_1$ implies that $\nu \times \nu$ must be supported on an H-orbit. From Fubini's theorem, one easily deduces that ν must be supported on an H_2 orbit in H/H_1 where H_2 is a conjugate of H_1, and in particular is compact. Thus, support(ν) is compact. Since λ is quasi-invariant under $\pi(\Gamma)$, support (ν) is $\pi(\Gamma)$-invariant, and since H_1 is also compact, it follows that $\pi(\Gamma)$ is contained in a compact set. This completes the proof of (i).

ii) Let $P' \subset H$ be a minimal parabolic K-subgroup, so that H_K/P'_K is compact, and P'_K contains no normal algebraic subgroup. We again wish to apply the same type of argument as in the proof of Theorem 6.5. The first step is to prove that analogue of Theorem 5.7 over K. In fact the proof in [52] shows that GL(n,K) acts smoothly on $M(\mathbf{P}^{n-1}(K))$. We can assume that we have a faithful rational representation of H_K on K^n so that H_K/P'_K is an orbit in $\mathbf{P}^{n-1}(K)$. By amenability of the Γ-action on G/P, there is a Γ-map $\varphi: G/P \to M(H_K/P'_K) \subset M(\mathbf{P}^{n-1}(K))$. By smoothness of the GL(n,K)-action on $M(\mathbf{P}^{n-1}(K))$, we can view φ as a map $\varphi: G/P \to [\mu \cdot GL(n,K)] \cap M(H_K/P'_K)$ where $\mu \in M(H_K/P'_n)$. Let S be the stabilizer of μ in GL(n,K). If S is compact in PGL(n,K), then by the argument in part (i), $\overline{\pi(\Gamma)}$ will be compact in PGL(n,K), and so $\overline{\pi(\Gamma)}$ will also be compact in H. If not, then using an argument as in Furstenberg's Lemma (5.9), we can, as in part (i), assume the Zariski closure L of S is a proper algebraic subgroup. Furthermore, we can clearly assume from the construction of L as in part (i), that for any $g \in GL(n,K)$, $\dim(H \cap gLg^{-1}) < \dim H$. By the condition of Zariski density, it therefore suffices to see that $\pi(\Gamma) \subset gL_K g^{-1}$ for some $g \in GL(n,K)$. We have $\varphi: G/P \to GL(n,K)/L_K$ a measurable Γ-map. In the real case we showed φ was rational by showing it could be built from homomorphisms of unipotent subgroups of G which had to be rational. In the present situation, we can construct the same type of homomorphism using the argument of Theorem 6.5, but now, since the image group is totally disconnected, these maps must be constant. We thus conclude that $\varphi: G/P \to GL(n,K)/L_K$ is essentially constant. Since $\pi(\Gamma)$ leaves $\varphi(G/P)$ fixed, this implies $\pi(\Gamma)$ is contained in a $GL(n,K)$ - conjugate of L_K, and this completes the proof.

We now turn to the proof of theorem 9.6 itself. We may take the semisimple Lie group to be $G^0_{\mathbf{R}}$, where now $G \subset GL(n, \mathbf{C})$ represents an algebraic group defined over Q. Thus $\Gamma \subset G^0_{\mathbf{R}}$ is an irreducible lattice. The following lemma is classical, and follows for example from an argument of Selberg [42] (see also [40 ,Prop. 6.6] for the same argument.) This argument is based on expressing the embeddings of Γ into G as an algebraic variety, and then choosing a real algebraic point of this variety. However, with theorem 9.7 at hand, we present an alternative argument due to Margulis [29].

Lemma 9.8. There is a real algebraic number field k and a rational faithful representation of G such that, identifying G with its image under this representation, $\Gamma \subset G_k$.

Proof. The first step is to show that for K the field of real algebraic numbers we have $Tr(Ad(\gamma)) \in K$ for all γ. Following Margulis, we let σ be an automorphism of $\mathbf{C}$. Then σ acts on matrices with entries in $\mathbf{C}$ by taking $(z_{ij}) \to (\sigma(z_{ij}))$, and since G is defined over $\mathbf{Q}$, σ induces an automorphism of G. (Of course this is an automorphism of G only as an abstract group, and will in general not be measurable.) Let us assume for the moment that G is simple. Then $\sigma|\Gamma$ satisfies either i) $\overline{\sigma(\Gamma)}$ is compact; or ii) $\sigma|\Gamma$ extends to a rational automorphism of G. This follows from Theorem 9.7. In the first case, all eigenvalues of $Ad(\sigma(\gamma))$ will have absolute value one, and in the second case, these eigenvalues coincide with those of $Ad(\gamma)$. (We remark that if $A:G \to G$ is an automorphism we have $dA \circ Ad(A(g)) \circ (dA)^{-1} = Ad(g)$, so $Tr(Ad\ A(g)) = Tr(Ad\ g)$.) Hence for each $\gamma \in \Gamma$, $\{\sigma(Tr(Ad(\gamma)))|\sigma \in Aut(\mathbf{C})\}$ is bounded. The same can easily be seen if G is semisimple by examining the composition of $\sigma|\Gamma$ with projection on the simple factors. However, since $Aut(\mathbf{C})$ is transitive on the transcendental numbers, it follows that $Tr(Ad(\gamma))$ is algebraic for all $\gamma \in \Gamma$.

Thus, identifying G with Ad(G), we have $\Gamma \subset G$ with $Tr(\gamma) \in K$ for all $\gamma \in \Gamma$. The next step, which is classical, is to observe that this implies that there is a faithful rational representation of G, defined over K, such that $\Gamma \subset G_K$, once again, identifying G with its image under this representation. We recall the construction. Consider $Tr:G \to \mathbf{C}$, and let $V = \mathbf{C}$-linear span of Γ-translates of Tr. By the Borel density theorem V will also be G-invariant. Choose a basis of V of the form $\gamma_i \cdot Tr$. Then one can verify in a straightforward manner that with respect to this basis, the matrix elements of $\gamma \in \Gamma$ acting on this space are all in K. Since Γ is finitely generated (in the property (T) case this follows easily [9]) we can find an algebraic number field k with $\Gamma \subset G_k$.

We now recall the basic operation of restriction of scalars. Suppose G is an algebraic group defined over an algebraic number field k. Then there exists an algebraic group $\tilde{G}$ defined over $\mathbf{Q}$ such that

i) There is an injective map $\alpha:G_k \to \tilde{G}_{\mathbf{Q}}$; and

ii) There is a surjective rational homomorphism $p:\tilde{G} \to G$ defined over k such that $p(\tilde{G}_{\mathbf{Q}}) = G_k$ and $p \circ \alpha:G_k \to G_k$ is the identity.

We recall here two ways of describing this construction. We can take $\tilde{G} = \prod_{\sigma} \sigma(G)$ where σ runs through the distinct embeddings of k in $\mathbf{C}$. Then $\alpha: G_k \to \tilde{G}$ is the map $\alpha(g) = (\sigma_1(g),\ldots,\sigma_r(g))$, and $p: \tilde{G} \to G$ is projection onto the factor corresponding to $\sigma = \mathrm{id}$. Alternatively, let $[k:Q] = r$, and choose an identification $k^n \longleftrightarrow Q^{nr}$. Then we have a map $\alpha: G_k \to GL(nr,\mathbf{Q})$, and we let $\tilde{G}$ be the Zariski closure of $\alpha(G_k)$ in $GL(nr,\mathbf{C})$. In this formulation, p arises from the fact that the entries of $g \in G_k$ are described by k-linear combinations of the entries of $\alpha(g)$. These linear expressions allow us to define a linear map from $nr \times nr$ Q-matrices to $n \times n$ k-matrices, and thus a map $\tilde{G} \to G$. (Recall that G is the Zariski closure of G_k.) Since this map is clearly a homomorphism on $\alpha(G_k)$, (being the inverse of α), it is also a homomorphism on its Zariski-closure, $\tilde{G}$.

Completion of proof of Theorem 9.6.

We let k be as in Lemma 9.8, and $\tilde{G}$, p, α as above. We have $\Gamma \subset G_k$, and we let $H \subset \tilde{G}$ be the Zariski closure of Γ. We still have p(H) = G by Zariski density of Γ in G. Since G is semisimple, p trivial on the radical of H, and, replacing H by the quotient of H by its radical, we can assume H is a semisimple group defined over $\mathbf{Q}$, and with trivial center.

We now claim that $(\ker p)_{\mathbf{R}}$ is compact. Let F be a simple factor of ker p. Then as algebraic groups defined over **R**, we can write $H \cong G \times F \times F'$ where F' is the product of the remaining simple factors. Since $\alpha(\Gamma)$ is Zariski dense in H, $(q \circ \alpha)(\Gamma)$ is Zariski dense in F where q is projection of H onto F. We claim $F_{\mathbf{R}}$ must be compact. If not, then $(q \circ \alpha)(\Gamma)$ cannot have compact (topological) closure since compact real matrix groups are real points of algebraic groups. This would imply by Theorem 9.7 that $q \circ \alpha$ extended to a rational homomorphism $h: G \to F$. But then $\{(g, h(g), f') \mid g \in G, f' \in F'\}$ would be a proper algebraic subgroup containing $\alpha(\Gamma)$, contradicting Zariski density of $\alpha(\Gamma)$ in H. This verifies compactness of $F_{\mathbf{R}}$, and doing the same for each factor, compactness of $(\ker p)_{\mathbf{R}}$.

Now consider $\alpha: \Gamma \to H_{\mathbf{Q}}$. For each prime a, the image of $\alpha(\Gamma)$ in $H_{\mathbf{Q}_a}$ must be bounded by Theorem 9.7(ii). This means that the powers of each prime appearing in the denominators of matrix entries of $\alpha(\gamma) \in H_{\mathbf{Q}}$ are bounded uniformly over $\gamma \in \Gamma$. But Γ is finitely generated, and hence only finitely many primes will appear at all. This is readily seen to imply that $\alpha(\Gamma) \cap H_Z$ is of finite index in $\alpha(\Gamma)$, and hence, applying p, that $\Gamma \cap p(H_Z)$ is of finite index in Γ. This in turn implies that $\Gamma \cap p(H_Z)$ is a lattice in $G_{\mathbf{R}}$. By Theorem 9.1 and Proposition 9.4, $p(H_Z)$ is a lattice in $G_{\mathbf{R}}$, and since $(\Gamma \cap p(H_Z)) \subset p(H_Z)$ is an inclusion of lattices, we also have $[p(H_Z) : \Gamma \cap p(H_Z)] < \infty$. This shows commensurability of Γ and $p(H_Z)$, completing the proof.

References

1. L. Auslander, L. Green, F. Hahn, Flows on Homogeneous Spaces, Annals of Math. Studies, no.53, Princeton, 1963

2. L. Auslander, An Exposition of the Structure of Solvmanifolds, Bull. Amer. Math. Soc., 79(1973), 227-285.

3. A. Borel, Linear Algebraic Groups, Benjamin, New York, 1969.

4. A. Borel, Density Properties for Certain Subgroups of Semisimple Lie Groups Without Compact Factors, Annals of Math., 72(1960), 179-188.

5. A. Borel, Harish-Chandra, Arithmetic Subgroups of Algebraic Groups, Annals of Math., 75(1962), 485-535.

6. A. Borel, J.P. Serre, Theorèmes de Finitude en Cohomologie Galoisienne, Comm. Math. Helv., 39(1964), 111-164.

7. J. Brezin, CC. Moore, Flows on Homogeneous Spaces: A New Look, preprint.

8. A. Connes. J. Feldman, B. Weiss, Amenable equivalence relations are generated by a single transformation, preprint.

9. C. Delaroche, A Kirillov, Sur Les Relations Entre L'Espace Dual d'un Groupe et la Structure de ses Sous-Groupes Fermés, Seminaire Bourbaki, no.343, 1967/68.

10. H.A. Dye, On Groups of Measure Preserving Transformations, I, Amer. J. Math., 81(1959), 119-159.

11. H.A. Dye, On Groups of Measure Preserving Transformations, II, Amer. J. Math., 85(1963), 551-576.

12. E.G. Effros, Transformation Groups and C^*-Algebras, Annals of Math., 81(1965), 38-55.

13. J. Feldman, P. Hahn, C.C. Moore, Orbit Structure and Countable Sections for Actions of Continuous Groups, Advances in Math., 28(1978), 186-230.

14. H. Freudenthal, Topologische Gruppen mit Genugend Vielen Fastperiodishen Funktionen, Annals of Math., 37(1936), 57-77.

15. H. Furstenberg, A Poisson Formula for Semisimple Lie Groups, Annals of Math., 77(1963), 335-383.

16. H. Furstenberg, The Structure of Distal Flows, Amer. J. Math., 85(1963), 477-515.

17. H. Furstenberg, Ergodic Behavior of Diagonal Measures and a Theorem of Szemeredi on Arithmetic Progressions, J. Analyse Math., 31(1977), 204-256.

18. F.P. Greenleaf, Invariant Means on Toplogical Groups, Van Nostrand, New York, 1969.

19. Y. Guivarc'h, Croissance Polynomiale et Periodes des Fonctions Harmonique, Bull. Math. Soc. France, 101(1973), 333-379

20. G. Hedlund, The Dynamics of Geodesic Flows, Bull. Amer. Math. Soc., 45(1939), 241-260.

21. E. Hopf, Statistik der Lösungen geodätischer probleme vom unstabilen typus, Math. Ann. 117(1940), 590-608.

22. R. Howe and C.C. Moore, Asymptotic Properties of Unitary Representations, J. Func, Anal.,32(1979), 72-96.

23. J.W. Jenkins, Growth of Connected Locally Compact Groups, J. Funct. Anal., 12(1973), 113-127.

24. D. Kazhdan, Connection of the Dual Space of a Group with the Structure of its Closed Subgroups, Funct. Anal. Appl., 1(1967), 63-65.

25. W. Krieger, On Ergodic Flows and the Isomorphism of Factors, Math. Ann. 223(1976), 19-70.

26. G.W. Mackey, Ergodic Transformation Groups with a Pure Point Spectrum, Illinois J. Math., 8(1964), 593-600.

27. G.A. Margulis, Non-uniform lattices in Semisimple Algebraic Groups, in Lie Groups and their Representations, ed. I.M. Gelfand, Wiley, New York.

28. G.A. Margulis, Discrete Groups of Motions of Manifolds of Non-Positive Curvature, Amer. Math. Soc. Translations, 109(1977), 33-45.

29. G.A. Margulis, Arithmeticity of Irreducible Lattices in Semisimple Groups of Rank Greater than 1, Appendix to Russian Translation of M. Ragunathan, Discrete Subgroups of Lie Groups, Mir, Moscow, 1977(in Russian).

30. G.A. Margulis, Factor Groups of Discrete Subgroups, Soviet Math. Dokl. 19(1978), 1145-1149.

31, G.A. Margulis, Quotient Groups of Discrete Subgroups and Measure Theory, Funct. Anal. Appl., 12(1978), 295-305.

32. C.C. Moore, Ergodicity of Flows on Homogeneous Spaces, Amer. J. Math., 88(1966), 154-178.

33. C.C. Moore, Amenable Subgroups of Semisimple Groups and Proximal Flows, Israel J. Math., 34(1979), 121-138.

34. C.C. Moore and R.J. Zimmer, Groups Admitting Ergodic Actions with Generalized Discrete Spectrum, Invent. Math., 51(1979), 171-188.

35. G.D. Mostow, Quasi-Conformal Mappings in n - Space and the Rigidity of Hyperbolic Space Forms, Publ. Math. I.H.E.S., (1967), 53-104.

36. G.D. Mostow, Strong Rigidity of Locally Symmetric Spaces, Annals of Math. Studies, no.78, Princeton Univ. Press, Princeton, N.J. 1973.

37. D. Ornstein and B. Weiss, to appear.

38. W. Parry, Zero Entropy of Distal and Related Transformations, in Topological Dynamics, eds. J. Auslander, W. Gottschalk, Benjamin, New York, 1968.

39. G. Prasad, Strong Rigidity of Q-rank 1 Lattices, Invent. Math., 21(1973), 255-286.

40. M. Ragunathan, Discrete Subgroups of Lie Groups, Springer-Verlag, New York, 1972.

41. A. Ramsay, Virtual Groups and Group Actions, Advances in Math., 6(1971), 253-322.

42. A. Selberg, On Discontinuous Groups in Higher Dimensional Symmetric Spaces, Int. Colloquium on Function Theory, Tata Institute, Bombay, 1960.

43. T. Sherman, A Weight Theory for Unitary Representations, Canadian J. Math., 18(1966), 159-168.

44. C. Sutherland, Orbit Equivalence: Lectures on Kriegers Theorem, University of Oslo Lecture Notes.

45. V.S. Varadarajan, Geometry of Quantum Theory, vol. II., Van Nostrand, Princeton, N.J. 1970.

46. S.P. Wang, On Isolated Points in the Dual Spaces of Locally Compact Groups, Math. Ann., 218(1975), 19-34.

47. R.J. Zimmer, Extensions of Ergodic Group Actions, Illinois J. Math., 20(1976), 373-409.

48. R.J. Zimmer, Ergodic Actions with Generalized Discrete Spectrum, Illinois J. Math., 20(1976), 555-588.

49. R.J. Zimmer, Orbit Spaces of Unitary Representations, Ergodic Theory, and Simple Lie Groups, Annals of Math., 106(1977), 573-588.

50. R.J. Zimmer, Uniform Subgroups and Ergodic Actions of Exponential Lie Groups, Pac. J. Math., 78(1978), 267-272.

51. R.J. Zimmer, Amenable Ergodic Group Actions and an Application to Poisson Boundaries of Random Walks, J. Funct. Anal., 27(1978), 350-372.

52. R.J. Zimmer, Induced and Amenable Ergodic Actions of Lie Groups, Ann. Sci. Ec. Norm. Sup., 11(1978), 407-428.

53. R.J. Zimmer, Algebraic Topology of Ergodic Lie Group Action and Measurable Foliations, preprint.

54. R.J. Zimmer, An Algebraic Group Associated to an Ergodic Diffeomorphism, Comp. Math., to appear.

55. R.J. Zimmer, Strong Rigidity for Ergodic Actions of Semisimple Lie Groups, Annals of Math., to appear.

56. R.J. Zimmer, Orbit Equivalence and Rigidity of Ergodic Actions of Lie Groups, preprint.

57. R.J. Zimmer, On the Cohomology of Ergodic Actions of Semisimple lie Groups and Discrete subgroups, Amer. J. Math., to appear.

58. R.J. Zimmer, On the Mostow Rigidity Theorem and Measurable Foliations by Hyperbolic Space, preprint.